MILK: BEYOND THE DAIRY

PROCEEDINGS OF THE OXFORD SYMPOSIUM ON FOOD AND COOKERY 1999

Frontispiece. Jatte téton avec son trépied, *soft-paste porcelain bowl on stand, designed by Jean-Jacques Lagrénée, heads modelled by Godin, Sèvres, 1788, made for the dairy at Rambouillet. (Height of stand: 7.4cm, diameter of bowl: 14.2cm). Courtesy of the Musée National de Céramique, Sèvres, MNC 23399. See the article by Carolin C. Young on* La Laiterie de la Reine, Rambouillet.

MILK:
BEYOND THE DAIRY

PROCEEDINGS OF THE OXFORD SYMPOSIUM
ON FOOD AND COOKERY 1999

EDITED
BY
HARLAN WALKER

PROSPECT BOOKS
2000

Published in 2000 by Prospect Books, Allaleigh House, Blackawton, Totnes, Devon TQ9 7DL, England

©2000 as a collection Prospect Books (but ©2000 in individual articles rests with the authors)

Designed and typeset by Tom Jaine.
Printed by the Cromwell Press, Trowbridge, Wiltshire.

ISBN 1 903018 06 4

The cover illustrations are reproduced thanks to the generous support of Robin Weir. The painting is a gouache on paper by Savario Xavier della Gatta, Naples 1820, entitled *Venditore di sorbetto a minuto — Franfelliccaro Napolitano*. Reproduction is by kind permission of Patricia Rogers, Head Librarian, The Juliette K. and Leonard S. Rakow Research Library, The Corning Museum of Glass, Corning, NY, USA. The whole painting is reproduced in black and white on page 348.

Contents

Milk and its Uses in Assyrian Folklore 9
   Michael Abdalla

Milk: Nutritious and Dangerous 19
   Ken Albala

Milk and its Products in Ancient Rome 31
   Joan P. Alcock

The Cheeses of Hokkaido and other Milky Issues in a Ricist Society 39
   Michael Ashkenazi

How the Bengalis Discovered *Chhana* and its Delightful Offspring 48
   Chitrita Banerji

Touloumotyro: Centuries Old and About to Die 60
   Rosemary Barron

Milk and its By-products in Ancient Persia and Modern Iran 64
   Najmieh Batmanglij

Rabbits, Fondues and Physics 74
   Tony Blake

Milk-borne Diseases: An Historic Overview and Status Report 81
   Fritz Blank

Hawking Milk: The Public Health Profession, Pure Milk,
and the Rise of Advertising in Early Twentieth-century America 86
   Daniel Ralston Block

The Hierarchy of Milk in the Renaissance, and
Marsilio Ficino on the Rewards of Old Age 93
   Phyllis Pray Bober

New York Milk Culture: Some History, Facts and Concerns 98
   Una Bray

Farmhouse Gouda: A Dutch Family Business 107
   Janny de Moor

Milk and Dairy Products in the Roman Period 117
   Carol A. Déry

Carabao Milk in Philippine Life 126
    Doreen G. Fernandez

A Spring-house in Pennsylvania: Design and Use 137
    Rebecca Fitzjohn and Harlan Walker

How Old is Old Cheese? *Gamalost* in Coffin-shaped
Boxes and Eccentric Jars 144
    Ove Fosså

The Origins of Taste in Milk, Cream, Butter and Cheese 161
    Sarah Freeman and Silvija Davidson

Cato's Roman Cheesecakes: The Baking Techniques 168
    Sally Grainger

Dairy Food in the UAE 178
    Philip Iddison

What's in a Name? Some Thoughts on the Origins,
Evolution and Sad Demise of Béchamel Sauce 193
    Cathy K. Kaufman

Low-temperature Cheese-making: Ancient Wisdom not Outdated 206
    Lidia Kitrilakis and Sotiris Kitrilakis

The Artisanal and Regional Cheeses of Greece 209
    Diane Kochilas

Fresh From the Cow's Nest:
Condensed Milk and Culinary Innovation 216
    Rachel Laudan

The Rise of the Cream Sauce, 1660–1760 225
    Gilly Lehmann

Finnish-American Milk Products in the Northwoods 232
    Yvonne R. Lockwood & William G. Lockwood

Names for Milk and Milk Products 241
    Jenny Macarthur

The Milk-tie 248
    Jeremy MacClancy

The Health Hazards of Milk 259
    H. Morrow Brown

The Art of Making Brie de Meaux Fermier 268
    Lizabeth Nicol

Medieval Arab Dairy Products 275
    Charles Perry

Images of Progress: Milk Advertisements in Greece 278
    Elia Petridou

Cheese in Art 287
    Gillian Riley

Animal Husbandry and Other Issues in the Dairy Industry
at the End of the Twentieth Century 293
    Cherry Ripe

Sandesh: An Emblem of Bengaliness 300
    Colleen Taylor Sen

Yoghurt in Iran 309
    Margaret Shaida

The Origins of the New York Dairy Industry 315
    Andrew F. Smith

The Wet-nurse 330
    Raymond Sokolov

Milky Medicine and Magic 337
    Layinka M. Swinburne

More on the Origin and History of the Ice-cream Cone 345
    Robin Weir

Use of Almonds in Late-medieval English Cookery 352
    Caroline Yeldham

La Laiterie de la Reine at Rambouillet 361
    Carolin C. Young

Other Papers Given at the Symposium 379

List of those attending the Symposium 381

Introduction

This volume of papers presented at the Oxford Symposium on Food and Cookery follows the pattern of previous collections. The Symposium entitled *Milk: its Uses, Properties and Substitutes* was held in September 1999 at Saint Antony's College, Oxford under the joint chairmanship of Alan Davidson and Dr Theodore Zeldin.

Again the success of the symposium resulted in the number of papers being such that it has not been possible to publish them all. A list of the papers given in Oxford but which are not in this volume is given on page 379 below.

As usual the College gave continuing support to our varied needs and whims. We are very grateful, particularly to Lynne Docherty.

On Saturday evening we enjoyed a dairy-inspired dinner at the Oxford University Press. This was planned by Hugo Arnold and managed by Jack Butler, who stepped in at the last minute on the resignation of his boss. We thank them both.

The special heroes of the weekend were Lidia and Sotiris Kotrilakis, Gareth Spencer Jones and Joy Davies who sourced, supplied, prepared and served a most delicious Sunday lunch with unusual and surprising Greek ingredients.

We are always dependent on fellow symposiasts not only for coming and giving their views in the form of papers and otherwise, but for much general and generous help. We particularly thank Emma Roberts, Hawys Pritchard and Ben Walker for their continuing attention to details. Finally we all know that without the efficient, endless and apparently tireless labours of our organizer Jane Levi our enjoyable and fascinating weekend would not have happened.

Harlan Walker
July 2000

Milk and its Uses in Assyrian Folklore

Michael Abdalla

At the Ethnological Food Research Conference on milk held in Ireland in 1991 I presented a paper on mostly traditional milk processing methods used by modern Assyrians in the Middle East, which was sketched against historical background and supplemented with culinary review.[1] A more detailed presentation of the subject was published in the *Journal of the Assyrian Academic Society*.[2] The present paper is not a copy of either, although it includes several parts thereof, but it is rather a considerably modified version of them both. It stresses mainly the narrative folklore tradition which developed around the milk motif. Such segments of folk heritage are worthy of being recorded and of comment since this has developed in the particular living conditions of the Middle Eastern villages. In recent years these conditions have been changing rapidly as a result of migration from the countryside to urban areas and from the Middle East to Europe.

It is less and less frequent to encounter mountain-dwelling Assyrian families, who commonly travelled long distances to reach a monastery often carrying their sick on a stretcher. On their arrival, they would spend the night in prayer to the saints pleading for the healing of their sick. They would also give the monks a sheep. The traditions of offering a fattened sheep to the local church or giving to charity the milk from the first milking or the first produced butter or cheese are also becoming extinct, yet at the end of the nineteenth century it was a common practice of the Eastern Assyrians from the area of Urmia (in Iran).[3] The high regard for livestock is apparent in folk magic. There were entreaties asking that the livestock maintain a favourable disposition toward their owner, that they retain good health, and that they be protected from theft. A book published in 1976 features 95 charms (presented in Assyrian and English), five of which deal with livestock protection, and the yield and quality of milk.[4]

Milk processing in outline
Fresh milk quickly turns sour due to the relatively high temperatures in that region. The process is also accelerated by insufficient observance of milking hygiene conditions, a difficult thing to maintain in the circumstances of the

Middle Eastern countryside. It is then considered a spoiled product (*ḥalwo ḥārīwo*). Therefore, after milking, it is quickly boiled. The residue of the boiled milk in the pot (sometimes even slightly burnt) is often a bone of contention among younger children – each of them asks his or her mother for this delicacy. It is usually given to boys and they add breadcrumbs to the sediment layer and eat the mixture from the pot.[5]

There is an opinion, based on some scientific grounds, that both the sediment layer and the 'skin' comprise milk's most nutritious elements, the remaining part being more sterile compared with milk before boiling.[6]

Yoghurt culture (*rawbe*) is added to milk, which is boiled in the afternoon and cooled to a temperature of about 40°C. The pot is wrapped in a warm garment and left indoors to cool slowly. The slow rate at which the milk cools has a positive effect on its coagulation. In the evening the pot is taken outdoors and put in a place inaccessible to animals, especially cats who are great lovers of yoghurt, so that it can slowly set in the coolness of the night. The yoghurt has a gelatinous texture, and the more solid the texture the better the yoghurt. The magnificent and refreshing taste and scent bouquet of yoghurt obtained through milk fermentation can be compared in its originality only to the aroma of freshly baked bread. This delicacy of Eastern origin has been popularized in Europe as recently as the 1950s.[7]

Butter is obtained from yoghurt by shaking it in a goatskin. It is not used as a spread on bread as this custom is unknown in the Middle East; instead, pieces of bread are dipped in butter and eaten. Butter is mainly used to make non-perishable milk-fat (*m:šḫo*, Arabic *samn*) which is added to dishes served on holidays throughout the year. The butter is salted and put into a container. When the container is full, the butter is taken out, put into a thick-walled, tin-plated tub, and then boiled. When boiled the milk falls into layers – pure fat forms on the top and whey and sediments collect underneath. When the milk has cooled, the fat is separated and boiled again, this time with a piece of bread. The boiling is stopped when the bread becomes crisp and light brown in colour. Butter is a particularly indispensable ingredient for the preparation of cereal dishes: it enhances their taste by its slight modification, and by giving a shine to individual grains it improves their colour and granularity. Butter should be warmed up and spread over the dish. The desired effect is achieved when, upon the contact of butter and cereal, a characteristic noise can be heard, described as *q:zz*, which is similar to the sound effect of submerging a red-hot metal rod into water. Housewives who are generous with fat enjoy the esteem of being hospitable and solicitous about their families' health. The phrase *muklo ḥāyyiro* refers to dishes which have insufficient amounts of fat (dishes with insufficient amounts of salt are referred to as *muklo šāhuwo*).[8]

Rhythmic songs about the traditional ceremony of butter-making have been written. Of two such songs in my collection, one is entitled *gūdī mayyānne b-dranāni* ('I am shaking the goatskin with the strength of my arms'), and is performed by a woman, while the other, performed by a man, describes the path along which a man's sweetheart can go, and on which she finds a goatskin.[9]

When the butter has been separated, buttermilk (*dawĝe*) remains in the goatskin, and this is drunk, often instead of water. In the summertime it is as popular as beer or Pepsi advertised in Europe. When there is a lot, it is given to farm animals, or poured into containers and sold in nearby cities. It is mainly bought by those who have recently migrated from villages to towns, who drink it with ice, or put pieces of stale bread into it and then eat it. It makes a dish for the poor. Lumps of butter in buttermilk increase its value and attract buyers. In the countryside the technology of full fat separation from butter-milk was not known.

Milk in folk medicine

Assyrian children brought up in Middle Eastern villages may be the best authority to question on whether milk can be used effectively. I (or, more correctly, my relations) recall a time when, little more than a baby, my eyes were so swollen as to conceal my eye lashes from view. Having difficulty in opening her son's eyelids, my mother during feeding time squeezed a few drops of milk from her breast onto the eyes. This was a common practice by mothers when their babies had sore eyes: if the mother was unable to supply the milk herself, she would ask a neighbour to do this for her. There was an abiding belief that a mother's milk was the best cure for such sores. Mother's milk was also used to wash the eyelids of a child whose eyes had become bunged up by gum during the night, especially in summer when families slept on terraces. There was a general belief not only in the healing effect of mother's milk on the eyes, but also on its prophylactic and even cosmetic properties.[10]

Milk was considered effective in the case of poisoning, and in scorpion and snake bites. In the case of such bites, the victim was bandaged above the bite in order to stop the flow of blood to the body. The locus of the bite would be cut and, with some milk in his mouth, the victim or, if he was not alone, a companion, would rapidly suck the blood and venom and proceed to spit it out. It was supposed that, after several repetitions, the milk would assimilate or dissolve the venom. Buttermilk was spread over the body to treat eczema caused by the sun's rays. The buttermilk was sometimes applied in the form of a poultice using as dressing the inner side of watermelon skin.

The advice of Assyrian folk medicine from early medieval times was collectively published in London in 1913, and then reprinted in 1976 in Amsterdam as a two-volume compendium (*The Syriac Book of Medicines*, see note 11 below). The first part of this work consists of an introduction with original Assyrian texts (collected in the nineteenth century around Mosul, Iraq). The other part includes translations of the texts and an index. Along with the great number of herbs and little-known medicinal plants and mineral substances which are listed in this compendium, there is also mention of milk as an ingredient in many medicinal mixtures. The practice of washing a baby's eyes with the mother's milk is reiterated. According to the compendium's author mother's milk was considered a suitable substitute for warm donkey's milk.[11]

The compendium includes a long list of diseases which were treated with the use of milk as a solvent for various extracts. For example a tongue ulceration could be healed by washing the mouth with a solution including an extract from dried rose leaves, pomegranate skin and wheat flour diluted in sour milk. Inflammation of the uvula was relieved by rinsing out the mouth with a solution including whey, grape juice and a bit of salt. Nervous disorders were treated with a milk solution including resin extract, pepper seed, rock salt, red raisins and garlic. Pulmonary diseases were treated with a dill seed extract dissolved in goat's milk, or with a drink made from sheep's cream, eggs and pepper seeds. Camel's milk mixed with a free fungi extract (probably *Polypori*) was indicated for liver and stomach problems. In the case of jaundice, Assyrian doctors prescribed goat's milk mixed with an extract from roasted convolvulus and orach or spinach or, as an alternative, with a solution including pomegranate skin, yellow arsenic and donkey's milk. An extract of sorrel, well leached in milk, was also used. It is interesting to note the ingredients of pills used in the treatment of the spleen: jackal's milk, roasted squill, cardamom, gum ammoniac, pyrethrum, rape seeds and soapwort.

Medicine to be ingested for colon discomfort included a drink made of iris petals and whey or an enema of diluted aloe paste, myrrh, opium, goat's cream, and castoreum (the secretion of the anal glands of a beaver), thinned in sweet wine and taken with dill oil or with rue oil; or, alternatively, an enema of aloe, opium, cow's cream, castoreum, soft tar and incense. Worms could be removed with a syrup made from convolvulus, copper, fern, fennel flower, garlic and goat's milk; or, better yet, with milk mixed with copper, convolvulus, fern, Indian salt and costmary. This mixture was customarily drunk three days after the milk treatment. Neck pains could be relieved by retaining salted milk in the mouth. A compress made of barley flour boiled in donkey's milk helped relieve the pain of frozen hands or feet. For the maintenance of healthy breasts, nursing mothers were advised to drink the milk of a red cow, or a drink combining fresh milk with liquoriceroots, figs and dates. People with cardiac problems were offered an extract of incense, pomegranate skin and wild mustard, all of which were diluted in sour milk and wine vinegar. Insomnia was cured by applying on the head a compress consisting of lettuce seeds, rose oil, and sour milk. Fever from travel fatigue was attended by a mixture of rock parsley ground in milk. Thin persons seeking weight gain were advised to consume galbanum with goat's milk. Scabies could be cured with a balm including fenugreek flour, wine vinegar and mother's milk.

Early medieval Assyrian doctors believed that milk taken orally could dissolve poison. A common antidote for poison was ingestion of gazelle's bile diluted in goat's milk. It should be noted that the belief that milk is an antidote to swallowed poison remains widespread among Assyrians today. To deal with a child's diarrhoea, the youngster was made to eat a piece of suet from a sheep's tail, boiled in fresh milk and powdered with sumach.[12] To stop abdominal bleeding, one drank broth made from hare and sheep, cooked with cow's milk and honey. Even flatus could be curtailed by use of a compress of barley paste diluted in wine, milk, wine vinegar and baby's urine.

Mother's milk – something better?

Much more importance is attached to mother's milk in the folk culture of the Assyrians than to any other food. The metaphysical symbols of mother's milk are present at different periods and situations in one's life. One often hears children swear an oath 'on mother's milk' to convince others of the truthfulness of what they have said. Very often a doubtful child will demand that his/her friend swears an oath 'on mother's milk', which is a guarantee that the information or tales are true; it is considered a mortal sin to abuse this oath.

A baby is breast-fed for a fairly long period of time, from one to three years. Breast-feeding is interrupted only when natural contra-indications occur.[13] It sometimes happens that one woman breast-feeds two babies at the same time, only one her own. There would be nothing strange about this if it were not for the belief that it is not advisable for a boy and a girl fed on the same breast to marry in the future, despite the fact that they are completely unrelated. This means that mother's milk shared by unrelated babies is believed to make them siblings. The same law, which forbids 'milk siblings' marrying, is also binding for Muslims.

An honest, truthful, loyal and well brought-up man is said to have 'sucked true, uncontaminated, legal milk'. What is meant here is that a man's conduct reflects the permanence of his mother's love which was given to him as a baby during breast-feeding, and loyalty acquired during the first months of life. Irresponsible, treacherous people are said to have 'sucked false, contaminated, illegal milk'. When children have been extremely naughty, some mothers say to them: 'May my milk be to you *ḥarām!*' This is understood as a curse. Perhaps there is nothing more forceful, or at the same time more painful, for a child's psyche than his or her mother's disapproval, expressed so firmly. When a child hears this, he or she usually cries, apologizes and remains submissive until the mother forgives. So the expression: 'May my milk be to you *ḥalāl*' sounds like the best blessing and wish for a success.[14]

Milk in verbal folklore

Among the Assyrians, like the majority of other Middle Eastern peoples, milk-related folklore does not seem to be very rich. From my memory I could only recreate the first three of the proverbs quoted below. When I enquired about that, my elder compatriots, who – as I do – live nowadays in Europe, could not remember many more. It is evidence that over a relatively short time the majority of emigrants ceased using this kind of phrase. Therefore the two collections of Assyrian proverbs, which have been recently published in Syria, are extremely valuable. One of the volumes is the source of four sayings (numbered 4 to 7),[15] and in the other I have found one which is quoted last.[16] The ninth and tenth sayings were sent to me by Simon Atto, an Assyrian immigrant in Holland.

1. *Yāq:ḏ lešone bū-ḥalwo, np:ḥle bad-dawġe.*
   (His tongue got burnt by milk; he blew air into buttermilk.)
   It is used metaphorically in at least two situations of conflict: a person unjustly suffers a wrong and is unable to pay back the wrongdoer; he vents his anger towards a completely impartial person; a person's quick and thoughtless acts lead to negative consequences; his attempts to redress the resulting damage are futile.

2. *Noše lo kōm:r 'u-qātirayḏi ḥāmuṣoyo.*
   (Nobody admits that his yoghurt is sour.)
   This saying, very popular among Western Assyrians, expresses the notion that people usually hide their weaknesses and mistakes, while they boast about what they have and what they do. It is not unlike the proverbs found in other nations – for example, a good Polish equivalent would be 'Every wagtail boasts of its tail' (to blow one's own trumpet).

3. *Kīto manṯo bu-qāṭiro.*
   (There is a hair in the yoghurt.)
   Such phrases can be heard when the speakers give an impression of hiding something or when they want to mislead others on purpose.

4. *Ḥāwi mēḥe dawġe.*
(His brain turned into buttermilk.)
He is so tired and exhausted that nothing gets through to him. Having been given too much advice and instruction, he does not worry anymore and is not able to comprehend what he is told. He has heard so many statements and contradictory arguments that he cannot tell any more which are genuine and honest, and which are false and untrue. He has lost his ability to concentrate. He is in a bad condition, almost dazed. Alternatively, he made too much physical and mental effort. He is not aware anymore of what he is doing. Every move he makes is a bad one.

5. *Kōḫ:l mū dawšo, hul lū-ḥalwo daq-qāṭune.*
(He devours everything: from honey to cat's milk.)
He never refuses what he is offered. That is his nature: he is always eager to accept everything either from the poor or from the rich. He takes every opportunity to receive something. He lives on other people's work and at their expense. He is greedy and canny.

6. *'Uḥa d-yōq:ḍ bū-ḥalwo, g-zōy:' mū qāṭīro.*
(Who got burnt with hot milk is afraid of yoghurt.)
This saying shows an excessive deliberation before taking decisions or acting. The speaker, having his own painful experiences, warns and recommends that it is worthwhile to be particularly careful not to repeat a mistake.

7. *Ḥālōwo dī-bōqo.*
(Milkman of aphids.)
It is hard to find a greedier man. He never offers anything to others, and he would rather live in poverty himself than spend an extra penny. Its Polish equivalent can be *dusigrosz* (penny-pincher).

8. *Sāḥ:b š:qf:t ':gg:bne y:'r:f ḥālo.*
(The one from the piece of cheese knows what I mean.)
There is a special story accompanying this saying expressed in the Arabic dialect of the Assyrians originating from Azah (Bazebde, nowadays Idil in south-eastern Turkey). A single lady from Azah employed several farm workers to plough her field. Their salary was a loaf of bread a day. To prove that a woman can also be as enterprising as men, she decided to motivate the ploughmen to make maximum effort. She promised to each of them individually an additional payment provided that the others would not find

out about their deal. The bonus was to be a piece of cheese. Each worker treated it as a distinction of enjoying their employer's special favours. To show their gratitude each of them promised to work the most effectively.

When overseeing their work, the energetic lady shouted out loudly now and then: 'The one for the piece of cheese knows what I mean'. Each worker understood it as a remark addressed to him alone, demanding more effective work according to his promise. Ploughing the field, a task which usually took the whole week, was finished before three days had elapsed. Over the years, what may have been actual fact has turned into myth and the saying into a proverb. It is often said as a joke among friends and acquaintances at feasts.

9. *Sāwī ṭabb l-madīne. :r:b 'araq, ḥasabō šinīne.*
   (A Bedouin came to town. He drank arrack and thought it was buttermilk.)
   This saying in Arabic must have been created in urban centres on the frontier of Syria, Turkey and Iraq. It proves that the Bedouins are not familiar with alcoholic beverages, and buttermilk is extremely popular among them. Arrack is always drunk diluted with water. Then it turns white and may resemble buttermilk.

10. *Ḫdī t:rto d-kobayzo 'ū ḥalwayḏa.*
    (Like a cow which spills its own milk.)
    This remark is made about a person who does not appreciate the fruits of his own work and intentionally or by accident wastes them, or who does not allow others to share what he has got in surplus. He would rather let things be wasted than share them with the needy. In the Arabic dialect of the Azahs there is a similar saying: *Flān y:šbah baqara zare, t:ḥl:b w-tk:bb l-ḥalīb.* 'He is like a piebald cow, who gives milk and then spills it afterwards.

For several years I have been collecting folk sayings relating to food products from the Western Assyrians. Such forms of expression have their equivalents in all cultures. It is interesting to pay attention to this specific manifestation of spiritual folk creation, the inspiration for which are both traditional food products and the modern ones which are advertised in mass media. Contrary to artistic objects, which reflect this sphere of human life in a more or less concrete form, the orally transmitted sayings are threatened with becoming forgotten.

REFERENCES

[1] Abdalla, M. 'Milk in the Rural Culture of Contemporary Assyrians in the Middle East', *Milk and Milk Products from Medieval to Modern Times.*, ed. Patricia Lysaght, 27–40, Canongate Press, Edinburgh, 1994.

[2] Abdalla, M. 'Milk and its role in Assyrian Culture', *Journal of the Assyrian Academic Society*, vol. 6, no. 2, 12–36, 1994.

[3] Note, for example, Asahel Grant's personal recollections, *The Nestorians or Lost Tribes*, Philo Press, Amsterdam, 1973, 182–3 (reprint).

[4] Gollancz, H. *The Syriac Book of Protection*, APA-Philo Press, Amsterdam, 1976, Codex B: 10, LXXI (reprint). The texts were collected in 1802–1803 and come from the vicinity of Urmia. They were catalogued in 1901 for the Cambridge University Library by William Wright and published in London in 1912.

[5] Beestings (*'alwo*) is another rare delicacy for the children. After a cow has calved, the parents try to ensure that the beestings from the first milking is equitably distributed between the newly-born calf and their own children. It is boiled and then given to children.

[6] A common belief is that this fraction of boiled milk is the most delicious. On the other hand, some of those who pose to be rich or belong to the upper class, would be ashamed to collect the sediment from the pot bottom and walls for fear of being considered excessively thrifty or greedy. Children in Polish kindergartens are said to protest when they spot the smallest part of 'skin' in their milk. In the Middle East the protest would be raised by those children who weren't offered any of the 'skin'.

[7] Tamime, A.Y., Robinson, R.K. *Yoghurt – Science and Technology*, Pergamon Press, 1985, 234.

[8] When margarine imported from Holland appeared on the Syrian market, it was called *samn frangī* (from 'Frank' or European). Housewives used to say that even a multiple of a regular dose of margarine did not yield the same result as traditional fat. It had different sensory properties. Imported fat was mainly purchased by townsmen, since using products imported from Europe was perceived as a kind of ennoblement and a sign of prestige.

[9] There is a popular custom practised by Eastern Assyrians on a wedding day. Before the bride enters the bridegroom's house she stops for a while in front of the door, dips her finger in the *m:šḥo*, and makes the sign of the cross on the roof frame. This is said to bring happiness and wealth to the newlyweds.

[10] A widely-held belief among Assyrians is that teenagers and young women from wealthy families who want to have a lighter complexion bathe in animal milk. In fact I know from my personal experience as a young man that Assyrian mothers often try to persuade their children to wash their faces with spring dew in order to lighten their complexions.

[11] Budge, E.A.W. *The Syriac Book of Medicines*, APA-Philo Press, Amsterdam, 1976, vol. II, 85, 108, 153, 182, 277, 421, 455, 465, 505–7, 672, 674, 677, 682, 686, 689, 691, and 693 (reprint).

[12] Also in medicinal practices of ancient Mesopotamia there was a method for treating diarrhoea: 'If someone suffers from a diarrhoea which can not be stopped, a potion of sheep's milk boiled in a copper pot mixed with roasted barley flour should be pored into the person's anus. The diarrhoea will stop'. Cf. I.L. Finkel, *An Introduction to Babylonian Medicine* (Polish edn: *Zarys historii medycyny starożytnej Mezopotamii*, Instytut Historii UAM, Poznań, 1997, 68).

[13] To break the breast-feeding habit of a three- or four-year-old child, which was often at the expense of younger siblings, mothers would smear their nipples with a bitter substance. I remember that in the city of Kameshli it was available as a powder in a shop run by a sympathetic Jew, whose name was 'Izra.

[14] The expressions *ḥalāl* (allowed, not subject to any restrictions) and *ḥarām* (forbidden, prohibited, taboo) are Semitic words commonly used by the Muslim. Initially they had a

purely religious meaning, indicating not only food products which were allowed or forbidden by the Muslim law, but also referring to the attitude towards people of other denominations. Surprisingly, with time, if not in all Arabic dialects, then definitely in those spoken in the Asian part of the Arab world, the word ḥarām changed its meaning to 'mercy' or 'sin'. The word ḥarām is usually used to refer to aggressive behaviour towards animals or harm directed to another person.

[15] Asmar, G., Ḥikam az-zamān fī 'amṯāl as-Syrian al-'āmmiyye (Wisdom of Time in Assyrian Folk Proverbs), Damascus, 1991, 29, 38, 43, 97. Sayings in Assyrian with comments in Arabic. The second is Ishaq, L., 'Amṯāl min Bāzebde (Proverbs from Bazebde), Damascus, 1994, 61. The Assyrians coming from Bazebde use a specific sub-dialect of Arabic.

Milk: Nutritious and Dangerous

Ken Albala

In Western medical and dietary thought there is probably no food about which more contradiction and confusion has been generated than milk. On the one hand, it is obviously among the most nutritious substances. Mother's milk is the first food of infants and supplies them with all the necessary nutrients to sustain life. By logical analogy, animal milk should also be an appropriate food for humans of all ages. But on the other hand, an overriding obsession with putrefaction and fear of foods categorized humorally as cold and moist makes milk one of the more dangerous aliments. Deciding who should use milk, and in what context, thus became a major preoccupation among medical theorists in Western cultures.

Few other cultures have such difficulty deciding how to use milk, because few cultures think of milk as food for adults. The high incidence of lactose intolerance among many Asian peoples, Africans and native Americans is good evidence that these people historically had never made regular use of milk, and had never adapted biologically to digest it efficiently. In these cultures milk remained solely a food for children.

The situation was quite different in the West where cattle had been domesticated for around 10,000 years. This is confirmed not only archaeologically, but in the appearance of diseases associated with living in close proximity to cattle, for example small pox, measles and tuberculosis, which are probably mutated forms of animal pathogens.[1]

After many generations of using milk, with the gradual culling of the gene pool of those individuals with intolerance, the population would have evolved the ability to manufacture lactase into adulthood. Lactase is the enzyme that breaks down the sugar lactose; infants and children produce an abundance of it, but adults generally do not. Thus, Western cultures, and especially Northern Europeans, acquired the ability to consume milk in greater quantities without discomfort. This was, of course, biologically a relatively recent phenomenon.

That this process had not been complete, even by the time of classical antiquity, is evident in the comments of the Hippocratic authors and Galen. The Hippocratic books on *Regimen* mention milk only in the context of cheese, which 'is nourishing because the fleshy part of the milk remains in it'.[2] But the author does not suggest that milk should be a regular food. It usually only appears in medical compounds designed to cool and moisten the body, such as oxygala (oxygalacte or soured buttermilk). As an especially phlegmatic food according to humoral theory, it would also be dangerous in colder seasons, tending to throw off the internal balance of humours in the body and giving rise to numerable diseases associated with phlegm, such as rheums and catarrh, or what we would call a cold. A few centuries later, and with the full elaboration of the humoral theory by Galen of Pergamum, the pre-eminent physician of the Roman world, milk is clearly described as a dangerous food. In his *De alimentorum facultatibus*, Galen points out not only the abundant humidity which tends to corrupt within the body, but also the crass substance of milk's solid parts, and their tendency to obstruct the liver and cause kidney stones. Moreover, he comments that milk often causes flatus, which is a possible indication that much of the population was still lactose intolerant to some extent.[3]

In the classical world, fresh milk consumption was also associated with barbarity. As a simple natural food unimproved by art, it would be particularly appropriate for uncivilized nations, and may even have been suitable in past ages when people were more accustomed to living directly off the natural products of the earth. But for civilized peoples, a more civilized food is required. Milk must be made into cheese, grape juice must be fermented, olives preserved, and bread leavened. According to classical ideas, then, people gradually lost the ability to consume milk in adulthood, as their digestive systems became weaker and more delicate.

Of course the association of milk consumption with barbarians, and especially Northern Europeans, had a great deal to do with the fact that in hotter regions milk spoils very easily and must be made into cheese. In cheese, particularly when aged, lactose breaks down. Thus people around the Mediterranean and Middle East, who regularly consumed cheese rather than milk, would have retained lactose intolerance. In the North, and in mountainous regions, it could be consumed fresh, which is probably why lactose intolerance is to this very day a rarity in the north.

Despite this fact, it was classical nutritional theory that was inherited by medieval and early modern Europeans, first by way of Arabic interpreters such as Avicenna and Rhazes, but later via Galen and Hippocrates directly. What this means is that Europe accepted as orthodoxy a medical theory which extolled the importance of milk to the human species, but retained a serious ambivalence about how it should be used by adults. This central tension would only be accentuated when classical theory was inherited by Northern Europeans who were, by and large, well adapted to consume milk and would have to resort to more forceful arguments to dissuade their readers from using it. The example of milk thus provides a model of interaction and tension between material and cultural factors that influence food choices. That is, even among peoples who could freely consume milk, the fear wrought upon their consciences by physicians remained.

Despite turgid warnings, milk always forms a central part of all discussions of nutrition. According to theory, those substances which are most similar to the human body are most easily embraced, assimilated and incorporated into the flesh. Flesh itself, therefore receives pride of place in this theory. But so too does blood, since food must first be broken down and converted into blood in the liver before assimilation can take place. Only one step removed from blood, milk too is a powerfully nourishing substance. It is, in fact 'twice concocted' blood manufactured in the mammary glands from blood itself. It thus carries all the nutrients originally consumed by the mother in a highly refined form. At this stage, however, it has been deprived of much of its natural heat in nourishing the mother, and becomes a cold and moist aliment. It is ideally suited for nourishing infants who were also said to be relatively cold and moist humorally. But in youth, when the complexion becomes hot and moist, milk is no longer suitable, and in hotter and drier adult bodies, milk presents numerous difficulties

This theory was obviously designed to explain what were first probably empirically gathered observations. Infants have an easier time digesting milk than adults. Nature designed milk for specific ends, and provided the ideally suited aliment for infants, but taken out of context it poses problems for adults. This kind of teleological reasoning is found throughout Galenic thought, as well as in Aristotle, and was only accentuated when these ideas were grafted onto the Judeo-Christian idea of a purposeful and well-planned creation. It makes perfect sense that infants, who have no teeth, should be provided with liquid nourishment.

What is even more curious though, is that although milk remained indexed for adults, it was specifically recommended for the aged and infirm. The process of ageing was described as the gradual burning down of the radical moisture and vital heat in the body, much like a lamp running out of fuel. As the body ages it thus grows colder and drier. But often in the very old an accidental accumulation of phlegm alters the natural course of events and the complexion becomes accidentally cold and moist, and the digestive powers become weaker and unable to deal with solid foods.[4] Milk once again becomes an appropriate food for the aged, exactly as it was for infants – because old people are also often without teeth, nature again supplies the appropriate aliment.[5] Old age was often compared to a second infancy, not merely mentally, but physiologically. In this case it is theory alone that informs food use. Just as hot and moist wine, another analogue of blood, is appropriate and nourishing for adults, cold and moist milk is similar to, and suitable for, very elderly bodies.

The real difficulty with this idea, however, was deciding exactly when the aged body is healthy, and when distempered. For while foods similar to one's complexion nourish in health, opposites correct in infirmity. An excessively cold and moist body would only be pushed to pathological extremes by cold and moist foods, and in this case a hot and dry food would logically act as a corrective. Hot and dry spices and wines would be far more suitable than milk. Therefore milk does make sense as a nourishment for the aged, but not as a medicinal food. Why then many dietary writers remark that milk should only be used as medicine remains problematic. The malleability of this theory obviously leaves wide room for disagreement and confusion.

Apart from these irregular uses for milk, the real controversy surrounds the use of milk by adults. The predominant fear, as has been mentioned, was putrefaction. Exactly as milk goes sour and curdles outside the body, it can also corrupt when exposed to the digestive heat of the stomach. No doubt the prevailing metaphor of the stomach as a pot whose seething contents rest on the flame of the liver, also suggested that milk can scald and easily burn. Should this happen, sooty vapours rise into the head, and the chalky deposits left by separation and burning would also be forced into the veins, accumulating in the kidneys as stones. Faulty processing of milk thus has resounding negative effects throughout the entire system.

This is why it was always recommended to drink milk on an empty stomach, and never after other foods, because by the time the first food is thoroughly 'concocted' the milk will already have spoiled.[6] Spanish physician Francisco Nuñez de Oria suggests that it is specifically nobles and grandees who err on this very point, eating milk products at the end of a meal, and thereby ruining their digestion.[7] Because of its cold and lubricating nature, milk can also draw other foods down to the bottom of the stomach before being completely digested.[8] This still 'raw' chyle would pass into the liver, bloodstream and eventually the entire body, and never being properly refined, would also never be assimilated into the body, thus offering no nutritive value whatsoever.

Even more dangerous, according to some authors, was to drink wine in the same meal with milk. Together they coagulate, and can lead to strangulation, once the crass substance makes its way through the bloodstream toward the heart and lungs.[9] For this same reason, abstaining from all other meats when drinking milk is recommended, as is avoiding exercise. Violently shaking the stomach contents can accidentally separate the different parts of milk, hastening their corruption.[10] After milk one should remain perfectly still and upright.

Several authors also suggest that many individuals have a particular aversion to milk, and for such people it can even be venomous.[11] They cite the immediate gag response, and point out that many phlegmy foods cause nausea and loathing. This is a good indication that they are unsuitable for our constitutions, and can never offer suitable nourishment.

Nonetheless, if one can avoid these numerous dangers, milk is still a very nourishing food. The fifteenth-century authority on dairy products, Pantaleone da Confienza, points to some earlier confusion about its specific qualities. As a nourishing food, logically it should be hot and moist: and the sweet flavour of milk is an indication that it should be categorized among the blood-promoting foods. This makes perfect sense, since milk is itself generated from blood. Moffet even called it 'white blood'. But by the authority of Galen, milk is a cold and moist food, watery and tending to increase phlegm in the body. Faced with this contradiction, the Arabic author Rhazes declared milk to be temperate humorally. Avicenna corrected his predecessor by explaining that blood is indeed hot and moist, but after nourishing the body of the mother it emerges in the form of milk as cold and moist, but retaining the elements suitable for nourishing infants.[12]

This confusion seems, however, to have persisted in Western nutritional literature, because one of the most frequent uses cited by medical authors for milk was to counteract melancholy. It would indeed moisten the dry body, but its cold qualities would only damage the already cold melancholic, particularly since such people are already prone to generate phlegm accidentally. Thus when Marsilio Ficino recommends all milky foods to combat melancholy, he is bending the logic of humoral correction.[13] And many authors followed him in this prescription. Take, for example, Andrew Boorde, who claims that milk is not good for the sanguine, but good for melancholics, old men and children. His countryman Thomas Cogan recommends it specifically for students, for whom melancholia is a 'common calamity'.[14] This opinion runs counter to most physician's advice that milk is particularly harmful for students because it offends the head and eyes.[15]

One of the only authors who seems to have thought out logically exactly which complexion should use milk was Antonius Gazius. He specified that milk is definitely not for phlegmatics, but is really well suited for adolescents and hot and dry bodies who would be moderately corrected by a cold and moist food. The only real danger here is that in excessively choleric bodies the milk may burn up and turn acidic, exactly as it would in a pot. But at least he thinks out the humoral logic thoroughly.[16] What a confused reader might have thought about all this contradiction one can only guess, but it may well have ended in despair of ever being able to follow the physician's advice.

There is no less confusion within dietary literature about what kind of milk is best to use. Some prefer goat's milk, others cow. The only point of agreement on this topic is that human milk, because closest to the human body itself, is most easily assimilated, and therefore the most nourishing. Hiring a wet-nurse for infants was obviously a common practice for wealthier households, but finding willing subjects for adults must have posed greater problems, especially since all doctors insist that milk be imbibed at the source. Platina suggests that milk from a healthy, beautiful, young woman of temperate complexion is best.[17] Platina was the Vatican's first librarian, and one can easily imagine the long line of clerics waiting for the cure. In fact, he suggests that human milk should only be used in small quantities for medicinal purposes, implying that some people would have preferred to use it as regular food.

That it was in fact used frequently is evident from numerous comments throughout the literature. For example, Thomas Cogan remarks, 'Common experience prooveth that Womans mylke sucked from the brest, is without comparison best of all in a consumption'.[18] He also notes that the old Earl of Cumberland was cured this way and afterward engendered the present Earl, who was George Clifford, one of Queen Elizabeth's favourites. Of course, as with choosing wet-nurses, one had to be very circumspect regarding the character and habits of the donor. These are passed on directly into the humours of the consumer, and materially alter the emotions. The great medical humanist John Caius was to learn this first hand. 'What made Dr. Cajus in his last sickness so peevish and so full of frets at Cambridge when he suckt one women (whom I spare to name) froward of condition and of bad diet; and contrariwise so quiet and well, when he suckt another of contrary disposition? verily the diversity of their milks and conditions, which being contrary one to the other, wrought also in him that sucked them contrary effects.'[19]

Comparable medical advice is also offered by Ficino, 'Choose a young girl who is healthy, beautiful, cheerful and temperate, and when you are hungry and the Moon is waxing, suck her milk; immediately eat a little powder of sweet fennel mixed with sugar. The sugar will prevent the milk from curdling and putrefying in the stomach; and the fennel, since it is a fine friend of the milk, will spread the milk to the bodily parts.'[20] Note, he too emphasizes a cheery disposition. In one of the more bizarre and gruesome passages in the dietary literature, Ficino also suggests that in lieu of a girl, the blood of a young boy will suffice.

For those content to use animal milk, the dietary literature also abounds in contradiction. To begin with, it was considered important to understand the relationship of the various components of fresh milk. The buttery parts were usually considered hot and moist qualitatively, while the parts that are solid and cheesy are cold and dry, and most importantly in substance and texture are crass, earthy and difficult to digest on their own. The qualities of the watery part, or whey, were a matter of debate. Because whey is clearly nourishing on its own, the Arabic authority Mesue concluded that it is hot and moist. Avicenna's rejoinder placed it in the cold and humid category, which he argued, makes milk relatively tempered, because all three opposing components tend to balance each other out. Nonetheless, since the watery part predominates quantitatively, milk is ultimately a cold and moist food in total.[21]

The three components of milk could also be described in Paracelsian terms. The butterfat corresponds to the sulphurous element and is associated with flame, the cheese is saline comprising the earthy part, and the whey is mercurial. This scheme actually just substitutes chemical terms for older humoral ones, and obviously involves no actual chemical analysis.[22]

All this was of central importance in appraising the merits of milk from different animals, because clearly some contain a greater proportion of fat, and others are more serous and watery. Maimonides considered pig's milk to be best because of the animal's anatomical similarity to humans.[23] How the Jewish philosopher was able to test his theory, one can only guess. Most authors do agree that cow's milk contains the greatest proportion of fat and is therefore the most nourishing. Goat's milk is the most watery, and sheep's milk lies somewhere between the two. Deciding which of these is best for drinking fresh is less clear though. For robust bodies who are able to digest crass foods, cow's milk would be preferable. For delicate bodies goat's milk, even though less nourishing, is superior. Typically, this rationale is left unexplained though, so in some authors there is a simple statement claiming that goat's milk is always best, and cow's worst. For example, Hugo Fridaevallis offers a jingle to remember: 'Quod praestat? Caprae. post? ovis. inde? bovis.' (What's best? Goat. After? Sheep. Next? Cow.)[24] In other authors, exactly the opposite claim is made. Baldassare Pisanelli ranks them in the following order: human, cow, sheep, goat and lastly buffalo.[25] Many authors, following their Greek and Arabic sources, also mention ass's milk which is the most serous of all and is used to treat dry and emaciated bodies.[26] Averroës even claimed that next to human milk, ass's is best. Some authors also mention mare's milk and camel's milk, though few presumably had the opportunity to try the latter. Both were considered extremely watery and dangerously cold. At any rate, the subject is again fraught with controversy.

Apart from the differences among species, the dietary authors are also surprisingly attentive to the importance of quality pasturage, and even the mood of the animal. In a passage that has a remarkably modern ring to it, Pantaleone insists that happier cows produce more and better milk.[27] A cheery disposition actually warms the milk, tending to render it more temperate and nutritious than milk produced by melancholy cows. According to the Greek Aëtius, animals fed on lowland pastures and near water tend to be colder and moister.

Mountainous pastures are far preferable, and milk from such animals is converted more quickly in the body to laudable blood.[28] The mountainous herbs, being well aerated and warmed by the sun, are better suited for conversion into warm humours in the animal's body, and hence the milk they produce is also more tempered. Only mountainous regions, especially in the north, that are too cold are unsuitable for dairy herds. Milk produced in the spring tends to be more nourishing than other seasons as well because the animals themselves are better nourished.[29] Animals that are well exercised also produce warmer and better quality milk. All these considerations are obviously meant to counteract the naturally cold and moist faults inherent in milk.

For the very same reason, milk from younger animals was thought to be hotter and moister. Pantaleone even goes so far as to suggest that certain outward signs offer clues about the humoral make-up of the animal, just as in the art of physiognomy a person's character and complexion can be judged by the face, body size, and hair colour. White cattle tend to produce colder and more humid milk which is less nourishing. Milk from black cattle is naturally hotter. Red or brown cows and goats produce more tempered milk.[30] Larger animals tend to be fattier, and are thus more humid than small muscular ones. The odour of an animal's flesh can also be taken as an indication of the milk they produce. Pantaleone considers the milk of dogs, wolves and bears fetid for this very reason. Pigs, because they are such voracious and indiscriminate eaters, produce equally foul milk.[31]

Another major topic of concern was a proper way to correct milk's dangerous qualities, tempering it with hot and dry condiments to make it more suitable for human consumption. Following Galen, the most typical correction recommended was salt, sugar and mint. According to Gazius, the salt or honey is added to prevent coagulation. Salt also makes it descend quicker and as a preservative it helps the milk to be properly processed before putrefaction sets in.[32] Buttes counsels his readers that 'a litle afore you take it, put into it some salt, sugar, or hony, least it curdle in the stomacke'. This also renders the milk moist and temperately hot.[33] What is strange, however, is that often all of these correctives are recommended together. Sugar or honey alone appears to have been the preferred correction among nobles or at least that was the impression of Giorgius Pictorius,[34] but only one author, Menapius, suggests that salt ruins the flavour.[35] A far more thorough corrective was suggested by Paul of Aegina, and would have much the same effect as pasteurization: the milk should first be boiled with an easy fire, then seethed on a hotter fire. The milk should then be skimmed and any burned bits on the vessel should be sponged away. Lastly salt and sugar are added.[36] Tobias Venner takes this correction one step further by adding pepper as well.[37] Another common corrective was to cook the milk with millet or rice, which absorbs the superfluous humidity and thus reduces the tendency to cause gas, although it does also becomes harder to digest prepared this way.[38] As a particularly gross food, this kind of porridge can easily clog the body's passages causing kidney stones.

Other preventative measures might also be taken to avoid the dangers of milk. Frequently, dietary writers recommend that after drinking milk one should carefully wash out the mouth with wine or brush the teeth. For reasons which are not entirely clear, milk was thought to damage the teeth and gums.[39] Of course this runs counter to the modern idea that calcium is especially good for teeth and bones. The apprehension in this case appears to have been a concern that corrosive residues left in the mouth might rot and loosen the teeth. Venner, once again, takes extra precautions and warns his readers to gargle with wine or strong beer and rub the teeth thoroughly after using milk.[40]

Far more perplexing are two sets of recommendations which seem to be completely opposed. On one hand, there are often admonitions against using milk for people prone to headaches. Vaughan cites a common saying: 'Dare lac aut vinum febricitantibus et capite dolentibus, est dare venenum.'[41] (To give milk or wine to the feverish and headache prone is to give venom.) This makes perfect sense. In such a hot body, the milk would burn and send noxious fumes into the head. Presumably following this same logic, Hollings warns all studious people to avoid milk because it harms the head and eyes.[42] On the other hand, many authors say that milk is very good for the brain and eyes, and even for augmenting sperm production in men. The logic here is that a similarity of substance facilitates conversion. As Gratili explains, 'In milke and egges there is great abundance of fat and clammie moisture, the one appropriate to aliment and nouriture, the other to sperm or seede of generation.' He continues that milk is also especially good for the brain, and is thought to increase its volume marvellously. Thus milk is good for students. Similar comments are found throughout the literature.[43] While a distinction can be drawn initially between headaches and those hoping to increase brain size, after popularization and simplification these comments seem completely contrary. Once again, a lack of precision and consistency among the medical authorities probably left the reading public hopelessly confused.

In conclusion, although milk is cited in Western dietary theory as among the most nutritious foods, the strident warnings and numerous variables and conditions under which it should be consumed, probably left serious readers petrified for their lives at the very thought of drinking milk. Or, as is just as likely, those accustomed to it probably went on blithely ignoring the medical advice. That this was the case is evident from a few very interesting comments within the dietary literature. Spanish-born Ludovicus Nonnius, displaced in the Spanish Netherlands, remarks that 'in our age many Northerners consider milk a delicacy, as the Belgian people can abundantly testify'.[44] He seems to have found this strange and worthy of note. Similarly, Frenchman Charles Estienne associated milk drinking with the Scots and Irish and attributes their ferocity to this fact. Alsatian Melchior Sebizius thought milk drinking a peculiarity of the Swiss and Dutch.[45] The same distinction that opened this paper serves as its conclusion. Certain Northern European peoples, with acquired lactose tolerance, remained impervious to medical warnings. In the South, and even in England, authors were extremely ambivalent, and a certain proportion of lactose intolerance among these populations probably lent empirical support to theories inherited from antiquity. As these ideas were elaborated over centuries and eventually popularized, even corrupted over time, the confusion generated left milk with a reputation for danger. It may be no coincidence that milk drinking among adults remains a rarity in Southern Europe to this day.

REFERENCES

[1] Jared Diamond, *Guns, Germs and Steel* (New York: W.W. Norton, 1997), pp. 195–214.

[2] Hippocrates, *Regimen* II. LI, in *Hippocrates*, vol. IV, translated by W.H.S. Jones (Cambridge, MA: Harvard University Press, 1967), p. 325.

[3] Galen, *De alimentorum facultatibus*, in Opera omnia, vol. VI, edited by C.G. Kuhn (Hildesheim: Georg Olms Verlagsbuchhandlung, 1965), pp. 681–689.

[4] According to the Hippocratic *Regimen*, the last age of life was cold and moist and requires corrective hot and dry foods. In Galenic texts, following Aristotle, the body naturally grows colder and drier and requires hot and moist foods, but accidentally may become cold and moist and is then corrected with hot and dry ones. In either case it is not entirely clear why milk, which is cold and moist, would be a suitable corrective. See Richard Palmer, 'Health, Hygiene and Longevity' in *History of Hygiene*, edited by Yosio Kawakita, Shizu Sakai and Yasuo Otsuka (Ishiyaku EuroAmerica: 1991); Peter H. Niebyl 'Old Age, Fever, and the Lamp Metaphor' in *Journal of the History of Medicine*, Oct. 1971, pp. 351–368; Thomas S. Hall, 'Life, Death and the Radical Moisture' in *Clio Medica*, vol. 6, 1971, pp. 3–23.

[5] Thomas Moffett, *Health's Improvement* (London: Thomas Newcomb, 1655), p. 124. Commenting that milk is bad for the teeth: 'there is no greater enemy to them then milk itself, which therefore nature hath chiefly ordained for them, who never had or have lost their teeth.'

[6] Melchiore Sebizius, *De alimentorum facultatibus* (Strasbourg: Joannis Phillipi Mulbii & Josiae Stedelii, 1650), p. 701.

[7] Francisco Nuñez de Oria, *Regimiento y aviso de sanidad* (Medina del Campo: Fransisco del Canto, 1586), p. 162. Condemning continual use of milk and all things made from milk, he specifies 'y mucho hierran los nobles y grandes en las comer, especialmente comiendo las tales viandas a la postre de otros manjares'.

[8] Bartolomeo Sacchi, called Platina, *De honesta voluptate*, edited and translated by Mary Ella Milham (Tempe, Arizona: Medieval and Renaissance Texts and Studies, 1998), pp. 158 and 159.

9 Sebizius, p. 701. Antonius Gazius, *Corona florida medicinae* (Venice: Ioannes & Gregorius de Gregoriis, 1491), fol hvi. '... nec etiam post ipsum bibatur vinum quam alteret ipsum et coagulat et penetrare facit indigestum ut iuniores medici sentire videntur.'

10 Thomas Elyot, *The Castel of Helthe* (New York: Scholars Facsimile Reprints, 1937), p. 34; Symphorien Champier, *Rosa Gallica* (Paris: Iodoco Badio, 1518), fol. LVIII. According to Symphorien, Avicenna recommended rest after drinking milk, but never sleep. The stomach contents must remain upright to digest this particularly dangerous food.

11 Ioannes Domenico Sala, *De alimentis* (Padua: Ioannes Baptista Martinum, 1628), p. 129. 'Multos autem videmus, qui singulari quadam proprietate ita a quibusdam alimentis abhorrent, ut perinde ab ipsorum esu laedantur atque a venenis laethalibus, aliqui a caseo, alii a lacte...'; Moffett, p. 124, if one naturally loathes milk, 'it cannot possibly give him good nourishment but perhaps very much hurt in offending nature'.

12 Pantaleone da Confienza, *Summa lacticiniorum* (Lyon: Antonium Blanchard, 1525), fol. XVI. The first edition of this work appeared in Turin in 1477, where Pantaleone was physician to the Duke of Savoy. This was a region in which understanding the medical use of milk products would have been especially in demand.

13 Marsilio Ficino, *Three Books on Life*, edited and translated by Carole V. Kaske and John R. Clark (Binghamton, NY: Medieval and Renaissance Texts and Studies, 1989), pp. 134 and 135. He also suggests that 'It is especially good to drink milk mixed with sugar on an empty stomach, provided the stomach can tolerate it. These moist things are of great advantage to melancholics...' This, at least, is a more moderate compound, the milk being tempered by the sugar. Ficino, p. 159.

14 Andrew Boorde, *The wisdom of Andrew Boorde*, edited by Edmund Pool (Leicester, Edgar Backus, 1936), pp. 46 and 50 where he specifies cow's milk for melancholics; Thomas Cogan, *The Haven of Health, Chiefly made for the comfort of students* (London: Thomas Orwin, 1589), p. 154.

15 Edmund Hollings, *De salubri studiosorum victu* (Ingolstadt: Typis Ederianus, per Anream Angermarium, 1602), p. 35. Both Cogan and Hollings writing on the exact same topic offer completely opposite advice.

16 Gazius, fol hv[vo].

17 Platina, *Le grand cuisinier de B. Platine de Cremonne*, translated by M. Desdier Christol (Paris: Jean Ruelle, 1586), p. 38.

18 Cogan, p. 154.

19 From Moffett, as cited by Waverly Root, *Food* (New York: Simon and Schuster, 1980), p. 258.

20 Ficino, p. 197.

21 Pantaleone, fol. XVI[vo].

22 William Vaughan, *Directions for Health both Naturall and Artificial* (London: Printed by T.S. for Roger Jackson, 1617), p. 71.

23 Pantaleone, fol. XVII.

24 Platina, *Le grand cuisinier*, p. 38vo; Hugo Fridaevallis, *De tuenda sanitate* (Antwerp: Christopher Plantin, 1586), p. 146.

25 Baldassare Pisanelli, *Trattato della natura de' cibi et del bere* (Venice: Gio. Alberti, 1586), p.128–9

26 Joseph Duchesne, *Le pourtraict de la santé* (Paris: Claude Morel, 1606), p. 435–437. The author mentions that whey is used in Italy to purge adust or burnt humors and melancholy, and also that in mountainous regions it is consumed in place of water and made into soup.

27 Pantaleone, fol. XVI[vo]. 'Quis dubitat lac animalis iocundi laudibilius esse lacte animalis tristabilis...'

28 Gulielmo Menapius, *De ratione victus salubris* (Basel: Bartholomew Westheimer, 1541), p. 524.

29 Iacobus Sylvius, 'Consilium perutile adversus famem et victum penuriam' in Jean Liebault, *Thesaurus sanitatis paratu facilis* (Paris, 1577), p. 38 'Omnium delicatissime et citissime nutrit qui ex lacte verno.'

[30] Pantaleone, fol. XVIII. 'Varias animalium complexiones qui nobis maxime manifestat ex colore pillorum suorum.'
[31] Pantaleone, fol. XVII.
[32] Gazius, fol. hvi.
[33] Henry Buttes, *Dyets Dry Dinner* (London: Tho. Creede for William Wood, 1599), fol. N2^{vo}.
[34] Georgius Pictorius, *Dialoghi del modo del conservare la sanita* (Venice, Vicenzo Valgrisi, 1550), p. 37. 'Gli huomini grandi temperano la malitia del latte con il zuccaro, overo col mele, perche in questo modo vogliono che facilmente s'impedisca la sua corruttione.'
[35] Menapius, p. 525.
[36] This is related by Sir Thomas Elyot, p. 34.
[37] Tobias Venner, *Via recta ad vitam longam* (London: Edward Griffen for Richard Moore, 1620), p. 88.
[38] Gazius, fol. hv^{vo}. 'Multoque magis predicta faciet nocumenta sicut grossis cibariis permixtum fuerit.'
[39] Elius Eobanus Hessus, *De tuenda bona valetudine*, with commentary by Ioannes Placotomus (Frankfurt: Heirs of Christian Egenollf, 1556), p. 66vo. 'Dentibus et gingivis inimicum. Diligenter igitur post usum lactis abluendi.'
[40] Venner, p. 89.
[41] Vaughan, p. 72.
[42] Hollings, p. 35.
[43] Gulielmo Grataroli, *A direction for the health of magistrates and students*, translated by Thomas Newton (London: William How for Abraham Veale, 1574), fols. Iiv^{vo}–Ki^{vo}; Benedict of Nursia, *Opus ad sanitatis conservationem* (Bologna, Domenico de Lapis, 1477), fol. K4; Gazius, fol. hiii^{vo}, 'Lac...augmentum faciens in cerebro.'
[44] Ludovicus Nonnius (Alvares Nuñez), *Diaeteticon* (Antwerp, Petrus Bellerus, 1645), p. 206, 'et nostro aevo Septentrionalium plerique, lac in deliciis habent: quod Belgicae populi abunde testari possunt.'
[45] Charles Estienne (Carolus Stephanus), *De nutrimentis* (Paris: Robert Stephanus, 1550) p. 40, 'Scoti atque Hiberni obsui lactis usum ferocem, robustumque corporis habitum induere conspiciantur'; Sebizius, p. 751, 'Helvetiorum in alpibus degentium praecipua alimenta sunt lac, caseus et butyrum.'

Milk and its Products in Ancient Rome

Joan P. Alcock

An essential part of any modern diet is dairy produce, and the Elder Pliny, writing in the first century AD, implies that cow's milk, though fairly important nutritionally to the Romans (*Natural History* 28 33 124), was more productive in cheese-making. His comment may refer to the fact that milk goes sour very quickly in a hot country (hence the wide prevalence of homogenized milk in Italy today). But Varro *(De Res Rusticae* 2 11 2), writing in the previous century, says that of all liquids, milk is the most nourishing.

If Pliny were correct, the Romans were missing an essential part of their diet, because milk contains useful quantities of vitamins C and D, a high calcium content essential for strengthening bones, small amounts of phosphorous and potassium and a significant amount of protein. Retinol, or Vitamin A, essential to repair surface body tissue, is more present in summer milk than winter milk and this is when milk production in the Ancient World would be more prolific. Milk also supplies amino acid lysine which is missing from a diet based mainly on cereal products (Sheratt 1981, 176).

Celsus, in his first-century medical tract (*De Medicina* 2 18 11), does recommend milk in his lists of 'good' foods and 'strong' drinks. Varro (*De Re Rustica* 2 11 1) thought that sheep's milk was more nourishing, closely followed by goat's milk. Sheep and goat's milk were more highly regarded than cow's milk. Columella (*De Re Rustica* 7 2 2) says that for nomadic tribes which have no corn, sheep provide their diet, hence the Gaetae are called 'milk-drinkers'. Virgil (*Georgics* 3 16) esteemed goats for their plenteous and nutritious store of milk. Goats give a milk yield five times as much in proportion to their body weight as cattle, and four times as much as sheep; in modern France a goat yields about 700 litres a year, an average yield of 2.5–3 litres a day. Varro, in fact, recommends not milking dams after giving birth as they will then yield more wool and bear more lambs.

Milk had some value in the Roman diet. Anthimus (*De Observatione Ciborum* 75), writing in the sixth century AD, commends bread and milk as a treatment for dysentery. Varro (*De Res Rusticae* 2 11 2) speaks of the purging effect of milk. Pliny remarks on its medicinal properties (*Natural History* 28 33 124–9). He gives a list of problems which it could cure: some bizarre, but others not out of place today, such as stomach ulcers and bladder complaints. He recommends it for those who are convalescing or suffering from consumption and in a poor state of health. He also praised goat's milk for being the sweeter and more suited to the stomach than cow's. This may imply that the Romans had some knowledge of bovine lactic intolerance.

There were also milk drinks. Celsus (*De Medicina* 2 29) mentions *lactantia* in a list of foods which move the bowels and this could be a form of invalid food. Apicius (*De Re Coquinaria* 7 13 9) speaks of *melca*, a form of curdled milk, probably like a thin yoghurt. This is similar to Pliny's *oxygala* (*Natural History* 28 35 134), an astringent, coagulated milk, made by adding sour milk to fresh, which has also been suggested to be buttermilk. Paxamos in the *Geoponica* (18 21) mentions a somewhat similar product, but as this is placed in a cupboard to rest undisturbed, it seems more like a junket.

Martial (*Epigrams* 13 38) gives a present of beestings, the first milk drawn from the mother; this is also highly praised by Varro (*De Res Rusticae* 2 11 2). These are rich in nourishment and are part of the diet of nomadic communities as in present-day Mongolia. In that country milk drawn from mares is put into a bag made from leather or sheepskin and stirred with a stick by whoever passes it, often up to 2,000 times. The result is a fermented drink (*airag*). If a similar method was used in the ancient Roman countryside, it would leave no trace.

One problem connected with milk is that if it has not been pasteurized it can be a carrier transmitting disease-causing organisms, especially tuberculosis and brucellosis (McGee 1984,13). Several examples of tuberculosis of the bone have been found in skeletons excavated from Roman sites; three were noted at the Roman town of Poundbury (Dorset) (Greene 1987; Farwell and Molleson 1993, 185) which could have been caused by drinking infected milk. Contamination of milk is easily done. Homer in the *Iliad* (16 641–3) compares the Greeks swarming round the body of Sarpedon to flies on a spring day in a cattle yard buzzing round brimming pails. Milk soon picks up the smell from another substance and is highly perishable, becoming sour, especially in a hot country. Leaving it in direct sunlight causes anti-oxidation giving rise to a burnt flavour. Some of this would have been noted by the Romans, but even if drinking milk was linked to the cause of death, obviously no attempt could be made to identify the actual agent of destruction.

Milk was used as a cooking medium. Apicius (1 8) makes meat sweet by cooking it first in milk, then in water; he soaks liver in a mixture of honey-water, milk and eggs, which can get rid of a bitter taste. Soaking in milk is a common method used by cooks. Milk could also be used as a cosmetic. According to Pliny (*Natural History* 11 96 238), Poppaea, wife of the Emperor, Nero took 500 asses with her wherever she travelled so that she could soak her whole body in a tub of it. Cleopatra was reputed to do the same; likewise some women today, the belief being that it smoothes out wrinkles and keeps the skin supple. Both Pliny (*Natural History* 11 96 239) and Galen (*De Alimentorum Facultatibus* 3146 = 6 638K) mention its use as a skin moisturizer.

Butter

Pliny (*Natural History* 11 96 239) comments that the barbarians have lived on milk for centuries, but that they do not know the blessing of cheese. Among the barbarians must be included the Celts and evidence belies Pliny's words. An Irish poem written in the twelfth century AD, but believed to be part of an earlier tradition, describes a fort in culinary terms, being surrounded by a sea of new milk, and having thick breastworks of custard, fresh butter for a drawbridge, walls of curd cheese and pillars of ripe cheese.

Wooden casks of a fatty substance given the name 'bog butter', some containing as much as 18.15 kg (40 lb), have been found in Irish and Scottish peat bogs (Raftery 1942–26; Raftery 1971). Although some of this may be adipocere, a waxy material formed from animal fat (Thornton, Morgan and Celoria 1970), some is conceivably butter, possibly put into the bog to preserve it. The earliest dated find is from the fifth century AD, but other finds are suspected to be earlier and the custom was known to have existed in the Highlands of Scotland until the nineteenth century (Frazer 1891). The butter is salted, which helps preservation and prevents it from going rancid.

It was this emphasis on animal fats, rather than olive oil, much more easily obtained in, and exported from, the Mediterranean regions (Du Plat Taylor and Cleere 1978) that distinguished the Celtic from the Roman diet, imparting a completely different cultural food experience. The Romans regarded butter more as a base for an ointment. Columella (*De Re Rustica* 6 14 5) recommends treating animal injuries with an ointment which included butter ('goat's fat'), akin to the remedy still used of putting butter on burns to relieve pain.

Pliny maintains that the barbarians considered butter their choicest food, the differing quantities of which distinguished the wealthy from the lower orders. Cow's milk was most commonly used, but sheep's milk gave richer butter. He recommends a traditional method of putting warmed milk in a tall container with a hole pierced close to the stopped mouth. As the milk curdles it is scooped off and salted. There might have been other ways of making butter that have left no literary or archaeological evidence. In Tibet, for example, warmed milk is put into a yak-skin bag, which is swung from a tree branch until butter is formed. Two other methods are to let children kick the bag about or to stir the contents violently with a stick with a club end; in both cases the milk solidifies to butter.

Cheese

Cheese is highly nutritious, with a high energy and protein value, containing vitamins A, D and B_2 (riboflavin). There is 700 mg of calcium and 400 calories to be found in 100 g of cheese. It is also high in saturated fat and, like milk, does not agree with people who are lactose intolerant or who cannot digest it. This fact was known to Hippocrates. In *On Ancient Medicine* (200) he said that there were some people who could eat as much as they liked without any problem, whereas others suffered acutely if they ate even the smallest amount. Celsus (*De Medicina* 2 21) emphasized that hard cheese causes flatulence and constipation; Pliny reiterates this. Even today many people believe that cheese causes heartburn and that it is 'binding'.

Cheese was included in iron rations carried by soldiers. The Emperor Hadrian made a point (when with his army) of observing the conditions of life of the ordinary soldier: in camp he ate bacon and cheese, and drank sour wine. Rustius Barbarus, in a letter found at the fort of Wâdi Fawakhie in North Africa (Davies 1971, 135), thanks his friend, Pompeius, for gifts including bunches of cabbages and a cheese.

Barbarian tribes ate it with great enjoyment. Tacitus writes in the *Germania* (23) about AD 97–98, that the Germans eat plain food – wild fruit, fresh game and curdled milk. The term he uses, *lac concretum*, can be identified as the solid milk formed when milk is left to go sour. This type of milk-cheese, naturally formed by lactic acid, has been a staple product throughout Europe for centuries. Harder cheese is made by putting curds into a container to drain off whey, which has always been a natural drink for humans and, as Cato knew (*De Agricultura* 150 2), is often fed to pigs as part of a fattening diet.

The Romans seem to have preferred cheese made from full-cream sheep's and goat's milk. Columella (*De Re Rustica* 7 2 1), in about AD 65, says not only do sheep satisfy the hunger of country folk with milk and cheese in abundance, but they provide the tables of people of taste with a variety of agreeable dishes. Both he and Virgil esteem goat's milk and Cato (*De Agricultura* 76–81), writing in the second century BC, when commenting on cheese, refers almost exclusively to that made with sheep's milk; ten of his twelve recipes include this product.

Both sheep and goat's milk have a greater concentration of short-chain fatty acids (McGee 1986, 42) and cheese made from their milk is easier to digest because of its smaller milk particles. Goat's milk curdles in about half the time of cow's milk, but though mild when fresh it develops its full flavour in several stages over four to five weeks, but it is these highly odorous molecules which gives the cheese a stronger taste, which some people find unpleasant.

Virgil, in his first Eclogue (1 33–35) written *c.* 37 BC, implies that it is the shepherd's duty to make sure that this cheese must be sold in the town marketplace, with the added implication that it must be freshly produced. Cheese-making often took place in rural areas where it might be more inconvenient to take milk in pails to market. Columella (*De Re Rustica* 7 8 1) sternly directs the shepherd to make cheese in the best possible manner, warning that if cheese is made with a thin consistency it should be sold as soon as possible while it is still fresh. One difficulty might be the production of a seasonable supply of sheep's milk, because the highest yield comes at lambing time, when ewes produce more milk than is needed for their young.

The Roman appreciation of cheese can be noted from the fact that writers give details of cheese-making and discuss the differences between hard (or dry, salty) cheese and soft cheese, which was akin to the modern Italian ricotta cheese. Varro (*De Res Rusticae* 2 11 3), in particular, favours soft cheeses for being more nutritious and less constipating. Pliny (*Natural History* 11 97 240) names several cheeses, even pinpointing the locality of two villages near Nîmes famous for their cheeses.

In particular demand were cheeses from the Apennines, from Sarcina in Umbria and a Luni variety (from the borders of Etruria and Liguria) which weighed up to 1,000 lbs (453 kg). Martial (*Epigrams* 13 30) says it would provide lunches for a thousand slaves. Making a cheese of this weight would be feasible, if somewhat impractical. Modern weights of cheeses are not so great: Parmesan rounds weigh 90 lbs (40.8 kg), and Emmental 240 lbs (108.8 kg).

Athenaeus (*Deipnosophistae* 14 658b) makes the actor Philamon say that while Sicily produces fine goat's cheese, a Tromilic cheese from the city of Achaea, admits no comparison with any other. A Vestinian cheese made in an area close to Rome is the only named cheese in the recipes of Apicius (4 2). Martial (*Epigrams* 13 31) gives its shape as a *massa* (a lump). In another verse he mentions a *quadra* (square) from Tolosa (Toulouse) in Gallia Narbonensis. This may imply that the cheese was a hard one which could be cut into blocks. He refers to the Sarcina cheese as cone-shaped and later compares it to a rounded breast. In modern France the cone shape and pyramid shape are often chosen because these cheeses mature more quickly than round or log-shaped ones.

Cheese-making

Cheese-making in modern times can still be the concern of local producers and this would have been even more so in the Ancient World, although mention of cheese from different localities implies an export trade from some areas, especially to Rome. Celsus (*De Medicina* 2 30) indicates, in a list of foods which 'confine the bowels', cheese 'which is rather strong in taste, either from age or because of that change which can be noted in cheese from across the sea'.

Cheese squeezes found in Britain at the forts of Bainbridge (Yorkshire), Usk (Gwent), Holt (Clwyd) and Corbridge (Northumberland) and the civilian sites of Camulodunum (Essex), Lower Halstow (Kent), Silchester (Berkshire) and Leicester include flat ones with large holes acting as strainers. A bowl-shaped vessel from the Templeborough (Yorkshire) fort with small holes in the base could have strained either curds or honey from a honeycomb; similar vessels were found in Pompeii. Wicker baskets would also let the whey run off. Curd residue, which resembled the modern Italian ricotta cheese, was shaped into cakes or put into a press. Soldiers could do this in the many conical and disc-shaped presses found at the Longthorpe fort (Cambridgeshire). Conical presses, found at the Eccles (Kent) villa, suggest that the occupants were preparing cheese for marketing.

The best vessels for making cheese would be the heavy, rounded, flat *mortaria*, common on Roman sites, which often have grit fired into the base of the interior (Dore and Greene 1977). After the curds had been scraped out, some bacteria would remain on the gritted pieces, thus acting as a starter for the next cheese-making session and avoiding the need for adding fresh rennet. A wide spout on the rim allows easy pouring of the whey. Modern milk bowls in Northern France and the Savoy almost exactly reproduce the shape of *mortaria* and use the same technique. Milk left overnight also acts as a starter, but Columella stresses (*De Re Rustica* 7 1) that milk must not be left to stand for a long time because it can turn sour.

A common process of natural souring is by the addition of rennet, the digestive juices secreted in the stomach of certain animals; this contains an enzyme which helps the milk to clot. Columella (*De Re Rustica* 7 8 1) regarded lamb's rennet as being the best – a small piece the size of a silver denarius was all that was needed. Varro (*De Res Rusticae* 2 11 4) favours that of hare or kid, utilizing a piece the size of an olive. The juice from some plants can also be used, in particular wild thistle, nettles and fig. Palladius (*De Agricultura* 6 9), writing in the fourth century AD, recommends a small piece of skin from a chicken's stomach; presumably the bacteria would instigate curdling, though it might also be a source of disease. Once rennet is added to the milk, it can be left in a warm place to thicken. An ideal spot would be on a shelf in a warm outhouse or even near the fire kindled on the hearth.

To make hard cheese the curds are placed in wicker baskets or a *ficus* (a flexible leather bag pierced for straining) to drain the whey, then pressed with weights in a mould for further drainage. Columella (*De Re Rustica* 7 8 4) says the cheese should be taken out of the mould, sprinkled with salt, and put into a cool, shady place to drain off further liquid and harden. It is pressed again for nine days, then washed with fresh water and set in cool rooms to dry; it is vital to follow this method, otherwise the cheese becomes full of holes, a result of not being pressed sufficiently, and is too salty and dry. Cheese made in this way can be exported.

If a cheese is to be eaten quickly, then it is merely drained, dipped in salt and dried in the sun, which would seemingly produce a creamy cheese with a stiff rind, somewhat like a Camembert. Columella also mentioned a cheese that could be hardened in brine and Pliny directs that a cheese should be steeped in vinegar to preserve it. This would help to retard the mould. Columella suggests that cheese should be sliced, put into a vessel treated with pitch and filled with must (grape juice or thickened wine), then covered for twenty days. He is not too sure about the taste as he adds that although seasoning may be added the product is not too unpleasant by itself. Cheese can be flavoured with crushed pine kernels, thyme and other herbs; in fact Columella says that adding any chosen seasoning can give cheese any flavour. Smoking imparts its own special flavour according to the type of fuel used and Pliny (*Natural History* 11 97 242) says that it increases the flavour of goat's cheese. Some woods and charcoal give an aromatic flavour; Columella (*De Re Rustica* 7 8 7) suggests the use of applewood.

Smoked cheese was produced commercially in Rome. Martial (13 32), whose epigrams were written about AD 84, mentions one street in Rome, the Velabran, on the west side of the Palatine, as famous for producing it. He included eggs baked with some of this cheese in the menu which he had prepared for his friend, Julius Cerealis. He also mentioned that Trebula, in the country of the Sabines, had cheeses 'which were tamed by a moderate fire'.

Varro points out that pasturage on which cows are fed affects their milk, a fact known to Pliny (*Natural History* 11 97 242) who comments that in Bithynia cows which eat on salty pasture produce a salty cheese. This is well known to modern dairy farmers in Cheshire because the salty pastures round the Northwich and Middlewich areas produce some of the best Cheshire cheese.

Large emporia were producing cheese. One villa in Italy, at Gragnano, between Pompeii and Stabia, was mainly concerned with wine production, but the discovery in one room of a large bronze vessel, which could hold a considerable quantity of milk, has been suggested as evidence for a cheese factory. Milk would have been obtained from Mons Lactarius (Monte Lattare); the milk of this area was famous for its richness and medicinal properties. Slaves ran the emporium managed by a *viticus*, but ear-rings and bracelets found in one room indicated the presence of a woman, possibly the wife, who could have been in charge of cheese production, for women seem always to have been traditionally connected with this occupation (Della Corte 1923).

This huge emporium might not be unique. Given the popularity of cheese and the organizing ability of the Romans, commercial cheese-making could have been a viable proposition throughout the empire.

BIBLIOGRAPHY
Classical texts
Athenaeus *Deipnosophistae,* London. Loeb Classical Library, Heinemann, 1928–41.
Anthimus *De Observatione Ciborum,* translated by Grant, M., Totnes, Prospect Books, 1996.
Apicius *De Re Coquinaria,* translated by Flower, B. and Rosenbaum, E. as *The Roman Cookery Book: a critical translation of the Art of Cooking by Apicius,* London, Harrap, 1958.
Cassianus Bassus *Geoponica,* Leipzig, Teubner 1895.
Cato *De Agricultura,* London, Loeb Classical Library, Heinemann, 1934.
Celsus *De Medicina,* 3 vols., London, Loeb Classical Library, Heinemann, 1935–38.
Columella *De Re Rustica,* 3 vols., London, Loeb Classical Library, Heinemann, 1951–55.
Galen *De Alimentorum Facultatibus,* Leipzig, Teubner 1923.
Hippocrates *Hippocrates,* 4 vols., London, Loeb Classical Library, Heinemann, 1923–31.
Homer *The Iliad,* translated by Rieu E. V., London, Penguin Books, 1950.
Martial *Epigrams,* 3 vols., London, Loeb Classical Library, Heinemann, 1919.
Palladius *Opus Agriculturae,* Leipzig. Teubner 1975.
Pliny *Natural History,* 10 vols., London, Loeb Classical Library, Heinemann, 1938–62.
Tacitus *Germania,* London, Loeb Classical Library, Heinemann, 1970.
Varro *De Res Rusticae,* Loeb Classical Library, London, Heinemann, 1934.
Vergil *Eclogues,* London, Loeb Classical Library, Heinemann, 1932.
Vergil *Georgics,* London, Loeb Classical Library, Heinemann, 1932.

Secondary texts
Davies, R. W. (1971), 'The Roman Military Diet', *Britannia,* vol. 2, 1971, pp. 122–142.
Della Corte, M. (1923), 'Villa Rustica esplorata dal Signor Cav. Carlo Rossi-Filangieri in un fondo del commune Agnello Machetti', *Notizie Degli Scavi di Antichità,* 20, pp. 278–280.
Dore, J. and Greene, J .P. (eds) (1977), *Roman Military Pottery Studies in Britain and Beyond, British Archaeological Reports,* No. S30, Oxford.
Du Plat Taylor, J. and Cleere, H. (eds) (1978), *Roman Shipping and Roman Trade. Britain and the Rhine Provinces,* Council for British Archaeology Research Report, No. 24.
Farwell, D. E., and Molleson, T. I. (1993), *Excavations at Poundbury 1966–8), vol. II The Cemeteries, Dorset Natural History and Archaeological Society,* Monograph Series No 11, Dorchester, Dorset Archaeological Society.
Frazer, W. (1891), 'Bog Butter: its history with observations', *Journal of the Royal Society of Antiquaries of Ireland,* vol.1, pp. 583–586.
Green, C. E. (1987), *Excavations at Poundbury, vol. I The Settlement. Dorset Natural History and Archaeological Society,* Monograph Series No.7, Dorchester, Dorset Archaeological Society.
McGee, H. (1986), *On Food and Cooking. The Science and Lore of the Kitchen,* London, Allen and Unwin.
Raftery, J. (1942–43), 'A bog butter vessel from near Tuam, Co. Galway', *Journal of the Galway Archaeological and Historical Society,* vol. 20, pp. 31–38.
——, (1971), 'A bog butter find', *Journal of the County Kildare Archaeological Society,* vol. 15, pp. 17–18.
Sherratt, A. (1981), 'Plough and pastoralism: aspects of secondary products revolution ' in Hodder, I., Isaac, G. and Hammond, N. (eds), *Patterns of the Past. Studies in Honour of David Clark,* Cambridge, Cambridge University Press, pp. 261–306.
Thornton, M. D., Morgan, E. D. and Celoria, F. (1970), 'The composition of bog butter', *Science and Archaeology,* vol. 20, pp. 2–3.

The Cheeses of Hokkaido and other Milky Issues in a Ricist Society

Michael Ashkenazi

With apologies to Augustin Berque.

Japan and the modernization of food

The story of Japan has been told a number of times: how a small, feudal society rose through grit and effort to political and economic prominence. How after several wrong turnings it finally found its way, by emulating its betters, manufacturing excellence and managerial innovation, to the head of the world's premier league. Like all stories, this story is true. To a degree. And it, of course, obscures more than it says.

To put this global story into a more personal perspective, it's useful to illuminate this with an anecdote. Several years ago I was collecting material on changes in food consumption in Hokkaido, Japan's northernmost island. I found myself in an exhibition of Hokkaido foodstuffs, put on by the Hokkaido Agricultural Federation. Among the delicacies (there's hardly a point to going to a food exhibition unless one can *taste* the food) – dried squid, fish roe, hairy crab, smoked salmon, beef – was a display of local Hokkaido cheeses. Among the cheeses were several prominently displayed examples of local Hokkaido Camembert. Now, perhaps the French government would object (in fact, it probably has) to the name, but the cheese itself was excellent: creamy, ripe, with a hint of acid and a measured saltiness. I asked the person at the counter if this cheese sold well, and he was emphatic that it did.

Intrigued, I asked, 'Why? What's the attraction?'

His answer is the first thread that I shall pursue throughout this paper: '*Yawarakai dakara. Sono ue, nihonjin wa minna gurume desu.*' (It's soft. And Japanese are all gourmets.)

East Asian agricultural economies, and thus their kitchens, are rarely based on pastoralism. China, Japan, and Korea's agricultural economies have, historically, been based mainly on grain production, usually rice. This is, to a degree, even true of areas not truly suited to rice culture such as Hokkaido (Berque 1979). And yet, in less than 50 years, the Japanese, at least, have transformed themselves culinarily into a nation that prides itself on its manufacture and consumption of milk-based products. This raises an intriguing question: how does such a change come about?

I will argue two parallel issues in this paper: that the process of modernization opens up 'food niches' in a food culture, into which new products can enter, many (but not all) borrowed from other cultures; and that the way into these niches is a complex dance between consumers, manufacturers, food culture, and taste.

The milk products of Japan

The consumption of milk products in Japan has grown steadily. Japan's milk production in 1997 was over 8,000,000 metric tons. 200,000 tons of cheese were consumed, almost 90 per cent imported (176,862 tons in 1998, part of a steady growth of about 5 per cent annually. [Agra Europe 1999]). Much of the milk is consumed as processed milk: yoghurts (of which more anon), chocolate, cake, cocoa, and ready-made milk tea accounting for much of the production. This means that the average Japanese consumed some seven kilograms of milk or milk products per year. This statistic is of course misleading, both because I've used extremely crude measurements, *and* because this consumption is skewed in various ways, which I shall point out as we proceed.

Hokkaido cheeses

Hokkaido boasts 30-odd producers of cheese. This northernmost island, colonized by the Japanese in the nineteenth century, is largely agricultural. And because of the environment (rice only grows on the southern half of the island) it is much devoted to 'non-traditional' agriculture: milk and meat production, and cold-climate fruit

The cheese producers are generally small, slightly larger than cottage industry size. They sell over the Internet and by mail. They produce Camembert, Emmenthal, and Edam-type cheeses.

Several things are significant: mode of sale, type of cheese, and location. The use of new marketing mechanisms such as the Internet is related to the nature of Hokkaido. The island has long been the haunt of Japanese social fringes. It has something of the aura of Hi Brasil about it: a land where, as one informant put it, even Japanese can be individuals. In other words, even though, to non-Japanese eyes, the differences in conduct, interest, and environment are minor, to Japanese they are quite significant. It is not surprising that much of Japan's milk production is concentrated in two areas: Chiba, near Tokyo, and Hokkaido. Chiba is dominated by the large commercial firms, whereas in the conceptually remote Hokkaido, individualism can flourish.

The types of cheeses made are, in a European sense, quite conventional. Emmenthal and Edam cheeses predominate. The one that attracts my attention, however, is the Camembert-type cheese. On first impression, that would not strike me as a likely cheese for several reasons. It has a strong and distinctive flavour and an unusual textural characteristic (rough mushroomy crust and gooey interior).

There are several features in this cheese's favour from the Japanese point of view, however. First, it is made, the ads claim, by small farmers doing their best to provide naturally-made farming products. As Knight (1994) has noted, this approach has a powerful appeal for the Japanese public, seeking desperately for their gradually loosening attachment to their countryside. It is also made in remotest Hokkaido: most of the farms are in the extreme north and east of the island – Japan's last frontier. The cheeses thus breath adventure and iconoclasm.

At the sensory level too, these cheeses are, surprisingly, well within the Japanese canon of taste.

'The texture is wonderful,' enthused one informant, 'soft and gooey.'

And he went on to enthuse somewhat in the vein of Benjamin Franklin's comment on the lubricious quality of good cheese.

The cheeses are also small, and fit the Japanese preference for small elegant packages. The fermentation and reprocessing of the cheese make it analogous to many preferred Japanese foods: the ubiquitous *miso*, fermented *natto*, and other foods familiar to the average Japanese.

The significance of the cheeses of Hokkaido can be seen at a number of levels: as a marketing success (the Hokkaido Milk Marketing Board is behind many of the marketing campaigns); as a significant change in Japanese diet (though the penetration of these cheeses into the national diet is marginal at best); as a repetition of the Japanese success at imitation; and as a tale of individual success in a society conditioned to collective, rather than individual, effort.

The amazing story of Yakult

The success of Yakult, a yoghurt-type drink, was the first case of an indigenous milk product taking off in Japan. Much is due to the innovative ideas of the founder of the company, Shirota Minoru, who, in 1930, isolated a lactobacillus (*Lactobacillus casei*) which, when used to ferment milk, became the basis of Yakult. In the 1960s, Yakult ladies circulated on their bikes, the back-carriers loaded with tiny (90ml) bottles of the pink-orange coloured, slightly sour drink. These would be delivered to households early in the morning, in packs of five. Children took them to school. Japan probably has more vending-machines in service than any other country (Ashkenazi 1997), and Yakult is available from them too.

Yakult (and its emulators) straddles the boundary between the health drinks popular in Japan and milk-product drinks. Advertisements in Japan, as in the UK (Yakult 1998), make much of the natural process of the 'beneficent' organisms that ferment the milk. As in Hokkaido cheeses, Yakult also benefits from the association of the drink with the plains and mountains of mysterious northeastern Asia, where the Japanese Empire was at its zenith (for older people) and where hardy pioneers could survive individual challenge (for younger ones).

One must ask why this drink (and others, such as Calpis, another milk extract with a similar flavour) took off in a country that did not consume milk? Clearly the nationalistic and mystical played a part. No less, however, were marketing, serving size, and the nature of the drink. Yakult's success was originally almost exclusively due to household delivery. The deliverers themselves were housewives, who took some time off every morning to make their rounds, chat with their neighbours, and emphasize the qualities of the drink. This was pitched, initially, at women and their children: in the lean post-war years particularly, mothers were not unnaturally concerned about their children's nutrition. I would assume that delivery by a comfortable, perhaps even familiar neighbour tended to reassure these housewives and relax the prejudices against milk products: a strategy that may have been copied from the government's similar marketing of condoms in the 1950s.

Part of the reason for the popularity is the association of the drink with health (Hentzepeter 1998). Many Japanese, particularly males, drink a tonic of some sort at least once a day. These may be extracts of ginseng, vitamin C, algae drinks, or caffeine. The choice is large and growing. Whether or not Yakult is as beneficial to health as is claimed is a moot point. What is undeniable is the large market in Japan (as in other East Asian countries) for 'elixirs'. I deliberately use the historical term here. The Daoist search for immortality and sexual potency (the two issues are related in Chinese, and therefore in Japanese ideology) has always involved a search for the elixir of life. 'Health drinks' sold in small (90ml) bottles are a feature of every kiosk and many vending machines in Japan and Korea. Yakult fits this 'health drink' niche very well. Company marketing sheds an interesting light on sexual relations as well: the Yakult company also manufactures Toughman, a health drink presumably designed to attract men,[1] to complement Yakult which, as its advertising in Japan and its mode of sale indicate, is aimed primarily at women and children.

Japanese cultural preference in meals is for a multitude of small, distinctively shaped and coloured items. Yakult – sold in small bottles of distinctive shape and colour – fits this preference. Yakult, indeed, does feature in a number of *ekiben* (railway packed lunches) for children. It is this combination of small size and distinctive shape which helps make it desirable.

Finally, is the issue of taste. One of the major flavour issues is that of fats. I'm not aware of any research on the subject, but from a subjective impression, the (im)proper use of fats in cooking and in foods is a major issue to which we should turn our attention. A liking for fats is, presumably, a biological preference in the species *Homo*. But which fats? The smell of a fat cooking or frying in one culture can make people from another culture feel queasy. The difference between the fat in a British cut of beef and a Japanese one is prominent: a good cut of British beef must have a solid layer of thick fat, and red, fat-free interior, whereas for Japanese, the best beef is marbled, or laced throughout with a thin tracery of fat. Men in Japan will eat certain fatty *getemono* (odd meats, such as boar and bear) as an energy pick-me-up. Yakult, though a milk product, is, or appears to be, very low on fat. The slightly sour taste cuts the impression of fat even further. Thus there is little resistance, even among traditional Japanese (who would not normally drink milk, for both cultural and physiological reasons) to the drink, notwithstanding its milky origins.

Raisin butter

Butter has been a staple of the Middle East, South Asia and Europe virtually since the domestication of sheep and cattle. It is a convenient way of preserving milk products, high in energy, and useful in a number of proto-industrial contexts. It is, however, not a component of East Asian, let alone the Japanese, diet. In fact a term of derision for foreigners was (and perhaps still is) *battakusai* (stinking of butter).[2]

I was introduced to Japanese butter by a friend at his home. The butter itself came in the shape of a small log, about 4cm in diameter and 15cm long. As butters go, it was reasonably good butter I suppose. It was served in slices, on a plate, with toothpicks to serve oneself. The butter was dotted with raisins, somewhat like a miniature roly-poly pudding. We drank beer with it. Later I was introduced, in the same way, to strawberry-flavoured butter, butter with nuts, and I have heard of, but not seen, green tea-flavoured butter.

Clearly, in the context of Japanese cuisine, butter represents something different from the European foodstuff. Of course, butter is used for a variety of purposes in different societies: to spread on bread (with or without Marmite or jam) in Britain, to cook with in the Middle East and India, even as a beauty aid. So there is no reason to view the Japanese use as peculiar. I am, however, concerned here with the *context* of use.

Rather than a spread, or a cooking medium, butter is seen as a food in itself. It is, or may be, consumed in the same way as a slice of fruit, a piece of confectionery, or a piece of dried seaweed. This is a completely different 'interpretation' of what butter is than one most Europeans or South Asians are familiar with. The fact that it may appear as *otsumami* (a snack eaten with alcoholic drinks) offered at bars, is another cue that there is a different perception of the foodstuff.

One can, of course find butter served in the 'conventional' British way in Japanese cuisine. In virtually any coffeeshop in Japan one can order toast: usually a thick slab of bread with a knob of butter on top. The bread appears to be viewed as a confection, and thus it is not the rather thin slice British people will recognize. The butter is the necessary accompaniment, as much a part of the toast as milk is in English tea. Butter is thus fitted in with the vast array of confections and side-dishes – *okazu* – which characterizes Japanese cuisine.

Principles of use: how milk became Japanese

How does this use of milk products come about? Of course, marketing and international relations (Australia and particularly the US have spent decades trying to force the Japanese to open their domestic markets, to, among other things, milk products, though the market is still considered 'undeveloped') play a major part. But we are less interested here in marketing mechanisms than we are in the cultural issue: what makes a strange, unfamiliar set of foods acceptable?

I would argue that several processes within Japanese society have contributed to the success of some forms of milk culture, notwithstanding long-held prejudices within Japanese food culture to the contrary. Each of the examples above appears to be, from a European perspective, an anomaly. The uses, the flavours chosen, and the qualities that are associated with them by the Japanese are not the ones Americans, Europeans, Middle Easterners or South Asians would have chosen.

Japanese are still huge consumers of rice, and the traditional Japanese meal of *ichiju-sansai* (one soup, three dishes) is still the core of Japanese food culture. But quite often one tends to forget that though the Japanese appear to outsiders to be strongly homogenous in various ways, there are strong heterogeneous currents hidden within. What this means in the realm of foods is that there is a vast array of desirable foods that are acceptable within Japanese cuisine. In fact, Japanese have been incorporating foreign and 'strange' foods since early in their history.

Milk became Japanese as a consequence of two major socio-historical features: the meeting with the West, and a rise in affluence.[3] Both of these enabled the Japanese to increase their culinary inventory by the addition of foods that had not only not been available, but were largely excoriated in Japanese society.

It was, I would argue, the positive qualities of these foods (among others) which made them regular items in Japanese diet. These positive qualities included their sensory features (texture, flavour, shape) as well as their social contexts. In other words, it is within the preferences of Japanese society, and the ability of these new foods to find a 'fit' that we should look in order to understand why new foods are accepted.

Niches in food culture

Because food is such a basic aspect of our lives, we tend to 'preserve' as it were, a division of foods into two classes. To use a popular term, we have 'comfort' foods, and other foods. Comfort foods are those we are familiar with, perhaps from childhood. They are the template against which we measure, perhaps unconsciously, the qualities of other foods. This template includes not only the foodstuffs themselves. For example, many East and South Asians have a 'rice hunger' which overrides physical satiety. The template also includes the order in which the foods are served, the preparation modes, and possibly the social ambience. Comfort foods have a wide variation, because, even though they are cultural constructs, they are interpreted differently within the individual household.

'Other' foods are indeed other. The range to which one is exposed, the social surroundings, which either encourage or discourage experimentation (by social pressure, by availability), and individual and collective ideology, create much specific variation. Sometimes broad personal or collective distaste can be overcome by the provision of a specific 'niche' for a particular strange or otherwise unknown/unappetizing food. Given specific circumstances – heavy drinking at a pub – young British men and women eat a variation on Indian food that has only relatively recently become acceptable. Within the confines of 'maleness' (aka laddish behaviour), one can indulge in heavily spiced food that is a far distance from the relatively bland diet of the English middle and lower classes since the Industrial Revolution. Later the same food may spread beyond its niche, due to the media-exposure, greater availability, marketing, and even genuine preference.

In Japan, these culinary niches are very numerous. Partly this has to do with the social and economic history of modern Japan. Rises in affluence, Goody (1982) reminds us, are accompanied by growing demand for exotica and the rare. Affluence has certainly been a feature of Japanese society since the early 1970s. Partly these niches have to do with the compartmentalized nature of Japanese cuisine. A niche, in this instance, is a particular set of flavours, shapes, textures, and contexts, which people find familiar and/or appealing. And these sets are by no means the same as will be found in other cultures.

The centrality of core 'comfort' foods is not challenged by the rise, or the huge number, of 'other' foods. No Japanese informant ever indicated that they felt the need for Yakult – unlike rice, Chinese noodles, cooked vegetable stew, or *sushi* – though younger people have been consuming the drink all their lives. Foods such as Yakult are still 'other' foods. And 'other' foods, however acceptable, even desirable they may be in specific social contexts – entertaining a guest, going drinking, having a party – are by their nature, dispensable with. Some other niche food will always become available.

Conclusions

East Asian society has been 'ricist' for centuries, because environmental economics, and choices made, perhaps, in the Neolithic age, created a situation that people needed to live in. Rice provides high caloric production, demands high human-power input, which leads both to rising population and greater reliance on rice. In practice, Japanese society, like all other monocrop societies, relied also on other crops, both for dietary variety, and to ensure against crop failure.

One would expect that the acceptance of new foods and foodstuffs would be accompanied by some disruption or rearrangement of food culture. As a society modernizes[4] it becomes more receptive to new ideas, including new foods. In the Japanese case, as elsewhere in East Asia, people who can afford it (and in Japan, virtually anyone can) positively seek for new culinary experiences. In Japan, however, they do so on their own terms, and from within their own experience. That is to say two things. First, the essential cultural food basis remains a strong factor in individual consumption: Japan, even its youth, is still a ricist society. On the other hand, due, I would suggest, to the presence, and ideological acceptance of 'niches' within native food culture, 'strange' products can easily make their way. From the European rather conservative view of food, these niches may look somewhat odd, but from the native Japanese point of view, they fit perfectly into the cultural matrix.

That Japanese resist raw milk consumption (which is 'low', according to the US Department of Agriculture), may be partly due to lactose intolerance among adults. But consumption of specifically Japanese forms of milk products is hardly credited because the best approximation in Western terms for such consumer items as Yakult is 'milk products'. It is safe to say that neither Camembert cheese, Yakult drink, nor butter is conceptually or ideologically the 'same' product as consumed in other food cultures.

BIBLIOGRAPHY

Agra Europe 1999, 'Record Japanese cheese imports in 1998/99', *Agra Europe* 1856: M/5.

Ashkenazi, Michael 1997, 'The Canonization of nature in Japanese culture: Machinery of the natural in food modernization', pp. 206–221 in Pamela Asquith and Arne Kalland (eds.), *Japanese Images of Nature: Cultural Perspectives.* Richmond: Curzon Press.

Berque, Augustin 1979, 'The Rice Fields of Hokkaido; Les Rizières de Hokkaido', *Sociologia Ruralis* 19(2-3): 148-159.

Goody, Jack 1982, *Cooking, Cuisine and class.* Cambridge: Cambridge University Press.

Hentzepeter, Vincent 1998, 'Functional foods: A global trend', *Elsevier Food International* 1: 76-77.

Knight, John 1994, 'The spirit of the village and the taste of the country: Aspects of rural revitalization in present-day Japan', *Asian Survey* 34(7): 634-46.

Yakult Honsha Co. Ltd 1998, *Yakult Company Profile.* Tokyo: Yakult Co. Ltd.

REFERENCES

[1] These health drinks are almost always sold in identical small brown glass 'medical' shaped bottles with a twist-off cap. Advertising heavily features men engaged in strenuous activity flipping off the cap.

[2] The only time I was actually accused of butter usage to my face was from a very proper elderly Japanese lady who was being either extremely, and politely, abusive, or extremely innocent! 'Don't foreigners always cook with butter?' she asked sweetly, when I offered to cook a dish. She was promptly rebuked by a shocked 'Grandma!' from her eight-year-old granddaughter.

[3] A similar phenomenon can be seen in Korea, albeit several decades after the Japanese. In 1976, foreign foods were almost not to be found in Korea. In 1994, almost two decades later, there was almost nowhere they could *not* be found.

[4] A much abused and disputed term. For the purposes of this paper, modernization is a situation in which caloric needs of an entire society are met and large caloric surpluses are available for all segments of the population; and in which there is great social differentiation and heterogeneity.

How the Bengalis Discovered *Chhana* and its Delightful Offspring

Chitrita Banerji

In most countries, food varies widely by region and each region tends to be famed for a particular element or category of food. In India, most people today would agree that the Bengal region (originally signifying the province of West Bengal and the country of Bangladesh) excels in the taste and variety of its sweets – in particular, milk-based sweets. Of these, the sweets made from *chhana* are unique to the region. Nowhere else in India does the confectioner work such magic through manipulating the substance derived by cutting milk with acid. And since both vegetarians (the majority of people in the Indian subcontinent) and fish- and meat-eaters can relish sweets, Bengal's *chhana*-based concoctions have long been famed outside the region. Two sweets, in particular, *sandesh* and *rosogolla*, are practically synonymous with the sweet-toothed Bengali, whose long-standing reputation as an indolent, easy-going, comfort-loving gourmet readily evokes the English phrase 'mild as milk'.

But what is the reason for the pre-eminence of *chhana* in this eastern corner of the Indian subcontinent? And why is it not associated with sweet-making in the rest of the country? The answer lies in an encounter between two races – historic, yet forgotten by most Bengalis today. It began 500 years ago, when the Portuguese explorer, Vasco da Gama, landed on the western coast of India in 1498.

From Portugal to Bengal in eastern India, it is a distance not only of several thousand miles, but also of climate, topography, and terrain. The connection between the two is neither obvious nor memorable. Britain – not Portugal, France, or Holland – became the dominant colonial power in India, once the East India Company had cemented its hold over Bengal following a decisive victory in the 1757 Battle of Plassey. Traders and fortune hunters from these other European countries, however, had been coming to India long before the establishment of Britain's control. Some formed small settlements that bear their imprint to this day – the Portuguese enclave of Goa on the western coast of India and the French enclaves of Chandannagar and Pondicherry on the east. But it was the Portuguese settlement in Bengal, so hazy in the regional memory, which made a revolutionary contribution to the region's food universe – as can be seen from trawling the byways of the past.

Although Portugal does not head the list of European countries in terms of gourmet cheeses, it does have several unique varieties. In the introduction to her book, *The Food of Spain and Portugal*, Elisabeth Lambert Ortiz talks about their excellence. Most are made of sheep or goat's milk, but cow's milk is also used. She describes the 'innumerable fresh cheeses, *queijos frescos* … made into little cakes about 9 centimetres (3 ½ inches) in diameter. When mature, they are firm with a strong flavour. When fresh, they are soft and spreadable… [They] are all good with bread or biscuits and fruit as a cheese course, or at any time as a snack.' The importance of *queijos frescos* in the Portuguese diet is demonstrated by the migration of the product. The cheese can be found in the refrigerated food section of many speciality shops in American cities with large Portuguese communities.

Shift the scene to modern Bengal. One of the curiosities available in Calcutta's New Market (Hogg Market in the early days of British colonialism) is 'bandel cheese'. It comes in the form of little cakes of fresh cow's milk cheese, remarkably similar to the kind mentioned by Elisabeth Lambert Ortiz. But ask the shopkeepers why the cheese is called 'bandel' or what its origins are, and they are likely to be stumped. Certainly most people buying the cheese are not aware of any possible connection between this product and the Portuguese traders who followed Vasco da Gama and settled in large numbers in Bengal during the sixteenth and seventeenth centuries. Nor does the average consumer realize that the numerous sweets made from *chhana* that she or he loves are, in a sense, the siblings or cousins of this same 'bandel cheese'.

Each time the Bengali rolls his or her tongue around the spongy juiciness of a *rosogolla*, or revels in the delicate graininess of a *sandesh*, the forgotten encounter between two races comes to life. For the Portuguese not only contributed the comparatively obscure 'bandel cheese' to the gourmet Bengali's platter, their distinctive way of processing milk also initiated a whole new flowering of the Bengali culinary imagination.

Milk: beliefs, myth, and lore

In medieval European physiology, the four humours – blood, phlegm, choler, and bile – were thought to be the constituents of the human body. Their respective proportions determined an individual's character, mood, and health. Similarly, in the ancient Indian system of Ayurveda, the body is dominated by one of three elements – *sattva, rajas,* and *tamas* – each imparting specific characteristics. And extending the notion that we are what we eat, the ancients ascribed those same characteristics to different foods. In this hierarchical universe, milk (including its derivatives, *ghee*, buttermilk, cream, yoghurt) is easily in the top bracket. It is the purest of edibles whose quality is *sattvika* (descriptive form of *sattva*) – nutritive, agreeable, conducive to serenity and spirituality. Sages and ascetics, who left all worldly ties behind and isolated themselves in the wilderness in search of higher metaphysical truths, subsisted on milk provided by local devotees. Milk was the one food that would not induce worldly desires or distractions in their minds. This belief in the semi-sacred quality of milk is also reflected in its consistent use as an offering to the gods. Rice pudding, for instance, is one of the commonest items offered to important household deities like Lakshmi, the goddess of wealth and prosperity. Even outside of Hinduism, milk retains its connotation of purity. The first food with which the Buddha broke his long fast after achieving Nirvana (enlightenment) was milk offered to him by a female devotee.

A famous turn-of-the century Bengali food-writer, Bipradas Mukhopadhyay, in his 1906 book, *Mishtanno Pak* (Making Sweets), documents the different types of milk and their specific qualities as set down by the ancients. Starting with the milk of cows, goats, and ewes, the list goes on to enumerate water buffaloes, camels, mares, female elephants (!) and women as acceptable sources of milk for human consumption. Cow's milk, as one would expect, is defined as second only to human milk in its wide-ranging benefits. Whatever the source, all milk was believed to have several properties in common: tasty, soothing, energizing, sperm-generating, cool, rich, reducing bile and gout, and conducive to phlegm.

Milk was also an important part of the diet of ordinary people. Unlike the ascetic, the householder looked on milk with infinite desire. It was not only health food par excellence, it had powerful symbolic value as an image of achievable prosperity. In agriculture-based Bengal, 'milk-and-rice' became synonymous with the sustenance of a comfortable life. Vegetarians and non-vegetarians alike considered milk a precious food. Both the folk-tales as well as the substantial body of orally transmitted nursery rhymes (*chharas*) of Bengal are replete with images of milk that connote plenty and prosperity. Kings are anointed with milk and butter before their coronation. Princesses bathe in copiously flowing milk. Young girls hope to improve their complexion by rubbing their faces with milk. Mothers who suddenly encounter lost children after many years find their breasts spouting milk even before the children have been identified. Rivers of milk, rippling waves of milk, lakes of milk, trembling layers of thickened milk, even oceans of milk recur with amazing frequency in myths, folk-tales, poems, and songs.

A famous Bengali folk-tale, recounting the adventures of two young princes named Sheet and Basanta (also the words for winter and spring, respectively), demonstrates the extraordinarily vivid presence of milk in the folk imagination. Separated from his older brother Sheet, Basanta was discovered in the woods by a holy man who raised him to be his disciple. One day, sitting under a tree, Basanta heard two parrots talking to each other about a fabulous gem that the king of the elephants carried on his head. Anyone who got hold of it would be able to marry the beautiful princess Rupabati. Immediately, Basanta decided to set off in search of this gem. After travelling for twelve years and thirteen days, he reached the realm of the royal elephant. But a huge, white mountain barred the way. Down its body flowed cascades of milk; its peak was smothered in quivering layers of *sar* (the thick skin skimmed off from the top as heated milk cools down). Basanta climbed to the top and saw on the other side an expansive ocean filled with thick, rich, rosy milk. In that ocean bloomed thousands of fragrant golden lotuses and in their midst frolicked a beautiful, milk-white elephant, on whose head glowed a gem that seemed brighter than all the jewels of the world put together. But as soon as Basanta jumped into the ocean of milk, it became a sandy desert.

A more ancient tale, from the Hindu epic, the *Mahabharata*, describes how the gods and demons got together to churn the ocean of milk in the hopes of obtaining ambrosia which would make them immortal. But after many wondrous treasures, including the sacred pitcher of ambrosia, had been hauled up, a virulent poison, distilled from the venom of all the serpents who ruled the underworld, rose to the surface. And immediately, the ocean of milk was transformed into the expanse of salt water that we mortals know today.

Both these stories illustrate a deep conviction about the fragility of milk. The introduction of an alien element destroys the very nature of the life-giving, life-sustaining fluid. Deliberately doing so, could, therefore, easily be deemed sacrilegious. For many centuries, the importance of that belief was instrumental in limiting the people of Bengal (as well as of the rest of India) in the uses they found for milk and the products they derived from it.

Sweets of yore

The use of milk and the general beliefs about its properties continued unchanged into the medieval period, as can be seen from literary evidence. Medieval Bengali literature consisted mostly of long narrative poems. Some of them are full of descriptions of food that highlight the connection between human temperament and the nature of food and here, too, milk and its derivatives – *ghee*, yoghurt, or buttermilk – exemplify the nobler qualities. In the sixteenth-century *Chandimangalkabya*, written by Mukundaram Chakrabarti, descriptions of meals include those prepared for Shiva the Destroyer and Vishnu the Preserver. Shiva, who is considered choleric and prone to violence, eats food cooked with pungent mustard oil, not *ghee* (clarified butter). Vishnu, on the other hand, is imagined as having a serene temperament and is offered *sattvika* foods including tender vegetables cooked in *ghee*, and a variety of desserts, all derived from milk.

These medieval poems refer to many desserts that ended elaborate meals, and milk was the basis for quite a few. Rice pudding – called *paramanno*, meaning the ultimate rice or best rice, in both Sanskrit and Bengali – was not only offered to the gods but also was a human favourite. It was a feature of most festive meals and many secular rituals. So was sweetened yoghurt, which (along with its unsweetened version) had highly auspicious connotations. But the commonest material for making sweets was *kheer*, that is, milk which has been boiled down and reduced until it is either a thick, viscous liquid (similar to what is sold as evaporated milk in western supermarkets), or a tight, slightly grainy solid. This latter version is often called *khoa kheer* and has the virtue of remaining unspoiled much longer than any other form of milk – an important consideration in a humid, tropical climate.

The evidence is bolstered by another famous medieval work of literature, which was not mythic or fictional, but a biography in narrative verse. The *Chaitanyacharitamrita* by Krishnadas Kabiraj recounts the life and times of the remarkable religious reformer known as Sri Chaitanya. Born in an orthodox Hindu Brahmin family of Vedic scholars, Chaitanya rejected the strict hierarchy and cruel discrimination of the Hindu caste system. Like Jesus, he preached a message of equality, brotherhood, love, and non-violence. The oppressed members of the lower castes followed him in droves and emulated his personal habits, which included strict vegetarianism – a practice not common until then, even among Brahmins, in fish-loving Bengal.

The importance of this biography to historians of Bengali food is immeasurable. When barely out of his teens, Krishnadas Kabiraj became a devout follower of Chaitanya. The latter took him everywhere he went, including the homes of other wealthy disciples. This provided Krishnadas with intimate knowledge of every aspect of Chaitanya's daily life. In the biography, he describes in fascinating detail the numerous elaborate vegetarian meals prepared for Chaitanya in the homes of admirers. A staggering variety of sweets are mentioned, indicating that despite his abjuration of human intimacy and worldly possessions, Chaitanya had a very Bengali predilection for sweets. Many of those were made with puffed, popped, or flaked rice, combined with white or brown sugar and/or *kheer*. Others were concocted from flour, coconut, ground legumes, or sesame seeds. Krishnadas also mentions an impressive array of purely milk-based sweets – *kheer* mixed with sliced mangos, sweet yoghurt, and items like *dugdha-laklaki, sarbhaja, sarpupee,* and *sandesh*.

For those unfamiliar with Bengali food, some explanation of these terms is required in order to appreciate the point made earlier – the taboo on making a deliberate, invasive change to the nature of milk which, clearly, still prevailed at this time. Three of the sweets served to Chaitanya – *dugdha-laklaki* (known today as *raabri*), *sarbhaja*, and *sarpupee* (known today as *sarpuria*) – are mutations of *sar* (mentioned above in the story of Basanta), which is as precious to the people of the Indian subcontinent as cream is to the people of the west. In a tropical region, before the advent of refrigeration, the only way to preserve milk (without making it into *kheer*) was to repeatedly boil it – the notion of pasteurization being still far into the future. A by-product of all this boiling was the transformation of the fatty top layer into a 'skin'. Each time the milk came to a boil, a new skin would form and it would be skimmed off, added to the previous layers and pressed together. These thick layers were used to make *sar*-based sweets. *Dugdha-laklaki* (*raabri*) is layers of *sar* cut into squares and floating in mildly sweetened milk, sometimes flavoured with saffron. *Sar* fried in *ghee* and soaked in syrup became *sarbhaja*. Fried in *ghee*, layered with crushed almonds, *khoa kheer*, and cardamom, and then soaked in sweetened milk, it became *sarpuria*. As for the *sandesh* mentioned by Krishnadas Kabiraj and other contemporary writers, it was sweetened pellets of *khoa kheer*.

What is notable in all these descriptions is that not a single sweet is made from *chhana* (cottage cheese or acid-curd cheese). Modern food historians like K.T. Achaya have discussed the Aryan taboo on cutting milk with acid. It is notable that in all the myths about the young Krishna (a later incarnation of Vishnu), who was brought up by foster parents among the dairy farmers of Brindaban (in the state of Uttar Pradesh), there are thousands of references to milk, butter, ghee, and yoghurt, but none to *chhana*. Even now, the practice of adding acid to make cheese is not to be seen in northern India. Sweets offered in the temples of modern Brindaban (sacred to Krishna worshippers) are invariably made of solidified *kheer*. In making sweets from milk, Chaitanya's medieval contemporaries were therefore adhering to the tradition prevalent throughout the Indian subcontinent at that point in time.

And yet, as mentioned earlier, whenever Bengali sweets are mentioned today, it is the *chhana*-based confections that everyone thinks of. The introduction of *chhana* into the Bengali (Indian) food universe in the centuries following Chaitanya – and its enthusiastic adoption – remains a wonderful metaphor for the enrichment of societies through encounters with the unknown.

Enter the invaders

Aryan Hinduism had its first major interaction with alien cultural norms from the waves of Muslim conquests of India, starting from the eleventh century.

Coming from the central Asian countries and Afghanistan, the conquerors (Pathans and Moghuls) made their presence felt – at least in terms of food – by introducing many kinds of meat dishes. But despite their prolonged domination over the subcontinent, they did not introduce too many new foods. They were not traders, looking for an exchange of goods. They were solely interested in exploiting the riches of India and establishing themselves in a land that was far more fertile and richly endowed than their rough native territories.

It was the advent of European traders which permanently changed many aspects of eating in the Indian subcontinent. Not only was it a case of East meeting West in terms of diet and cookery, it also meant a significant enlargement of the subcontinent's food repertoire. The Europeans who came in search of eastern spices, brought with them the vegetables they had discovered in the New World. The earliest and foremost traders were, of course, the Portuguese who discovered the direct sea-lane from Europe to Asia. For almost the entire sixteenth century, Portugal virtually monopolized this route.

During that time, the Portuguese spread their area of operations along both coasts of India. In the east, they settled in large numbers in Bengal, along the Hooghly River. Initially, they had a fearsome reputation in Bengal, since they used their navigational skills to commit daring acts of piracy along the coast, as well as in the interior where the numerous rivers served as primary conduits for goods and passengers. But unlike the British, who came later, the Portuguese also intermarried with the locals, thus paving the way for a more intimate exchange between the two races. Among the new crops they introduced were tobacco, potato, cashew, papaya, guava, and a host of other vegetables.

Modern compendiums on the cheeses of the world stress the paucity of cheeses in the cuisine of Asia – a fact attributed to the humid tropical climate, which made it difficult to apply the sophisticated preservation techniques needed for the famous cheeses of Europe. Among the few cheeses found in the Indian subcontinent today are the ubiquitous *paneer* (familiar to Westerners through the good offices of Indian restaurants serving dishes like *mattar-paneer* and *saag-paneer*), a couple of varieties from Gujarat, and two from Bengal.

Books like *The Simon and Schuster Pocket Guide to Cheese* and Geoffrey Campbell-Platt's *Fermented Foods of the World* refer to the two cheeses of Bengal as *chhana* and *bandal*. Both are described as 'acid-curd cheeses' made from cow's or buffalo's milk, although no mention is made of how they came into being.

Bandal, however, is pronounced *bandel* in Bengali and a little digging reveals that it is cheese that is made only in Bandel, a town situated 25 miles north of Calcutta on the banks of the Hooghly River. One British writer, in 1679, described Bandel as 'a large town' and 'very populous'. The name derives from the Bengali term *bandar*, meaning port. The Portuguese had originally chosen the nearby town of Hooghly (same name as the river) as their centre of operations. But in 1632, they suffered a serious defeat at the hands of the imperial Mughal army. They then retreated to Bandel, which was, at the time, the chief port on the Hooghly, and formed a second, more lasting establishment. The reason for Bandel's continued popularity as a settlement, among not only the Portuguese but also other Europeans, was its supposed salubrious qualities. Many of them went there to convalesce and recover from the trying effects of the local climate. A report in the *Calcutta Gazette* of September 3, 1799, for example, says, 'Sir Robert Chambers, Judge of the Supreme Court, had gone to spend the vacation at the pleasant and healthy settlement of Bandel'.

Today, the chief relic of this flourishing Portuguese colony is the Bandel Church, the oldest Christian church in Bengal. The present structure, according to some historians, 'replaced an older one built by the Portuguese in their fort in 1599, which was razed to the ground by the Mughal army on the capture of the town in 1632. The present church and monastery are said to have been built in 1660 by Gomez de Soto, who had managed to save the keystone of the old church, bearing the date 1599, during the sack of the town.'

K.T. Achaya documents the establishment of the Portuguese community in Bengal: 'By the second half of the seventeenth century, they [the Portuguese settlers] numbered 20,000 ... with some at Rajmahal. ... They loved cottage cheese, which they made by "breaking" milk with acidic materials. This routine technique may have lifted the Aryan taboo on deliberate milk curdling and given the traditional Bengali *moira* a new raw material to work with.'

Given this well-established presence, the influence of Portuguese cooking techniques on the eating habits of Bengal is not surprising. It was noted by at least one contemporary travel writer. François Bernier, a French doctor, spent seven years in India from 1659 to 1666. He mentions in detail the physical beauty of Bengal and its lush plenitude of grains, vegetables, fish, and meat. He also notes: 'Bengal likewise is celebrated for its sweetmeats, especially in places inhabited by the Portuguese, who are skilful in the art of preparing them and with whom they are an article of considerable trade.'

Although *bandal* or *bandel* cheese is now associated with West Bengal (and found only in a few speciality shops), the process of making acid-curd cheeses found another incarnation across the border in the eastern part of the Bengal region – the famous *Dhakai paneer*. Dhaka, the capital of modern Bangladesh, was known as Rajmahal during the sixteenth and seventeenth centuries when the Portuguese began settling in Bengal. *Dhakai paneer*, as described in *The Simon and Schuster Pocket Guide to Cheese*, is made from cow's or buffalo's milk, or a mixture of the two. It is drained in wicker or bamboo baskets, pressed, and dried for about two weeks, before being smoked. Wedges of salt, placed in the middle, help preserve it and lend sharpness to the taste. The cheese is eaten plain, or sliced and fried gently in clarified butter, or even added to legume and vegetable dishes. Both *Dhakai paneer* and *bandel* cheese, however, remain speciality products and not common items for regular consumption in any part of Bengal.

Neither, however, can be made without curdling milk with acid. It is the solid separated by curdling, that the Bengalis called *chhana*, which has found such wide application in Bengali sweet production and left an entire region indebted to the Portuguese. The etymology of the term is rather obscure, but according to several major Bengali dictionaries, it is a case of a verb becoming a noun. *Chhana* is related to another verb, *shana*. Both mean kneading vigorously by hand to create a fine paste or dough. The naming of *chhana* seems based on the fact that all *chhana*-based sweets require the curdled milk solid to be kneaded first. In fact, the excellence of many sweets depends on the right degree of kneading and often the reputation or status of a Bengali *moira*, or traditional confectioner, depended on his success in achieving the right consistency of kneaded *chhana* respective to the sweet being made from it. It is also of interest that the word *chhana* has a separate meaning in colloquial Bengali – children or offspring. And if one considers curdled milk to be the offspring of untreated milk, then this is indeed a serendipitous example of *double entendre*.

The moira *doesn't eat* sandesh

In one of the stories of *The Book of Thousand Nights and One Night*, a beautiful female slave called Sympathy the Learned is brought to the court of the Khalifa Harun-al-Rashid and quizzed by a series of scholars and wise men on different branches of knowledge. In answering the questions of a doctor about the treatment and prevention of disease, Sympathy says that gluttony is the cause of all disease. To avoid gluttony, one has to divide the belly into three parts – one to be filled with food, one with water, and one with nothing at all so that the body has room to breathe and the soul can lodge comfortably.

Whatever one may think of the efficacy of this charming formulation (which is not so far from modern directives of health), the image of the Bengali *moira*, or traditional confectioner, is that of a man at the other end of the spectrum. A mountainous figure with a ballooning middle, he gives the impression that all three parts of his belly are more than full. He's a famously sedentary character, sitting all day in front of his stove, surrounded by huge containers filled with *chhana* and *kheer* from which he concocts the infinite variety of sweets that are synonymous with gourmet eating in Bengal. Clad only in the traditional white *dhoti* from the waist down, he leaves his torso bare except for a red and white checked towel (*gaamchha*) flung over one shoulder and used frequently to mop his sweating face and neck. The aphorism about the *moira* never eating *sandesh* (Bengal's most favourite sweet) is supposed to indicate a gluttony that has resulted in absolute satiety.

It should be noted that the *sandesh* today is a totally different animal from the one offered to Chaitanya and his medieval contemporaries. The latter, as mentioned above, was made from sweetened, solidified *kheer*. Since the dryness of the *kheer* made it easy to preserve, Bengalis developed the custom of carrying some *sandesh* with them whenever they visited someone. The term *sandesh* also meant news, and the sweet therefore became the perfect offering for someone who comes bearing news and is interested in getting the host's news.

Chhana has different consistencies. As Achaya notes, 'mild precipitation of milk using whey yields a soft but perishable *chhana* product, while the use of lime juice yields a gritty one which sets to a hard, long-lasting product'. It is hard to determine exactly when the term *sandesh* came to indicate a sweet made with *chhana* rather than with *kheer*. But it is reasonable to assume that it had become common usage by the latter half of the nineteenth century. Today, the simplest Bengali *sandesh* is called the *kanchagolla* (literally, 'uncooked ball'), that is, hot, sweetened *chhana* formed into round balls. The term *kancha* (uncooked) does not indicate a lack of processing by heat after the milk has been curdled. The *chhana* is actually tossed lightly with sugar over low heat and the mastery of a *moira* is indicated by the complexity of texture he can achieve despite the shortness of the cooking/processing time. Some of the best *kanchagolla* is available from sweet shops around the famous Kali temple in Calcutta's Kalighat neighbourhood. They are generally given as offerings to the goddess. Once they have gone through the ritual of offering, the devotees are free to eat and enjoy these soft yet grainy rounded shapes served on disposable plates of *shaal* leaves, their milky flavour curiously enhanced by the slightly smoky odour of the leaf container.

In more elaborate incarnations, the *chhana* for *sandesh* can be pressed, dried, flavoured with fruit essences, coloured, cooked to many different consistencies, filled with syrup, blended with coconut or *kheer*, and moulded into a variety of shapes (including those of elephants, conch shells, and tigers). Fancy confectioners in Calcutta or Dhaka even take proprietary pride in a particular shape or flavour of *sandesh* as their particular invention. The Bengali obsession with this sweet is indicated in the flights of fancy displayed in the naming of different kinds of *sandesh*. Bipradas Mukhopadhyay, writing in 1906, lists more than twenty names, based on the form, content, consistency, and flavour of the *sandesh*. Among the most memorable are: *abaar khabo* (I want to eat it again), *pranohara* (robber of my soul), *manoranjan* (heart's delight), *nayantara* (pupil of my eye), and *ahladey putul* (pampered doll). In the heyday of colonial rule, the British, too, were honoured, as indicated by names such as 'good morning' and 'Lord Ripon'.

Chhana-based sweets in Bengal (including both West Bengal and Bangladesh) are too numerous to enumerate in full. But some of the most famous deserve mention. Next to *sandesh*, the *rosogolla* is the best known as the representative sweet from Bengal. The most obvious characteristic is that of being soaked in syrup (*ros*). The other is the exquisite smoothness of the *chhana*. There is no room for graininess in a good *rosogolla*. A variety referred to as the 'sponge *rosogolla*' (sic) is considered the best. Other syrup-soaked sweets made from *chhana* include *chamcham, pantua, chhanabora, chhanar jilipi, rosomundi, golapjaam,* and *kalojaam*. Sometimes the fame of a sweet is tied to the name of a place, as in the '*kanchagolla* of Natore', the '*monda* of Muktagachha', and the '*chamcham* of Tangail'. Sometimes it is the name of a shop that serves as a label of excellence, as in the *sandesh* from Bhim Chandra Nag or the *rosogolla* from K.C. Das.

One interesting point of speculation is why these sweets are rarely made at home, for the true subtlety and excellence of most Bengali foods (vegetables, legumes, fish, meat, and sweets not made from *chhana*) is generally to be found in home cooking. According to the eminent Bangladeshi historian, Professor Abdur Razzaque, it was the professional sweet-makers – many of them Muslims – rather than the home cook, who were originally responsible for the creation of all these *chhana*-based sweets. This has some credibility when one considers that Hindu households, with all their practices of ritual purity, might initially have resisted the invasion of the kitchen by a substance made by the Portuguese 'heathens'. It was thus left to the professionals, working in a shop and not a home, to experiment with a new medium of gustatory delight. The practice, once started, still prevails. And capitalizing on the enduring Bengali addiction for the varied, delightful offspring of *chhana*, sweet shops flourish in every corner of a city or town in West Bengal and Bangladesh. The home cook need only step out to buy any *chhana*-based sweet she or he fancies and therefore, has little incentive to invest time, energy, and skill in making them.

The arrival of Vasco da Gama in India signalled many pivotal developments in the history of Europe and Asia, including the sad scourge of colonialism. But today, in the combined glow of a post-millennial and pre-millennial light of enquiry, the Bengalis, at least, can savour a very special sweetness that is the gift of the Portuguese.

ACKNOWLEDGEMENTS
The paper benefited greatly because of the many stimulating conversations the author had with Mr. Radha Prasad Gupta, a noted expert on old Calcutta and Bengali food history, who died in March 2000.

BIBLIOGRAPHY
Achaya, K.T., *Indian Food, A Historical Companion.* Oxford University Press, Delhi, 1998.
Banerjee, Amiya Kumar, *West Bengal District Gazetteers: Hooghly.* Government of West Bengal, Calcutta, 1972.
Birmingham, David, *A Concise History of Portugal.* Cambridge University Press, Cambridge, 1993.
Campbell-Platt, Geoffrey, *Fermented Foods of the World: A Dictionary and Guide.* Butterworths, London, 1987.
Carr, Sandy, *The Simon and Schuster Pocket Guide to Cheese.* Simon and Schuster, New York, 1981.
Mathers, Powys, tr. from original French version by Dr. J.C. Madrus, *The Book of Thousand Nights and One Night.* Routledge, London, 1996.
Mitra, Sudhir Kumar, *Hooghly Jelar Itihash o Bangasamaj* (Bengali), Vol. 2. Calcutta, 1963.
Mitra-Majumdar, Dakhhinaranjan, *Thakurmar Jhuli* (Bengali). Mitra & Ghosh Publishers Pvt. Ltd., Calcutta, 1996.
Mukhopadhyay, Bipradas, *Mishtanno Pak* (Bengali). Ananda Publishers, Calcutta, (Reprint of original 1906 edition) 1981, 1991.
O'Flaherty, Wendy Doniger, tr. from the original Sanskrit, *Hindu Myths.* Penguin Books, London, 1975.
Ortiz, Elisabeth Lambert, *The Food of Spain and Portugal.* Atheneum, New York, 1989.

Touloumotyro: Centuries Old and About to Die

Rosemary Barron

Why, I wonder, should I, or anyone else, feel a sense of unease at the imminent disappearance of a cheese, in particular, a cheese that is by no means 'world-class'? Is it because I once took this cheese – *touloumotyro* – for granted, during the long periods I've spent on Crete; now, when I'd like one for us all to share, I've had to have it specially made? Or maybe it's because similarly-made cheeses were mentioned in the classical texts? Or that *touloumotyro* is, to many older Greeks, simply 'cheese' – one of our greatest foods? Perhaps I enjoy *touloumotyro*'s sharp tang because it demonstrates above all man's ingenuity – skin used as packaging? Or is it a fond food-memory of a time when I was younger, perhaps? A feeling of nervousness as to the future of our foods? Or is all this consideration just esoteric nonsense?

I do know that I have always been fascinated by the foods or dishes that have been with us since antiquity, and that I feel a sense of sadness that another generation will not have these links with the past; *touloumotyri* represent the oldest and most traditional of all the cheeses made in Greece, thus of those belonging to our own classical past. But perhaps future generations will have something far better to eat, and *touloumotyro* should be allowed to rest in peace?

Cheese, and a little boy's wish

Every week, just before the war, a little boy in Piraeus used to wish hard. There is, of course, nothing unusual in a little boy making wishes, but it was what this particular little boy – Manolis – wished for, that made me write this paper.

Manolis began wishing the moment his mother sent him to buy the week's cheese. As he walked to the cheese shop through the busy, noisy streets of Piraeus, he wished harder. By the time he reached his destination, Manolis was praying. For his instruction was to buy 'cheese' and in this instance – summer, in pre-war Greece – 'cheese' meant *touloumotyro*.

Among many Greek villagers, it is still the custom to refer to a cheese by its name only when it's a seasonal speciality. After all, 'everyone' knows which cheese is at its best at any given time, therefore the word 'cheese' is sufficient description. During the short autumn and spring seasons, however, when there may be several different fresh cheeses available, even Greeks recognize that it could be confusing to tell someone that you wanted 'cheese'. *Touloumotyro*, however, for older Greeks today, remains 'cheese', and according to descriptions of cheeses in texts of the past, *touloumotyro* has been 'cheese' for a long time.

My first introduction to this great food of classicism was literally accidental – I tripped over a low mound of stones straddled alongside a Cretan mountain path, knocking the stones, and what appeared to be an old and battered grey rugby ball underneath them, down the mountainside. A shepherd who had witnessed my clumsiness shouted awful insults at me; my friend and I thought it prudent to scramble down the mountainside and retrieve the old rugby ball, which was obviously the source of the problem. We suspected that it may have had some emotional value for the old man.

By the time we returned to the path, the rugby ball had begun to smell rather 'high' in the hot sun, and the shepherd had calmed down; he invited us to supper.

Later that evening, the magnificent centrepiece on the shepherd's table was the old rugby ball. Next to it was a bowl of shrivelled brown-black olives, a small clay flagon of deep-green olive oil, a mound of misshapen, green-streaked tomatoes, a huge round loaf of dark brown bread, and a crock of red wine.

It was what happened next that made me understand the story my friend, Manolis, told me years later. The shepherd (Vassilis) took his hunting knife and slashed the rugby ball. Inside the, now rather smelly, old leather pouch was a pure white mass of cheese of a texture I have seen neither before nor since; towards the centre, the cheese was soft and creamy; nearer the skin, lumpy within the softer mass. What caught my eye though were the tiny threadlike trails, mostly close to the outside skin. Manolis would have known exactly what they were… On the days he had had the misfortune to reach the cheese shop at the wrong moment, the piece of cheese he would be given would be lumpy – a texture he didn't like. If he were really unfortunate, he would have been given cheese from close to the pouch, a place more likely to hold any worms. His father told him that these worms were a sign of a good cheese – a story Manolis believed – as we all do when told such stories in times of hardship – right up until a few years ago.

Luckily for the spirit of the evening, it was not obvious to me exactly what caused the 'threads' in the magnificently transformed rugby ball. The cheese, as would be expected in high summer, was very salty, but its rather acidic bite was perfectly matched by the intense sweetness of the tomatoes, the chewy bread made from local mountain greens, and the earthy musty little olives. Few meals have tasted better.

Touloumotyro – *a description*

A type of *touloumotyro* has been made throughout Greece since pre-classical times, the choice of skin, or enclosing packaging, depending on custom and availability. The skin of black piglets is preferred in parts of Crete, goat elsewhere. I suspect, without proper evidence, that pigskin, being short-haired, is less likely to cause worms, and its distinctive colour means that it doesn't turn that rather unpleasant grey of my own first *touloumotyro*.

The skin is boiled, and 'cured' in a strong salt-brine, to clean it, then stretched out in the hot sun to dry. Before filling, it is soaked once more in brine, then turned inside out so its smooth side is outside; it remains porous.

Touloumotyro is made throughout winter and spring, when milk is plentiful. Its final flavour depends on the feed of the goats, sheep or cows contributing the milk, as well as the way each skin is cured and the length of time the cheese ripens. The flavour of a *touloumotyro* made in the mountains is very different to one made on the islands or in the fertile valleys, where cow's milk is used, or added to goat's milk, to give a generally milder cheese.

In the past, almost every village household made its own *touloumotyro*, and always with a natural rennet, sea salt and whole milk. Once the skin is filled, and sewn up, the 'packaged' cheese is either hung over the rafters of an outhouse or, in the case of 'my' cheese, stored in a stone 'cheese-house' in the field. Every so often the cheese-maker (either the man or woman of the house) would give the skin 'bag' a squeeze, thus helping to distribute throughout the cheese any brine remaining from the skin-cure, as the porous skin accelerates the ripening process.

After two or three months the cheese is ready for the more delicately-tongued among us – fresh, slightly salty, slightly acidic, and pleasantly pungent in flavour, a gorgeous creamy-white colour, and its creamy texture such that worm holes are less obvious. After four or five months, its flavour changes to tongue-tingly acidic and highly salty, its colour sometimes to off-white/greyish, and its texture to lumpy around the edges.

A mystery surrounds the successful processing of *touloumotyro* as, unlike many other simple 'farmhouse' cheeses, it keeps for several months, a fact that undoubtedly explains its longevity on Greek tables. One possible explanation is the ritual squeezing of the cheese whilst ripening – this could distribute some useful enzymes (and the salt) contained in the skin throughout the cheese.

Touloumotyro – *an enjoyment*

The best *touloumotyro* has the texture of a spread, and is eaten this way on strong, flavourful bread. Bread and fresh tomatoes detract from its saltiness, and *throumbes* olives (the olives which fall naturally from the trees, and which a farmer keeps for himself) provide a powerful taste-sensation. *Touloumotyro* is often 'watered-down' with good yoghurt, to create a dish my neighbour on Crete always referred to as 'cheese in a temper'. She said it took a lot of beating to make it taste the way she liked it. It is a versatile cooking ingredient – delicious crumbled over bakes, wonderful as a filling in pies or as an ingredient in cheese bread.

Milk and its By-products in Ancient Persia and Modern Iran

Najmieh Batmanglij

The use of milk and its by-products have become so prevalent that we often forget that milk is a liquid secreted by the mammary glands of female mammals for nourishing their newborns. In traditional Iran, there are many expressions defining a man's character based on what kind of milk was used to raise him. At one end of the spectrum, a good man is said to have been raised on his mother's milk, while an aggressive man is said to have been raised on wolf's milk. Indeed, the legend of Rome's founders, Romulus and Remus, would seem to bear out this characterization of she-wolf's milk!

Use of animal milk for human nourishment in Persia must have come very soon after the domestication of the sheep and goat, east and south of the Caspian Sea, around 9,000 BC.[1] Cows weren't domesticated until much later, around 6,000 BC. The Persian epic *Book of Kings* tells the tale of how the evil King Zahhak discovers that his downfall is destined to come at the hands of Fereydoun, a hero being raised by a shepherd on cow's milk. Zahhak does not find the infant Fereydoun but does locate and kill the poor cow Barmayeh! The milk of other animals were also popular: the Greek historian Herodotus referred to the milk of horses in his *Histories* (*c.* 430 BC) and the Roman savant Pliny called camel's milk the sweetest in his *Natural History* (AD 77). Much later, the Islamic prophet Mohammad is said to have meditated in the wilderness for 40 days living on goat's milk and dates.[2]

A relief on a sarcophagus at Ur from about 2,500 BC shows an early milking scene. In it, a man is milking a cow while another is seated, rocking a large narrow-necked clay jar as two men strain the liquid from it. The narrow-necked clay jar pictured is known as a *kuzeh* in Persian and is still used in some parts of Iran for making butter and yoghurt.[3]

Ancient Babylonian cuneiform texts on clay tablets deciphered by Jean Bottéro, the French Assyriologist and Cordon Bleu chef, reveal the existence of highly developed culinary arts by at least the time of Hamurabi, around 1750 BC.[4] These tablets described recipes and ingredients for preparation of several meat and vegetable dishes. Many of these, such as the ones for goat-kid stew, tarru-bird stew, and bird pie, used milk or sour milk (yoghurt) in their preparation. Milk and cheese were served to 47,050 guests at a feast inaugurating Ashurbanipal's palace at Nimrud in 668–627 BC.[5]

The Greek writer Polyaenus recounts the quantities of ingredients necessary for a king's dinner in Achaemenid Persia (500–300 BC), which he transcribed from an inscribed bronze pillar looted by Alexander the Great:[6] fresh milk, 10 maries;[7] sweetened buttermilk, 10 maries; rams, 400; oxen, 100; horses, 30; fattened geese, 400; turtle doves, 300; miscellaneous small birds, 600; hares, 300; goslings, 100; gazelles, 30.

Though the ancients had many types of cheese, the modern Persian taste seems to be for the more basic white goat, sheep, or cow's milk cheeses similar to Bulgarian feta. In Iran, the snack of bread, cheese and herbs are a favourite of kings and beggars alike. The Qajar king Nasir al-din Shah not only enjoyed bread and cheese himself, but insisted that everyone waiting to see him also be offered it.[8] C. J. Wills, an English doctor of Her Majesty's Telegraph Company of Persia who spent fifteen years in Iran from 1866–81 tells an interesting story about cheese:

> Cheese too is much eaten for the morning meal, with a little mint or a few onions. The banker at Shiraz, to whom the government moneys were entrusted – a rich man – told me that he or any other merchant never thought of any more elaborate breakfast than these named above. This same man, when *giving a breakfast,* would give his guests twenty courses of spiced seasoned *plats.* It is said of a merchant in Isphahan, where they are notoriously stingy, that he purchased a small piece of cheese at the New Year, but could not make up his mind to the extravagance of eating it. So, instead of dividing the morsel with his apprentice, as that youth had fondly hoped, he carefully placed it in a clear glass jar, and, sealing it down, instructed the boy to rub his bread on the bottle and *fancy* the taste of cheese. This the pair did every morning. One day the merchant, being invited to breakfast with a friend, gave his apprentice the key of his office and a halfpenny to buy a loaf of bread; but the apprentice returned, saying he could not get the door open, and though he had bought his bread, could not eat it without the usual flavour of cheese.[9]

Whereas milk, butter and cheese are the most popular dairy products in the West, in Iran – although milk is given to infants and is used in cooking – it is yoghurt which is the preferred milk by-product. Yoghurt, which must have quickly followed the use of milk, has been a staple of the Persian diet for at least 4,000 years. Yoghurt was no doubt discovered by accident. A well-known story goes: a desert nomad, possibly a Persian somewhere on the Iranian plateau, was carrying some milk in his goatskin canteen. During his journey, heat and the right bacteria (one of several *lactobacilli*) transformed the milk into yoghurt. The nomad probably took his chances and drank the sour mixture. To his surprise, he found a creamy refreshing taste. When he did not get sick, he shared his discovery with his friends. It is interesting to note that yoghurt did not become popular in Western Europe until the 1960s, and then only after it was flavoured with fruit.

The Sassanid Queen Boran (r. AD 629–630) was extremely fond of yoghurt, thus her royal chefs somewhat wisely devised many dishes combining yoghurt and vegetables. These yoghurt-based dishes are still known as *boranis* to this day. A very popular *borani* is that of eggplant and yoghurt.[10] A well-known story by the fourteenth century Persian poet, Obeyd-e Zakani goes:

> One day the king (Sultan Mahmud), who loved eggplant *borani* was eating it to excess, all the while saying how tasty it was. His courtier gave a long lecture about its qualities. Later when the king got a stomach-ache and cursed the eggplant *borani*, the courtier agreed and started to say how bad and harmful it was. 'What are you talking about, you wretch?' said the king. 'You were just singing its praises.'
>
> 'But of course,' said the courtier, 'I am *your* servant, not the eggplant *borani*'s.'

Persian cuisine also uses several yoghurt by-products. The most common is called *kashk*. *Kashk* is drained, salted, sun-dried yoghurt commonly used with eggplants and in soups as a souring agent. Less common but also used as a souring agent is *qaraqorut* which is whipped and cooked whey.

In addition to its various applications in Persian cuisine, from marinades for kebabs to cakes, yoghurt also features prominently in the Persian lexicon. The English name for yoghurt comes from the Turkish word *yoghurmak*, which means 'to thicken'. In Persian the word is *mast*, from the Sanskrit *mastu*. The word is deeply embedded in Persian culture and is used not only in some of the oldest existing recipes, but also in Persian poetry and common idioms. In poetry some of the references are:

I was milk but he trained me and turned me into yoghurt.
—*Boshaq al-Atameh, who wrote a whole divan on food (fourteenth century)*
If a stranger brings you yoghurt, be sure that it will be two-thirds water.
—*Sa'adi's* Golestan *(twelfth century)*
We have had much with yoghurt, not once but many times
—*Sana'I (eleventh century)*

Expressions:
- *Boro mastet o bezan* (go beat your yoghurt): Mind your own business.
- *Moo ro as mast keshidan* (pulling hair from yoghurt): Someone precise.
- *Mast mali kardan* (spreading yoghurt): Confounding and confusing an issue, or covering up something with mischief and trickery.
- *Masta ro kiseh kardan* (bagging the yoghurt): Packing up something naughty upon the arrival of authority.
- *As sefidi-ye mast ta siai-ye zoghal* (from the white of yoghurt to the black of coal): Every which way.
- *Mast-i keh torsh ast as tagharsh peydast* (it is apparent from the jug that the yoghurt is sour): Such a beginning, such an end.
- *Mast-o darvazeh ra mibandand* (closing the yoghurt and the gate): This is a play on words as the word for making yoghurt and closing the gate is the same. Hence the expression means to ensure that the first step is taken to do something properly.

Yoghurt production vocabulary:
- *mastban* (yoghurt maker)
- *mastkesh* (yoghurt distributor)
- *mastabeh* (the liquid drained off from making yoghurt)
- *mastdan* (large clay bowl in which yoghurt is made)
- *mastkhory* (small glass, or china bowls used at table for eating yoghurt)

Regional records of ancient civilizations from Egypt to Iran refer to yoghurt's healthful properties. Indian yogis in 500 BC called yoghurt and honey the 'food of gods.' In Iran, yoghurt's reputed benefits range from prolonging life and remedying baldness, to calming the nervous system, curing skin diseases and soothing gastro-intestinal ailments. It is always recommended for the very young and old alike due to its ease of digestion. Indeed, yoghurt is considered a virtually essential dietary item. The Persian equivalent for the Western expression 'bread and water' (as in 'the prisoners received only bread and water') is 'bread and yoghurt'.

Yoghurt is considered a 'cold' food in the traditional Persian food classification system.[11] In this system, food is categorized into hot and cold, dry and humid, and neutral. 'Hot' food is high in energy and 'cold' food is low in it. These categories do not refer to measurements of temperature or spice content but to energy and specifically defined culinary properties. Often in Persian cuisine one finds that 'hot' and 'cold' ingredients are combined to produce a balanced meal. For example, the popular Persian beverage *dugh* combines yoghurt (cold), with mint and pepper (both hot), to create a refreshing and 'balanced' drink.

This culinary balancing act stems from the broad dualistic philosophy of the ancient Persians which dates back to at least 1500 BC, when the semi-nomadic Aryan tribes descended into Iran and India.[12] The belief is that everything that affects an individual, including climate, seasons, age, and gender, can be counterbalanced through a dietary program. For example, hot and dry foods generally have a robust quality and give strength, while cold and humid foods have calming, enfeebling qualities – and also serve to inhibit the libido. Indeed, in the Caspian region, where much yoghurt and rice (another cold food) is eaten, the men are said to be lacking in passion! People also have hot and cold natures. People with hot natures should eat cold foods while people with cold natures should eat hot foods in order to achieve a good balance.

Whether due to its reputed health benefits, or the attempt to balance hot and cold foods, or even its taste, yoghurt is the most prevalent dairy product in Iran and can be found throughout Persian cuisine. As its popularity increases in the West, people will be seeking new ways to incorporate it in their diets.

RECIPES

Mast-e khanegi. Home-made Yoghurt
Makes 6 servings: Preparation time, 20 minutes plus 12 hours setting time.

2 quarts whole milk (8 cups)
1 cup plain yoghurt

1. Bring the milk to a boil over medium heat in a very clean, non-reactive pot. Dirty or greasy utensils will not produce the desired results.

2. Remove the milk from heat and let stand until cool but not completely cold. The temperature of the milk is very important at this stage. It should not be too cold or too hot. If the milk is too cool, the culture will not grow; if it is too warm the heat will kill the bacteria in the culture. If you are using a thermometer, the temperature should be 115°F/45°C, or with some experience you can test by hand. Put your little finger in the milk and count to 20. The temperature is correct if you can just tolerate the heat.

3. Pour the milk into a Pyrex dish. Pull out the rack in the centre shelf of the oven and place the dish on it.

4. Add the yoghurt to the four corners of the Pyrex dish. Gently push the rack back inside the oven, close the oven door and turn on the inside light. Do not heat the oven. Allow to rest undisturbed at least overnight. (Yoghurt must be stored in a draft-free, protected spot and must not be moved or touched during this period, and your oven is the ideal place for this. As an alternative, cover and wrap the dish with a large towel or blanket and let it rest undisturbed

in a corner of your kitchen for at least 24 hours.) Keep the yoghurt in the refrigerator and use as needed.

VARIATION: Drained or thick yoghurt (*mast-e kisei*). Pour the yoghurt into 3 layers of cheesecloth or a cotton bag, pull the ends together and then hang the bag for 15 to 20 minutes over a large pot to catch the yoghurt drips. All the liquid will slowly drain out, leaving behind a thick and creamy yoghurt.

Panir. Persian Cheese

Makes 6 servings: Preparation time, 20 minutes plus 4 hours to settle and set.

> ½ gallon whole milk
> 1 cup plain yoghurt
> 2 tablespoons salt (optional)
> ¼ cup fresh lime juice
> Cheesecloth for straining the mixture
> ½ tsp salt
> 1 tablespoon plain yoghurt

1. Pour the milk into a large pot. Bring to a boil over low heat.
2. Add the yoghurt, 2 tablespoons salt, and lime juice to the pot of milk. Mix well and simmer over low heat for 3–5 minutes, or until its colour turns yellowish.
3. Line a strainer with three layers of cheesecloth and place the strainer in a large container. Remove the pot from the heat and immediately pour the milk-yoghurt mixture over the cheesecloth. Let drain for several minutes. Save the liquid in the large container for step 5.
4. To remove excess water from the milk-yoghurt mixture, bundle the free ends of cheesecloth together over the mixture to enclose it. Place the cheesecloth in the centre of the strainer and place a heavy weight on top of the cheesecloth. Allow to stand for about 2 hours, then remove the weight. Place it in a container and refrigerate for 1 hour to set thoroughly.
5. Remove the cheesecloth and place the cheese in a sterilized jar. Fill the jar with the strained liquid from step 3 and add ½ teaspoon salt and 1 tablespoon yoghurt. Refrigerate until ready to serve. Serve the cheese with bread and fresh vegetables and herbs.

Kashk.

Kashk is a drained, salted and sun-dried yoghurt.
This recipe was given to me by Mr Parsa.
Makes ½ pint: Preparation time, 2 days: Cooking time, 4 days in a warm oven.

8 cups non-fat yoghurt
½ tablespoon salt

1. Make yoghurt as for the yoghurt recipe or use commercial plain yoghurt. Store yoghurt for 2 days at room temperature for it to become sour. Place the soured yoghurt in a blender and mix for 5 minutes, until it reaches a smooth consistency. Boil the yoghurt until thick, then drain through two layers of cheesecloth. Tie up the corners and allow the liquid to drain from the yoghurt into a bowl.

2. Transfer the yoghurt to a Pyrex dish. Cover it with another glass and place it in a warm oven (140–150°F/60–65°C). After few hours check the amount of condensation on the lid. No condensation indicates that the temperature is too high. Slight condensation means the temperature is correct. Adjust the oven setting as needed.

3. Leave the dish covered in the warm oven for 3–4 days; do not disturb it.

4. Remove the dish from the oven and uncover it. Observe the colour and scent: the contents should be yellowish and have a distinctive smell.

5. Line a strainer with 2 layers of cheesecloth. Pour the contents of the container over the cheesecloth. Tie the ends and allow to drain for several hours so that the excess water drains into a large pot. See *Qaraqorat* below.

6. Remove the contents from the cheesecloth and place in a food processor. Add ½ tablespoon salt and mix in food processor for another 3 minutes to obtain *Kashk*.

7. Place *Kashk* in an airtight jar and keep in the refrigerator for 2 weeks, or store in separate small plastic bags in the freezer and use as needed.

TRADITIONAL METHOD: Add water to diluted whole yoghurt. Then pour yoghurt into a goatskin canteen. Shake for a long time to separate the fat from the yoghurt solids. Boil the yoghurt solids, drain and then dry in the sun to create *Kashk*.

BLACK WHEY (*Qaraqorut*): Cook the drained liquid from step 1 over high heat for 20 minutes. Whip and continue cooking until it reaches a consistency of thick, black paste, which is extremely sour and has many nutritional properties. It is used for soup, fish, and sometimes sauces.

Borani-e bademjan. Yoghurt and Eggplant Dip

Makes 4 servings: Preparation time, 15 minutes: Cooking time: 1 hour.

2 large or 3 small seedless eggplants
¼ cup olive oil
1 onion, peeled and thinly sliced • 4 cloves garlic, peeled and crushed
1 tsp salt • ¼ tsp freshly ground black pepper
1 cup yoghurt, drained in cheesecloth for 3 hours
1 tablespoon chopped fresh mint, or 1 tsp dried mint
¼ tsp saffron dissolved in 1 tablespoon hot water
½ cup chopped walnuts (optional)

1. Preheat oven to 350°F/175°C. Wash eggplant and prick in several places with a fork to prevent bursting. Place whole eggplants on oven rack and bake for 1 hour. Be sure to put a tray under the eggplant to catch drips.

2. Remove eggplant from oven, let cool, then peel and mash.

3. Heat the oil in a non-stick skillet and cook the onion and garlic until golden. Add the eggplant and mix well. Cover and cook for five minutes over low heat. Add salt and pepper. Remove from heat and let cool.

4. Transfer to a serving dish, mix with drained yoghurt and garnish with mint, saffron water, and walnuts.

5. Place the eggplant dish on a platter. Cut up *lavash* bread and arrange around the dish on the platter. Serve as an appetizer.

Borani-e karafs. Yoghurt and Celery Dip

Makes 2 servings: Preparation time, 10 minutes plus 1 hour's refrigeration.

4 or 5 celery stalks (the inner, more tender stalks)
2 cups drained plain yoghurt
½ tsp salt
¼ tsp freshly ground black pepper
2 tablespoons chopped fresh mint

1. Chop the celery into one-inch pieces.

2. Place celery, yoghurt, salt, pepper, and mint in a salad bowl. Toss well. Refrigerate for an hour.

3. Serve with bread as an appetizer.

Tah chin-e barreh. **Baked Saffron Yoghurt Rice with Lamb**
Makes 6 servings: Preparation time, 70 minutes, plus 2 hours' marinating:
Cooking time, 3–3 ½ hours.

> 2 large onions, peeled and thinly sliced
> 1 tsp salt • ¼ tsp freshly ground black pepper
> 2 pounds boned leg or shoulder of lamb, cut in 2-inch cubes
> 2 ½ cups plain yoghurt
> 2 cloves garlic, peeled and crushed • 2 egg yolks, beaten
> 1 tsp ground saffron dissolved in 4 tablespoons hot water
> 4 tablespoons slivered orange peel with bitterness removed
> 3 cups long-grain basmati rice
> ¾ cup clarified butter (*ghee*) or oil
> 1 ½ tsp ground cumin seed (optional) • 1 tsp *advieh* (Persian spice mix)

1. Preheat the oven to 350°F/175°C. Place one onion, salt, pepper, and meat in an ovenproof baking dish. Cover and bake for 1 hour, drain and allow to cool.

2. In a bowl, combine the yoghurt, garlic, saffron water, and the slivered orange peel. Add salt to taste and marinate the meat in this mixture for at least 2 hours and up to 24 hours.

3. Clean and wash 3 cups of rice 5 times in warm water.

4. Bring 8 cups water and 2 tablespoons salt to a boil in a large pot. Pour the washed and drained rice into the pot. Boil briskly for 6 to 10 minutes, gently stirring twice to loosen any grains that may have stuck to the bottom. Bite a few grains. If the rice feels soft, it is ready. Drain in a large, fine-mesh colander and rinse in 2 to 3 cups lukewarm water.

5. Heat 1 tablespoon oil in a non-stick skillet and sauté the other onion. Preheat oven to 350°F/175°C. Remove meat from the marinade. Combine the beaten egg yolks, 4 spatulas of rice, sautéed onion, the marinade, sprinkle with cumin and *advieh* and mix well.

6. Heat ½ cup butter or oil in a Pyrex baking dish in the oven. Add the mixture of rice and marinade, spreading it across the bottom and up the sides of the baking dish. Place the meat pieces on top, cover with layers of rice. Pour the rest of the butter over the rice. Pack firmly using a wooden spoon and cover with oiled aluminium foil.

7. Place baking dish in the oven and bake 1½ to 2 hours, until the bottom turns golden brown.

8. Remove baking dish from oven. Allow to cool on a damp surface for 10 to 15 minutes (do not uncover). Then loosen the rice around the edges of the baking dish with the point of a knife. Place a large serving dish over the baking dish. Hold both dishes firmly together with two hands and turn them upside down.

9. Serve hot with fresh herbs, yoghurt, and Persian pickles (*torshi*).

NOTE: Rice with chicken can be made the same way. Simply replace the cooked lamb with a cut up uncooked, deboned chicken, with the skin removed and follow the recipe as directed.

Dugh. Yoghurt Drink

Makes 2 servings
Preparation time: 10 minutes

1 cup whole-milk yoghurt
1 tsp chopped fresh mint or a dash of dried mint flakes, crushed
½ tsp salt • ¼ tsp freshly ground black pepper
1 ½ cups club soda or spring water, chilled

1. Pour yoghurt, mint, salt, and pepper into a pitcher. Stir well.

2. Add club soda or spring water gradually, stirring constantly. Add 3 or 4 ice cubes and mix again.

3. Serve chilled.

REFERENCES

[1] Reay Tannahill, *Food in History* (New York: Three Rivers Press, 1983), 26–7.
[2] J. B. Glubb, *Syria, Lebanon and Jordan* (London: Thames and Hudson, 1967)
[3] Shell and white limestone relief, inlaid in shale, from al Ubaid near Ur, circa 2500 BC
[4] Jean Bottéro, 'The Culinary Tablets at Yale', *Journal of the American Oriental Studies* 107 (1987), 11–19.
[5] From James B. Pritchard, ed. *The Ancient Near East: A New Anthology of Texts and Pictures*, vol. 2 (Princeton: Princeton University Press, 1975), 103–4.
[6] A. D. H. Bivar, 'Achaemenid Coins, Weights and Measures', *The Cambridge History of Iran*, ed. Ilya Gershevitch, vol. 2, The Median and Achaemenid Periods (Cambridge, and New York: Cambridge University Press, 1985), 638–9.
[7] A marie was 9.3 litres.
[8] Jakob Eduard Polok, *Persien, das Land und seine Bewohner* (Leipsik: Brookhaus, 1865), 90.
[9] C. J. Wills, *In the Land of the Lion and Sun* (London: Ward, Lock and Bowden, 1893), 172.
[10] See recipe in the appendix.
[11] While milk and yoghurt are considered cold foods, other by-products are not: feta cheese is considered neutral, and whey is a hot food.
[12] This Indo-Iranian philosophy was later developed by the prophet Zoroaster and influenced many other major religions including Hinduism, Buddhism, Taoism, Judaism, Christianity and Islam. Zoroastrianism was based on the dualistic concepts of the opposing principles of good and evil and light and darkness. Zoroaster believed that everything in nature and all events in life have a relationship to each other involving opposing forces. This dualistic concept probably spread to China via commercial contact in the third century BC and was a major influence in forming the entire school of Ying and Yang. It influenced every aspect of Chinese thought including astrology, divination, government, art, medicine and food. Later, in AD 1200, Dogen, the founder of Zen Buddhism, introduced the concept of Ying and Yang to Japan.

Rabbits, Fondues and Physics

Tony Blake

As I live in Switzerland, there were several milky themes I could have chosen for my talk. Since I was born in Wales, I have chosen a title which allows me to discuss not only what has been described as the national dish of Switzerland, cheese fondue, but also that oddity of British cuisine, the Welsh rabbit (rarebit). Since I am also a food scientist, I would like to share my interest in the underlying science of these two dishes.

Fondue has with some justification been described as the national dish of Switzerland and has become an essential feature of winter-sports holidays even beyond this country's borders. It is claimed that the first historic record of a 'fondue' is an incident which took place about the time of the battle of Kappel in 1531. During the Religious Wars the Catholic cantons fought the attempts of the Protestant Zurichois to impose their influence. After the battle, a truce was signed and religious tolerance guaranteed. At one point in the hostilities two groups of enemy soldiers met; one side had only milk and cheese, the other only bread. Agreement was eventually reached to put the saucepan of fondue in no-man's land and the opposing groups shared bread and fondue from the same dish, each telling the other to 'graze on your own land' (*broute sur ton terrain*).

While this historical anecdote refers to a dish of melted cheese, there is, however, no mention of wine. Whether or not this story is true, there is no doubt that today, throughout the Alps of Switzerland, the Haute-Savoie and Austria, groups of tired skiers sit around pans of melted cheese with bread poised on long forks and with glasses of chilled white wine close to hand. This is all a long way from that weary and hungry hill-farmer who first heated some cheese with his wine to make it more palatable. What prompted this culinary invention will probably never be known. We can only speculate that the cheese was hard and by melting it with hot wine it became edible. Maybe the evening was cold and the idea of eating cold cheese with cold wine had little appeal. Whatever the reason, that initial experiment has been repeated many times and probably laid the foundation of the processed cheese industry of today. There are many examples of melted-cheese dishes and possibly all of them originated as a way of using up waste scraps of cheese.

The recipe for fondue is essentially simple. Like many traditional recipes which have been handed down from generation to generation, it has numerous variations, many dos and don'ts and a wealth of advice for the inexperienced. Our Swiss neighbour, an expert on cheese who once owned a cheese shop in Geneva, insists on using a blend of at least four cheeses, together with a carefully selected cheap white wine. As he explains, a good quality wine doesn't necessarily make a good fondue. One might think that taking some grated cheese and melting it with hot wine was a simple process, but for the novice

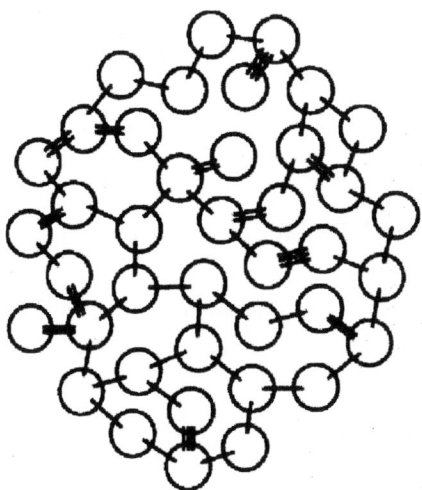

Figure 1. Sub-unit model of the casein micelle. The lines connecting the individual sub-micelles represent bridges of colloidal calcium phosphate. Unconnected surface regions of sub-micelles are rich in ë-casein. (From Dickinson, E., An Introduction to Food Colloids, *Oxford Univ. Press 1992.)*

fondue-maker it is full of pitfalls. The result of the first attempts to make fondue is usually a layer of tough, pasty cheese sitting at the bottom of a pan of hot greasy wine! The luckless experimenter will ask those with more experience. He will be given lots of tips about the variety of cheese, the choice of wine, whether or not to add some cornflour (frowned upon by the expert), or the merit of adding some sodium bicarbonate to lighten the texture. He may try again. More often than not, however, he will remain content only to eat fondue when prepared by someone else or bought from the supermarket (ready-made fondue is increasingly available in a packet). However, fondue-making can be mastered and as with many food products which are emulsions, success is helped by some knowledge of physics and chemistry.

A fondue is more complex than other more familiar emulsions in the kitchen. For example, margarine or butter are fine dispersions of water droplets in a continuous phase of semi-solid fat; and a mayonnaise is a dispersion of oil droplets in a small but continuous phase of water (containing in addition the salt, sugar, mustard, lemon juice or vinegar, etc). A fondue by comparison is a three-component emulsion consisting of the fat from the cheese, the water (and preferably also alcohol) from the wine and kirsch, and the proteins from the cheese. It is these latter which give fondue-makers problems.

The protein in cheese is predominantly casein. This evolved in milk, specifically to provide raw materials for the growing offspring to build flesh and bones; it would be surprising if by chance it also had the ideal characteristics for making fondue! Casein is present in milk as particles (often called micelles) which are roughly spherical and with a diameter of about 100 nanomillimetres (10,000 end to end would extend 1 mm). These casein micelles are in fact composed of four smaller proteins called: á-s1-casein, á-s2-casein, â-casein and ë-casein. These four proteins are bound together with calcium salts (mainly phosphate) to make up the porous sponge-like casein micelle which is not homogeneous; the á- and â-caseins lie within the micelle and it is predominantly the ë-casein which lies on the outside (see Figure 1).

ë-casein has a hydrophilic or water-loving region which protrudes from the surface of the casein micelle into the surrounding water and carries a negative electrical charge which causes the casein micelles to repel each other and prevents them from aggregating together. This is what gives casein its stability in milk. However, during digestion (or cheese-making) the hydrophilic ends of the ë-casein are removed from the surface of the micelle by the action of the enzyme rennin. It is this change which causes milk to form a curd.

The protein in cheese is thus essentially casein micelles – stripped of the hydrophilic, negative outer layer – which have been coagulated into a mass. The calcium is still present in the form of calcium phosphate. This protein/phosphate

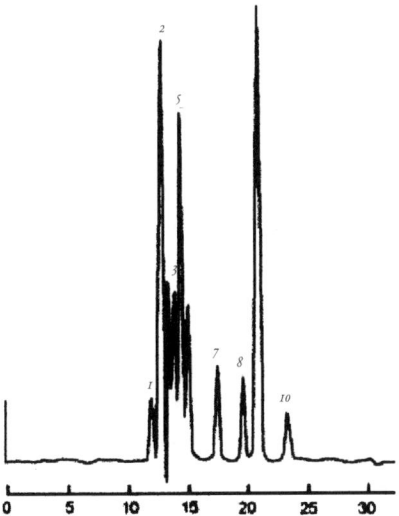

Figure 2. Separation of the acids in dry white wine. Ten acids are identified here. For our purposes the important peaks, reading from the left, are: (1) Citric acid; (2) Tartaric acid; (3) Phosphoric acid; (5) Malic acid; (7) Succinic acid; (8) Lactic acid; (10) Acetic acid. (From Frayne, R.F., Am. J. Enol. Vitic. 37, 181, 1986.)

complex exists in cheese in the form of a viscous gel encasing the dispersed fat. When heated the fat melts, the casein separates out from the excess wine and fat and settles to the bottom of the pan. The fondue has 'gone wrong'.

How does one ensure the protein is dispersed more evenly in a good fondue? There are three factors which are relevant:

- The degree to which the casein in cheese has been broken down and become more dispersible as a result of the action of enzymes from the bacteria which are present during the maturation of the cheese.
- The viscosity difference between the molten casein gel and the wine.
- The rubberiness of the casein gel, dependent on its calcium content.

In the light of these three factors, we can start to understand the reasons behind a good fondue – one that remains homogeneous during cooking – as well as those behind the unsatisfactory fondue – which separates.

By increasing the thickness (viscosity) of the aqueous phase one can physically help prevent casein separation. This is simply achieved by the addition of cornflour or potato starch. In addition to its own thickening effect, the starch also helps emulsify the fat. The increased viscosity of the overall system makes separation of the casein more difficult.

The experienced fondue-maker, however, rejects this technique and relies much more on his knowledge of cheese. My neighbour is very careful about his blend of cheeses and their maturity He is not only picking cheese for its flavour but, without realizing, he is also selecting one which has been subject to considerable bacterial action whereby the casein has been broken down to smaller protein fragments, more dispersible in the aqueous phase. In addition, the soluble proteins are effective emulsifying agents for the fat. This increases the thickness of the aqueous phase. The physical effects are similar to the addition of starch.

Not only is the cheese important, so is the choice of wine. Again, my expert neighbour insists on choosing one which would normally be regarded as relatively poor for drinking: dry to the point of over-acidic, preferably very fruity. What he is doing is picking a wine with a high content of fruit acids, mainly tartaric, malic and citric (see Figure 2).

The salts of these acids (particularly of citric acid) are very effective at removing calcium. The significance of my neighbour's choice is a wine that dissolves the calcium present in the viscous casein gel. Once the calcium is removed from the casein micelles, they break up and can more easily form a disperse and homogeneous mixture with the wine. We can also, incidentally, see some reason for the addition of sodium bicarbonate since this will neutralize the free acids and create the salts which are more effective at removing calcium. No doubt the liberated carbon dioxide will also give a thicker and more mousse-like texture to the product.

So now we have the answers for making a successful fondue: a correctly aged blend of cheese and a wine with a high level of fruit acids. Perhaps it is the difficulty of achieving these two desiderata that leads published recipes to advise the addition of starch.

An alternative strategy, which may even be preferable, is to add extra amounts of those components normally present in wine which dissolve the calcium in the cheese. The most effective choice is citric acid in the form of its sodium salt. Addition of 1–2 per cent trisodium citrate to a fondue allows the preparation of a homogenous fondue even when relatively young cheese is used. It is also possible to make many interesting variations in this way, even alcohol-free, if that is your preference.

A foolproof fondue

400 g cheese of choice • 180 ml very dry white wine
garlic cloves (one at least) • 25 ml kirsch
8 g trisodium citrate

A blend of Gruyère and Emmental or, for a stronger flavour, Gruyère with Vacherin Fribourgeois, gives authentic Swiss fondue. For an English version, equal parts of Cheddar and Cheshire may be substituted.

Cut the garlic clove in half and wipe the cut surface around the inside of a fondue pan. The cheese is grated and added to the wine and other ingredients including the cut garlic cloves. Stir while heating over a gentle flame until fully melted. Serve at simmering temperature with pieces of bread and a chilled white wine.

What is given above is a classic Swiss cheese fondue. There are many possible variants. One, from Provence, contains tomatoes.

Fondue provençale

8–10 tomatoes • 2 garlic cloves
tomato juice (see below)
300 g Comté cheese (a mild Cheddar makes a good substitute)
300 g Emmental cheese
4 teaspoons cornflour • 2 soup-spoons cream
1 soup-spoon capers • 1 chopped onion
pinch of pepper
pinch each of ground oregano, thyme and basil

Cut the peeled tomatoes and garlic and cook until soft. Pass through a blender, sieve and add tomato juice to make up to 330 ml. Slake the cornflour with 4 soup-spoons of the juice and mix with the rest of the juice and the grated cheese in a fondue pan. Cook while stirring until the cornflour thickens then add the chopped onions, herbs and capers. Finally stir in the cream and serve with bread or croûtons.

Let me now discuss a dish from the land of my fathers: Welsh rabbit, which in more recent cookery books is described as rarebit. Older books such as that by Hannah Glasse take the view that Welsh rabbit is simply a piece of toast with a grilled slice of cheese on top, sprinkled with mustard. More recent recipes, however, get closer to a type of fondue. The following is taken from Elizabeth Ray's *Good Housekeeping Country Cooking* (Ebury Press 1979).

Welsh rabbit

60 ml milk or beer • 225 g grated cheese
2.5 ml Worcestershire sauce • 5 ml dry mustard
1 egg yolk • toast

All the ingredients (except the toast and egg yolk) are mixed and stirred over a gentle heat until they melt together. The egg yolk is added and the mixture cooked slowly until it thickens then poured over the toast and browned under the grill.

You should be able to see the similarities with fondue. In this case the mustard provides the thickening which helps to prevent the casein separating from the mixture. The addition of the egg yolk is interesting since this not only helps thicken the sauce but contains lecithin (phospholipids) which are powerful emulsifying agents for the fat and act in exactly the same way as they do in a mayonnaise to disperse the fat in the water phase. I have experimented with the addition of a small quantity of sodium citrate in the mixture and can report that this gives a substantially more stable and smoother consistency.

One question often asked and to which I would like to have a good answer is: why 'Welsh rabbit'? Is the name, as has been suggested by Elizabeth Ray, a pseudo-meat dish as in 'mock turtle' or 'Scotch woodcock', or is it more subtle? The change from rabbit to rarebit is an additional area of confusion. It is clear that the name rabbit is the earlier: being used in an 1801 copy of *The Lady's Assistant* by Charlotte Mason (thanks here to Fritz Blank and his extensive library of food books). *Larousse Gastronomique* (1984 edition) includes the recipe for Welsh 'rarebit'; but adds the unhelpful comment that 'Lapin Gallois' is a translation of a play on words and that the correct name should be 'Morceau de Choix Gallois'. It does, however, add that this British speciality appeared on the Continent in the nineteenth century 'dans les tavernes anglaises'. This link with France is interesting and my wife proposes an alternative explanation for the name: somewhere in France, someone once referred to a piece of toast with cheese sauce on top as 'Le Pain Gallois'. A passing tourist, missing the subtlety of *masculine* bread, went on his way describing it as 'Lapin Gallois'. Whether this is true or not I don't know but what is clear is that our Welsh rabbit with a slice of ham is not very far away from the *croque monsieur* of France. There is an interesting sexist connotation too: a *croque monsieur* with a poached egg on it becomes feminine as a *croque madame* while a Welsh rabbit with an egg on top is made masculine as a buck rabbit or a golden buck.

There is a modern end to this story of fondue. In 1912 two Swiss technicians were working on the problem of making cheese sterile and keepable in hot countries. Up until that time all attempts to pasteurize cheese had essentially failed because of separation problems. Maybe the two were experienced at making fondue or simply enjoyed eating fondue but whatever the case they solved the pasteurization problem by adding sodium citrate to their cheese and were able to put the cheese through a heat treatment to inactivate the bacteria and enzymes. They had invented processed cheese. Independently, and a few years later, Mr Kraft did much the same thing in the USA, except that he used phosphate salts rather than citrate for emulsifying the casein. Thus were cheese slices and a new industry created.

Milk-borne Diseases: An Historic Overview and Status Report

Fritz Blank

'Child suspended and ready for the application of plaster-of-Paris bandage.' From, Tuberculosis of Bones and Joints, by N. Senn (1893).

Within the confines of a large community of human food-borne infectious diseases, only three are considered to be milk-borne: brucellosis, tuberculosis and listeriosis. The aetiological agents of all three of these diseases are totally destroyed by pasteurization. Thus, as we approach the third millennium, the incidence of disease epidemiologically linked to milk or milk-product consumption is small – especially in the United States, where raw dairy products are prohibited from sale by Federal law.

Unfortunately, milk and milk products are not the sole vehicles of the dissemination of these three diseases which, accordingly, still pose a serious public health threat. Note that all three can be transmitted either indirectly by vectors other than milk and milk products, or retrogress by means of the pasteurized (which is assumed safe) product being contaminated after pasteurization. Indeed, tuberculosis kills more people world-wide than any other infectious disease!

Brucellosis

Brucellosis is a disease affecting man, domestic cattle, goats, sheep, swine, and certain wild animals, including bison and caribou (aka reindeer). Brucellosis also can cause abortion in beagle dogs, especially in the US.

The causative agent is a bacteria belonging to the genus *Brucella* – which was named in 1887 for British Army surgeon David Bruce. Bruce isolated an organism from human cases of Malta fever, and classified it (incorrectly) as *Mirococcus melitensis*. Also in 1887, Dr O. Bang, a Danish veterinarian studying the cause of 'contagious bovine abortion', isolated a bacterium and appropriately named it *Bacillus abortus* (later reclassified as *Brucella abortus*). Henceforth, abortive brucellosis in cattle is called Bang's disease. There are several species of *Brucella*, all of which can be transmitted to man. *Br. suis* (swine), *Br. abortus* (cattle), *Br. melitensis* (goats), *Br. ovis* (sheep) and *Br. canis* (beagle hounds).

Other names given this disease in humans include undulant fever, 'intermediate' typhoid, typho-malarial fever, Mediterranean fever, Rock fever, Gibraltar fever, Malta fever, Neopolitan fever, Cyprus fever, and incorrectly as Bang's disease – which is a term reserved for the disease in cattle.

Human symptoms of brucellosis include weakness, fatigue, and an undulating type of fever resulting in a chronic and severely debilitating illness. At one time it was considered to cause more human disability than any other disease shared by man and animals. The pathology and course of brucellosis is similar to that of tuberculosis; the chief difference being that TB is more insidious and takes longer to kill those infected.

Suffice it to say that brucellosis is principally a disease of animals, and that human infections from any source are much less common. The few undulant fever cases reported each year are usually associated with a patient-history of close animal contact. Especially vulnerable are workers in abattoirs and meat-processing plants, although retail butchers, veterinarians, livestock producers and rural farm housewives and their children are also more susceptible than is the public at large. Wherever animal brucellosis is endemic, human disease occurs.

As a food-borne disease, brucellosis has been more often associated with milk and milk products. This is probably a scientific myth. While *Brucella* is totally eradicated by pasteurization, the rate of human infections has remained unchanged – notably in the United States, where, with few exceptions, pasteurization is mandated by law. Many studies indicate that most cases of undulent fever arise from sources other than the ingestion of dairy products.

Some interesting and unique epidemiological findings concerning brucellosis indicate that contaminated barnyard run-off water, soil manure, and urine are prime reservoirs of the microbe. Also noteworthy is that these organisms are especially tolerant of cold but are very susceptible to destruction even at ambient temperatures when coupled with lowered pH. Although it has been demonstrated that survival time and rate of *Brucella* is increased in frozen ice-cream and refrigerated dairy products, very few cases of human brucellosis have been linked to the ingestion of aged cheese, ice-cream, butter, or drinking water.

Surveillance of human undulant fever linked to drinking fresh cow's milk is most often associated with localized outbreaks or geographically confined mini-epidemics. Schools and institutions – including several tuberculosis sanatoria – which maintain(ed) their own dairy herds and used the unpasteurized milk for in-house private consumption are especially at risk.

In many parts of the world, unpasteurized fresh cheeses from raw sheep's and goat's milk have been identified as a major locus for *Brucella* infections. However, the ingestion of ripened, aged goat's- and sheep's-milk cheeses, made from raw, potentially *Brucella*-containing milk, seem to carry little hazard. Beware, however, that indirect contamination from leaves and earthenware crockery used to wrap and/or store such cheeses is still possible, and may be pathogenic.

All and all, it is quite obvious that contact with infected animals plays a much more important role than does the use of contaminated raw-milk supplies.

Listeriosis

Listeriosis (formally known as listerellosis) is caused by the bacterium *Listeria monocrytogenes* which is a relative newcomer to the list of pathogenic organisms which can cause disease in man and animals. Probably, the first reported incidence of this disease occurred in Sweden in 1910 and was described as a sporadic infection in domestic rabbits. The organism was first isolated and characterized by Murry, Webb, and Swann in 1926, and was also associated with rabbits as well as guinea pigs.

Listeriosis, like brucellosis, has been, and continues to be, principally a disease of animals, producing a variety of clinical symptoms. Especially susceptible are wild rabbits, hares and rodents – including voles, lemmings, field mice and squirrels, but not rats [!] – deer, racoons, moose, wood grouse, Arctic fox, and elephants. Domestic animals susceptible to listeriosis include chinchillas, silver fox and mink, while sheep and cattle are the most frequently infected barnyard animals, followed by goats, swine, and various species of fowl.

In humans, sporadic meningitis – with a mortality rate of up to 70 per cent – and granulomatusis infantiseptica are the most frequently encountered clinical presentations. It can also produce venereal disease and abortion. Epidemiologic surveillance is somewhat skewed, because diagnosis is predicated upon isolating and identifying the organism in the laboratory, which is not always easy.

Listeriosis may cause few or no symptoms in healthy people, but has been linked with serious illness in pregnant women, infants, the elderly, and individuals immunocompromised either by drugs – used in the treatment of AIDS, cancer and leukemia, and organ transplant recipients – or less frequently by underlying diseases such as diabetes, cirrhosis, chronic ulcerative colitis, and asthma.

Of recent interest and concern is the association of human listeriosis to the ingestion of milk and milk products – especially cheese – made from unpasteurized milk. However it should be noted that *Listeria* can be transmitted by other foods, including undercooked chicken and other poultry products, smoked and raw fish, vegetables, hot dogs, food purchased from delicatessen counters, and other ready-to-eat products.

Since the infective bolus for listeriosis is relatively low, so that as few as 1,000 CFUs per ml can result in clinical disease, it is important to know that *Listeria monocytogenes* – unlike *M. tuberculosis* and *Br. abortus* – is able to withstand extremes of heat and cold, surviving 80°C exposure for five minutes, and remaining viable at 3°C. It is thus able to multiply readily at refrigerator temperatures.

Tuberculosis

Tuberculosis historically has afflicted men since the beginning of human existence. In 1999, *Mycobacterium tuberculosis* continues to kill about 3,000,000 people every year – more than any other single infectious agent. It is currently by far the most serious infectious disease.

BIBLIOGRAPHY
General

Carmichael, Ann G., Ratzan, Richard M., 1991, *Medicine, A Treasury of Art and Literature.*
Cole, H.H. 1962, *Introduction to Livestock Production.*
Dubos, Rene and Hirsch, 1965, *Bacterial and Mycotic Infections of Man,* 4th edition.
Ewald, Paul W., 1994, *Evolution of Infectious Disease.*
Lennette, Edwin, Spaulding, Earle, Truant, Joseph (eds.), 1974. *Manual of Clinical Microbiology,* 2nd edition, A.S.M. publication.
Lyons, Albert S., Petrucelli, R. Joseph II, M.D., 1987, *Medicine, An Illustrated History.*
Morbidity and Mortality Weekly Report[s] published by the US Centers for Disease Control and Prevention (CDC), Atlanta, Georgia.
Rosen, George, 1958, *A History of Public Health.*
Spink, Wesley, 1958, *Infectious Diseases: Prevention and Treatment in the Nineteenth and Twentieth Centuries.*

Tuberculosis
Barnes, David S., 1995, *The Making of a Social Disease: Tuberculosis in Nineteenth-Century France.*
Huber, John Bessner, A.M., M.D., 1906, *Consumption. Its Relation to Man and His Civilization. Its Prevention and Cure.*
Bushnell, George, 1920, *A Study in the Epidemiology of Tuberculosis.*
Bargaining for life: a social history of tuberculosis 1876-1938.
Daniel, Thomas M., 1997, *Captain of Death: The Story of Tuberculosis.*
DuBois, René and Jean, 1951, *The White Plague Tuberculosis, Man and Society.*
Lowell, Anthony M., 1976, 'Tuberculosis in the World: trends in tuberculosis incidence, prevalence and mortality at the beginning of the third decade of the chemotherapeutic era' (USDA, CDC publication).
Lowrie, Douglas B. *et al*, 1999, *Nature*, vol.400.
Ott, Katherine, 1996, *Fevered Lives: Tuberculosis in American Culture since 1870.*
Rothman, Sheila M., 1994, *Living in the Shadow of Death: Tuberculosis and the Social Experience of Illness in American History.*
Senn, N., M.D., Ph.D., 1893, *Tuberculosis of Bones and Joints.*
Steindler, Arthur, M.D., F.A.C.S., 1929, *Diseases and Deformities of the Spine and Thorax.*

Listeriosis
Foodborne Pathogenic Micro-organisms and Natural Toxin Handbook published by the F.D.A. Center for Food Safety and Applied Nutrition.
Miller, A.J., Smith, J.L. and Somkuti, G.A. (ed), 1990, 'Foodborne Listeriosis' (Topics in Industrial Microbiology, 2)
Seeliger, H.P.R., 1958, *Listeriosis.*

Brucellosis
Huddleson, I. Forest, 1939, *Brucellosis in Man and Animals.*
Spink, Wesley W. (President), November 6-10, 1950, Proceedings of 'The Third Inter-American Congress on Brucellosis', W.H.O. publication.
Spink, Wesley W., 1956, *The Nature of Brucellosis.*

Hawking Milk: The Public Health Profession, Pure Milk, and the Rise of Advertising in Early Twentieth-century America

Daniel Ralston Block

Flashy advertisements and public health campaigns promoting the consumption of milk have become almost clichés in modern day America. The 'milk moustache' ads in particular are seen, imitated, and even parodied throughout the country. From the 1950s until the 1980s, the 'four food groups' idea, complete with a request for children to drink four glasses of milk a day, was constantly repeated at public schools.

It was not, however, always this way. The development of fluid milk advertising has its roots in a decision by early twentieth-century public health officials to choose milk as a staple and the concurrent development of the use of modern advertising techniques within public health. Rather than simply testing food in laboratories and producing informative brochures, beginning in about 1910, public health officials promoted milk consumption in an unabashedly direct style. This paper examines one of the most flamboyant examples of this promotion in the United States, by the Chicago Health Department.

Public health, milk protection and promotion

The field of public health in the early twentieth-century United States was in a period of quick development spurred by the discovery of the germ theory of disease and the development of the science of nutrition. Milk, both nutrient-rich and a vector of disease, became a focus of both public health protection and promotion. Protection focused around the production of clean milk at the farm and the promotion of pasteurization as a barrier between the farm and the consumer. Milk, which entered the city daily in a relatively unprocessed state, was one of the main vectors to disease discovered by public health officers looking out from their cities searching for incoming germs. The germ theory gave the officers both the power and impetus to do such searching. Bruno Latour writes of the previous situation in France: 'Since anything might cause illness, it was necessary to act upon everything at once, but to act everywhere is to act nowhere' (Latour, 1988, p. 21). With the development of the germ theory, public health officials were called on to set up barriers to germs and given regulatory powers to do so. In an age where refrigeration was either absent or primitive, the growth of bacteria in milk that sat out even a day, and the presence of dirt and disease brought from the farm, were constant concerns. In 1906, Dr George W. Goler, a Rochester, NY public health officer, went so far as to say in a speech to the City Club of Chicago, that 'bad milk is the cause of practically all the deaths of children under five years in big cities'. This was an obvious exaggeration, but since public health officials were also promoting milk consumption, the fact that milk was so dangerous made the protection of pure milk one of the main occupations of the early twentieth-century public health officer. Pasteurization was promoted through Health Department posters and brochures, but the main use of advertising in relation to milk was in the promotion of fluid milk consumption. This was part of a second prong of public health activity: nutrition.

Food chemists were taking food apart, and the discovery of essential, and mysterious, nutrients – which milk appeared to have in abundance – led to its christening as 'The Modern Atlas Supporting the World' by the Chicago Health Department. No other food except mother's milk, it was thought, provided as wide a range of nutrients as cheaply as milk. A leading dairy economist of the time stated:

> Our high standards of living can be kept without fear of ruinous competition in international trade only if we choose for our diet those foods that give the best nutrition for the least price. Among such foods dairy products easily take first place (King, 1920, p. 20).

Nutritionist E.V. McCollum of Johns Hopkins, the discoverer of vitamin A, in a speech to the City Club of Chicago stated in 1918:

> An adequate diet consists of proteins, sugar, fats, inorganic elements and certain unknown elements – the so called 'vitamines' [sic]…There are only two classes of what we may call protective foods – 'protective' because they contain the vital elements lacking in other foods. They are (1) milk and its products and (2) the leaves of plants. Milk is the most important. You may combine it with seeds, with roots or with meat and get a fairly satisfactory diet.

McCollum summarized his thoughts with a statement that invokes the cultural nature of the importance of milk, particularly its Western orientation:

There are some nations – for instance Japan and China – which use but little milk in the diet, but if they have maintained their vitality it is because they have eaten leafy foods in large quantities. It is not an exaggeration to say that only those peoples who use dairy products are strong and virile, with low infant mortality and a long span of life.

With these thoughts in mind, public health officials went out to promote fluid milk consumption. It was a duty of the public health officer to bring health to the people. In some cases this responsibility almost turned into a religious zeal. In a speech to the municipal health officers' section of the American Public Health Association, one officer stated:

I cannot conceive of any man in any profession ... where this sense (the sense of realizing that one has helped the community) is surer that in the public health officer, who had tried to do his duty and to help his time and improve the surroundings for little children. Who is more certain when he comes to the Last Judgement, the verdict of history, to hear from the Spirit of the Nation 'Inasmuch as ye have done it unto the least of these, my little ones, ye have done it unto Me.' (Hutchinson, 1913, p. 779).

Public health officials turned to the new science of advertising for instructions on how to popularize their messages. One charity officer compared publicity to medicine: 'It [publicity] should be entered into the Public Health Pharmacopoeia as an accredited remedy for human ills' (Moree, 1916, p. 97). The *American Journal of Public Health* from the time is filled with articles about publicity and modern advertising methods. A clinic was held at the American Public Health Association annual meeting in 1923 where an advertising agent, the director of an art school, and the head of a printing firm inspected Health Department literature and gave instruction on possible improvements. The chief weakness found was that many of the flyers were overly depressing, bland, or both. The advertising agent stated that while commercial advertisements might state, 'Eat our yeast, raisins and breakfast foods, be beautiful and youthful', public health advertisements often read, 'Avoid sickness, visit our clinics and use our facilities' (Routzahn, 1923, p. 248). A similar sentiment was related by the former director of the US Public Health Service about milk purity who commented that 'the dangers are real enough and bad enough without exaggeration…The statement is often made that danger "lurks" in milk. This is quite true, but every portion of milk is not a potion of poison' (Rosenau, 1913, p. 17). Rosenau recommended posters such as a *Chicago Health Bulletin* poster promoting pasteurization which 'is instructive yet excites no unpleasant sensations' (Rosenau, 1913, p. 16).

The Chicago Health Department

The Chicago Health Department was exemplary in learning these messages. Health Department bulletins and posters were often used as examples of good publicity. W. A. Evans, Health Department Commissioner from 1907 until 1911, began the trend. Evans converted the weekly bulletin of the Chicago School of Sanitary Instruction to a more easily-read *Chicago Health Department Bulletin*, where weekly and monthly tallies of various statistics including deaths, contagious diseases, and cows inspected were published, in addition to health hints. The bulletin also included one-line 'healthgrams', for instance: 'If your milk is not safe your life is not safe.' More importantly, Evans and later commissioners sponsored a series of posters showing various facts of nutrition. After his dismissal as Health Commissioner, Evans wrote a daily column in the *Chicago Tribune* covering health issues. He became something of an expert on publicity in newspapers. Evans, advising his fellow public health officials, stated: 'You have gotten the good from campaigns to control contagion. Further improvement necessitates changes in the habits of individuals' (Evans, 1916, p. 30). Newspapers, he goes on to say, are great methods of appealing to individuals. But how to make that appeal? Evans himself is somewhat cautious: 'News stories must be true in the main or the paper suffers' (Evans, 1916, p. 30). Evans thought of himself as a consulting-physician for the *Tribune*, giving practical and down-to-earth advice. A typical opinion given was, 'In the protection of food against infection the chain is not much stronger than the weakest link – that is, if the food is eaten raw. Cooking it or properly washing it will prevent most of the danger from the weak link.' Furthermore, Evans felt that use should be made of newspapers catering to immigrants and blacks, which was often overlooked but was 'the hardest problem which health officers have to solve'. Evans stated, in a surprisingly non-paternalistic way, that these communities should be addressed 'in their own language and by their own people' (Evans, 1916, p. 30).

In 1922, Herman N. Bundesen became Health Commissioner of Chicago. Except for a period in the late 1920s and early 1930s, he would remain Commissioner until the mid-1950s. Bundesen was a controversial figure. He was the health-huckster-extraordinaire, selling health to the people. In a presidential address to the American Public Health Association, Bundesen likened public health departments to corporations, selling a product: 'Every big business that has something to sell, spends vast sums of capital to train and send out live, alert and efficient salesmen. No other business in the world is bigger than that of keeping people well. Why then should we lack enthusiastic, efficient health salesmen?' asked Bundesen (Bundesen, 1928, p. 1452). The public health officer is advised to 'know his wares', 'have absolute confidence in his line', and 'have sound health himself and…surround himself with evidence of it.' Health officers were to be walking, talking, enthusiastic poster boards for the healthy life.

Bundesen transformed the weekly bulletins. He renamed the series *Chicago's Health* and featured frequent special issues devoted to particular topics and filled with flashy illustrations in multiple colours. A bulletin in 1924 was entirely devoted to milk. Another, entitled 'The Growing Child,' featured multi-coloured printing and was widely distributed. It is from the bulletin that 'The Atlas Supporting the World' graphic was taken. A booklet entitled 'Our Babies' was published in its entirety by *The Chicago Daily News*, which paid $5,000 for the privilege (Bundesen, 1927, p. 444). A series detailing the growth of a baby between particular ages was mailed free to a list of all new mothers in Chicago. Bundesen patterned the new bulletins, he said, after the Lord's Prayer and the Gettysburg Address. Both, he stated, are 'short, simple, and made up of one or two syllable words' (Bundesen, 1927, p. 442). He also stated that bulletins should, 'be understood by one with the mental age of about 12 years', contain lots of pictures, and refrain from wrong or misleading messages like showing a cute picture of a baby sucking his thumb while at the same time discouraging thumb-sucking in the written text (Bundesen, 1927, p. 442–3). In one easily-read graphic, milk is depicted as the ideal food. In a brochure promoting milk consumption, milk is connected with growth in children, puppies, and chickens, and promoted for strength in boys and beauty in girls. Another graphic, actually completed during a brief lapse in Bundesen's tenure as Health Commissioner, and which followed the same philosophy, depicts a boy, a girl, and a smiling milk bottle staring towards the Temple of Health. The milk bottle states: 'I'll take you there.'

Bundesen suggested that public health officials follow the example of health oriented advertisements, which appeal to health and beauty while suggesting the use of a product. He also was not above pure hucksterism. In a famous stunt in 1924, the department fuelled a locomotive with dried milk and had it pull a five-car train carrying 200 orphans about ten miles. Over 100 newspapers nationwide covered the event which was, 'perhaps the most successful single stroke of health publicity in the decade' (Kirschner, 1986, p. 70). *The Chicago Tribune* reported: 'When the skeptical crowd reached the station Dr Bundesen was at the throttle in the engine cab and…the fireman was shoveling chunks of dried cow's milk into the firebox.' *The New York Times* stated: 'Dr Bundensen's (*sic*) object in conducting the demonstration was not primarily to see if milk could furnish enough power to run a train, but to dramatize the idea that milk has energy.' Bundesen probably could have made this point in many cheaper ways, but he wouldn't have gotten it on the front page of *The Chicago Tribune* and *The New York Times*. In addition, it should be obvious that just because milk can power a locomotive does not mean that it is necessarily good for people. Eating coal, the normal fuel, would be a mistake. However, from now on, Bundesen and others could relate milk in your body to coal in a locomotive. Everybody who remembered the locomotive that ran on milk would associate milk with power.

Bundesen had many detractors. He used his position for profit by sponsoring various products, and occasionally sponsored strange items. He, for instance, authored a booklet in which he stated that candy, if eaten between meals, burns fat. Many doctors disapproved of the manner in which Bundesen and earlier health commissioners simplified health rules and placed what they felt was undue importance on particular products. By concentrating on extreme purity, however, many in the medical establishment had isolated themselves. When doctors stated that, 'cow's milk has been overestimated as a food and underestimated as a carrier or cause of disease', in *The Chicago Medical Recorder*, it was unlikely to be heard. However, when public health officials concentrated on the positives in multi-coloured pamphlets sent out to almost every new mother in the city, an impact was made.

Aftermath and conclusion

Bundesen was exemplary of the new public health movement to which publicity was central. By declaring milk 'the single most important food', Bundesen, and other public health officers like him, had placed themselves in a situation where they almost forced themselves to support further milk regulation. Public health departments had grown with the rise of the germ theory of disease, and milk was admitted to be an important vector of disease. Controlling diseases passed through the milk supply might have involved taking emphasis off cow's milk as an important food. Instead, milk was promoted as the most important food, primarily because it was economical. Placing such emphasis on a product that was historically often a vector for disease was risky, since political disaster could have resulted from a milk-based epidemic. Health officers accordingly pushed for stricter regulation of the product they emphasized. Increasingly, the regulations reached back away from the city and towards the farms in the surrounding countryside. This brought health departments in direct contact with farmers, often creating conflicts the resolution of which led to the development of the New Deal milk-marketing programmes.

The connection between advertising and the promotion of milk and other foods by public health groups continues to this day. In the post-war era, an alliance between dairy promotion organizations and public health groups, based in the United States around the 'four food groups' idea, blurred the borderline between advertisement and public health information. (Even more difficult to allocate to their proper source are the popular milk moustache advertisements, in which milk's nutritional value is ignored. Early twentieth-century public health officers such as Bundesen, who concentrated promoting foods in the simplest possible manner and performed such stunts as running a locomotive on dried milk, were the pioneers in leading towards this separation of the physical and nutritional characteristics of milk and its depiction in advertisements.

BIBLIOGRAPHY

Bundesen, Herman N., 1928. 'Selling health – a vital duty', *American Journal of Public Health* 18(12): 1451–1453.

Bundesen, Herman N., 1927. 'Value and use of health department bulletins', *American Journal of Public Health* 27(5): 442–447.

Evans, W.A., 1916. 'The role of the newspaper in the dissemination of public health news', *American Journal of Public Health* 6(1): 28–31.

Hutchinson, Woods, 1913. 'Publicity in public health work', *American Journal of Public Health* 3(8): 777–779.

King, Clyde L. 1920. *The Price of Milk*. Philadelphia: John C. Winston.

Kirschner, Don S., 1986. *The Paradox of Professionalism, Contributions in American History* 119. New York: Greenwood.

Latour, Bruno, 1988. *The Pasteurization of France*. Cambridge, MA: Harvard University Press.

Moree, Edward A., 1916. 'Public health publicity: the art of stimulating and focusing public opinion: I. The scope of publicity', *American Journal of Public Health* 6(2): 97–108.

Rosenau, M.J., 1913. *The Milk Question*. London: Constable.

Routzahn, E.G., 1923. 'Health education and publicity', *American Journal of Public Health* 13(3): 248–250.

The Hierarchy of Milk in the Renaissance, and Marsilio Ficino on the Rewards of Old Age

Phyllis Pray Bober

Save in the realm of cheese, modern consumption of dairy products in the developed countries (a signal that I do not intend to consider mare's, camel's or yak's milk) offers relatively little choice as to mammalian sources. Although goat's milk is now regularly available in well-stocked markets, it is the bovine variety that is taken for granted in recipe directions or in references to 'a drink of milk'. 'Twas not ever thus. Particularly was it not the case in the period of my focus in this paper – the fifteenth and sixteenth centuries in Italy. Aside from the universal heritage of *maigre* cookery using 'milk' made from almonds or other nuts, there were important distinctions to be made among animal and human *latticini*.

With the splendid Renaissance ardour for explicating theoretical principles that underlay any traditional practice, authors of alimentary treatises turned to scholarly humanist analysis of entrenched dietary, that is, humoural dicta. To my knowledge, the earliest to concentrate upon milk and milk products is Pantaleone de Confienza's *Summa la(c)ticinioru(m)*, printed at Turin in 1477, which I have used in the beautiful Vatican manuscript (Vat.lat. 4479), a presentation copy to its dedicatee, Pope Sixtus IV.

Pantaleo, a Piedmontese doctor who flourished in the first half of the Quattrocento, taught medicine for many years at Vercelli. Service as chief physician to the Duke of Savoy brought him experience at the French court and ultimate retirement in the Touraine. This means that his treatise can speak authoritatively about many more cheeses than those from Italy. He writes in detail solely of those personally savored throughout the French provinces, including the area still in English hands until 1453, as well as from Flanders and parts of Germany. But the material of interest in the present instance is found in the first portion of Pantaleo's work as he sets down the 'natural history' of milk, '*volens in hac su(m)mum lacticiniorum Aristotilis imita(n)ti doctrina(m).*'

In addition to Aristotle, a multitude of sources both ancient and 'modern' are cited, with special reference to the Arab scientists Averroës and Avicenna and to contemporaries like the Bolognese doctor Michele Savonarola, whose

Libreto de tutte le cosse che se magnano was written about 1450. Received authority on the qualities of different milks, stemming from Aristotelian/Galenic tradition, revealed conflicting opinions. Medicine taught that milk was nothing other than blood 'twice cooked' by the body,[1] its savour and whiteness brought about in breasts or teats. One tradition in the humoral system held it, bloodborne, to be warm and humid; it was impossibe to negate the moisture, but contrary findings emphasized its coldness.[2]

Renaissance rationality was able to reconcile the apparent discord in the sources. One could think of milk as an essence itself composed of four substances: '*caseale* [the cheese potential, curds], *butirosa, aquosa* and the *serosà* [the serio or whey].[3] Even if one more commonly confined the division to three, omitting the *aquosa*, it was still possible to distribute the degree of warmth or cold among components with secondary allowance for whether each was fresh or aged. The spectrum ranged from coolness[4] – with accompanying humidity – of the whey and its derivatives like ricotta, to the heat and dryness of old, salted cheese, with butter standing moderately between.

The result of this system can be read in warnings of health advisories against eating aged, hard cheese, save in small quantities and at the end of a meal when it may act as a 'seal' or weight to compress other foods for digestion. Savonarola quotes a poetic adage: '*caseus est sanus quem dat avara manus*';[5] and the poet-physician Battista Fiera, who set down in Latin verse his compendium of agronomic and medical/dietary lore associated with every comestible, writes: '*si calet antiquus siccescens caseus, humet/ Et gelidus novus est, inter urumq(ue) placet./ Si novus impinguat, siccat vetus, attamen ambo/ Et capiti & stomacho, pectorius-q(ue) nocent:/ Mensa ferat medium, vel nullum, renibus illi/ Sunt nocui & iecori, nil habet ista boni.*'[6]

How did all this affect actual cookery and service at formal dinners? The scholarly portion of the first printed cookbook, Platina's *De honesta voluptate et valetudine*, which precedes six chapters of recipes translated from Maestro Martino's vernacular manuscripts,[7] follows *lac*'s inclination to coolness, recommending that one drink it warm before eating other food, or mitigate any harmful quality by taking it as curds in the first (*credenza*) course at a meal, not at the end when milk will either spoil (*putrescit*) or draw other food down with it, undigested. In this he echoes medical advice, agreeing also that one must not exercise after imbibing to avoid curdling it by shaking![8] Again, in the section on cheese he iterates the medical consensus that, in contrast to nutritious fresh, cold cheese (only to be avoided by the phlegmatic), old cheese, hot and dry, should be accepted solely in the final course of a meal, although in small, sealing quantity. The possible injuries are those consistently cited by others, including gout, kidney stones and damage to the liver.

All of the doctrine sketched above explains Renaissance banquet practice in which the initial servings that we would call hors d'oeuvre, served usually with *malvasis* (malmsey wine), featured unctuous milk preparations such as *latte miele, capo di latte, neve di latte, fiorita* (akin to modern *fiore di latte* or *mascarpone*). In keeping with this taste for the unctuous, Pantaleo recommends *marzolino* as the favoured antipasto cheese, the Tuscan ovine speciality that is still enjoyed today, but which Platina terms 'rotted'.

Butter joins the first-course specialities thanks to its temperate quality, but moderate though it is, and highly nutritious because of its fat content, care must be taken to be certain that it comes from young animals, fed on the most salubrious pasturage, and preferably churned from May to July.[9] A favourite way of serving it was softened, mixed with sugar and rosewater, then piped through a syringe to make a *faux* pasta surprise. The surprise for us in reading recipes from the early Renaissance is the frequent use of butter as an ingredient (not to mention lard or *strutto*) in place of olive oil in areas of central and southern Italy where one would not later find it favoured above olive oil.

Unlike other writers on alimentary topics, Platina does not consider human milk in a hierarchic evaluation of different types of milk and their derivatives, stating that women's milk is used medicinally alone. Otherwise, in the period in question there is general agreement on ranking milk from different sources. In a change from the Salerno/ *Tacuinum sanitatis* tradition in which sheep's milk was deemed superior to that of goats,[10] the hierarchy in nutrition and quality reads: from woman, ass (though these do not make cheese), goat, sheep and, lastly, cow (although the best for butter). Perhaps the shift to preference for goat's over sheep's milk owes something to Pliny's evaluation (*Nat.Hist.*, XXVIII, xxxiii), likewise acknowledging the classical legend of Jupiter nursed as a baby by the goat Amalthea. Another factor may be the strong influence of Catalan/Aragonese cookery on fourteenth- and fifteenth-century Italian recipe books; note the specified use of goat's milk in Maestro Martino's *biancomangiare alla catalana*. In any case, the experts knew that it was less viscous and fatty than cow's milk, though less sweet than that of a sheep; one need merely be certain that the goats had not been eating hellebore and other *nocivi* shrubs and remember that their cheese lost humidity very rapidly.[11]

The perfect nutrient because of its temperate quality, human milk was not confined to the feeding of infants. It was not only the sweetest and most delicate (*dulcissimum, mollissimum*), but ancient precedent for its medical use stressed its efficacy in persistent fever or coeliac disease; in lung infections; expressed straight into bloodshot eyes following a blow or the 'flux'; and mixed with a little oil for earache; it even had a preventive role against rabies![12] Indeed, Conrad Gesner, the great natural historian, stresses that its best results in restoring bodily health come from sucking it directly from the breast, just as Eurython, Prodicus and Herodotus advised those wasting away from consumption to be nursed by their wives.[13] Several doctors mention that Aristotle had written (*Historia animalium*, III, xx, 522b) that milk from a black woman was the healthiest.

Literati in the Renaissance surely knew the story told by Valerius Maximus (V, iv, 1) concerning Pero, who had nursed her imprisoned father Cimon to keep him from starvation, although this did not become a favoured subject for painters until the seventeenth century. Such an icon of piety and compassion takes on a different cast when 'Il Panunto' writes that the nurse should be healthy, young (others specify between 25 and 35 years-old), and beautiful, as well as of temperate nature ('complexion' in humoral terms) and generate good spirits.[14] Not all these characteristics are likely to have defined a wet-nurse for an infant.

But the most explicit recommendations for therapeutic use of human milk – and splendid evidence for the rewards of old age, at least for one gender – comes from Marsilio Ficino's *De triplici vita*. The great Florentine philosopher prescribes (Bk. II, cap. XI): 'Immediately after the age of seventy and sometimes after the age of sixty-three, since the moisture has gradually dried up, the tree of the human body decays. Then for the first time the human tree must be moistened by a human, youthful liquid in order that it may revive. Therefore choose a young girl who is healthy, beautiful, cheerful, and temperate, and when you are hungry and the Moon is waxing, suck her milk. Immediately eat a little powder of sweet fennel mixed with sugar. The sugar will prevent the milk from curdling and putrefying in the stomach, and the fennel, since it is fine and a friend of the milk, fermenting will spread the milk to the bodily parts.'[15]

We can all believe in the efficacy of the therapy and the justice of having the human rule the Renaissance hierarchy of milks.

REFERENCES

[1] See for example, Michele Savonarola (Jane Nystedt edition, Acta Universitatis Stockholmiensis, Romanica Stockholmensia, 13, 1988, rather than her earlier version under the same imprint, 1982, which did not use the best manuscript, p. 146), citing the authoritative Mesue [Ibn Yuhanna Masawayh, d. 857]. Mesue's *Opera* were popular in the Quattrocento, editions being published in Venice and Florence in 1471. For a later use of the same phrase by Domenico Romoli, 'Il Panunto,' see his *La singolar dottrina dell'ufficio dello scalco...*, Venice, 1560, c.233 verso: '*Il latte non è altro che sangue due volte cotto.*'

[2] Savonarola, *loc.cit.*, takes notice of the alternate tradition, referring to Rasis, *De simplicibus*, although he comes down on the side of Isaac Judaeus and Mesue, while noting how the Milanese seemingly ascribe to the cold and humid nature of milk, using it to refresh themselves in hot weather.

[3] Savonarola, p. 147, Nystedt.

⁴ Note, however, a persistence in some texts of the generally temperately warm quality ascribed to milk itself in the *Tacuinum sanitatis* of Ibn Butlan, so widely diffused in the fourteenth century. Illogically, ricotta and curds (*lac coagulatum*) were in its manuscripts deemed cold and humid, while fresh cheese varied from cold to warm – cf. Luisa Cogliati Arano, *The Medieval Health Handbook*, New York, 1976, Liège 15–19, Paris 128.

⁵ *Op. cit.*, p. 148 Nystedt, 'Cheese is healthful served by a miser's hand.'

⁶ *Coena Batistae Fierae de Herbarum Virtutibus, & ea Medicae Artis parte, quae in Victus ratione consistit. Columella de cultu Hortorum. De generibus morborum, ex Imprecatoria Satyra Petri Montani*, Strassburg, n.d., f. 127 [p. 10]. The work is dedicated to the leading Roman humanist Pomponius Laetus. I hesitate to translate when I cannot maintain the metre – or the wit – but the sense is clear

⁷ The recognized 1475 Venetian edition follows an undated Roman one, *ca.* 1470, probably published by Han, but the work was written *ca.* 1465–67; see now the authoritative critical edition and translation by Mary Ellen Milham, *Platina: On Right Pleasure and Good Health*, Tempe, AZ, 1998. Sacchi knew Martino, as his comparison of him to Carneades affirms (the Greek philosopher was famous for being able to argue for or against any position as mental exercise), and, good humanist that he was, acknowledges his borrowing – cf. Emilio Montorfano, 'L'opera di Maestro Martino alla luce delle ultime ricerche storiche,' in *Maestro Martino da Como e la cultura gastronomica del Rinascimento* (Convegno internazionale di studi, Como, June, 1989), Milan, 1991, p. 33

⁸ Milk, Bk.II, cap,16; Cheese, cap. 17; Ricotta, cap. 18; and Butter, cap. 19. It is interesting that Platina speaks of the practice of discarding the beestings or colostrum when animals give birth, since it is considered harmful to the newborn. Despite his learning, he seems not to be aware that it was prized in antiquity, especially by Attic pastry chefs – see my *Art, Culture and Cuisine.* Chicago University Press, 1999, ch. 'The Hellenic Experience', note 10.

⁹ In addition to other sources already mentioned, see Carol Stefan [Stephanus], *De nutrimentis*, Paris, 1550, p. 41f., or Conrad Gesner, *Libellus de lacte, et operibus lactariis...*, Zurich, 1541, c. 21 verso, c. 23ff. (with paraphrases of Dioscorides and Galen; this work includes recipes copied from Platina).

¹⁰ I would think this the heritage of Varro's privilege accorded sheep's milk (*de re rust.*, II, xi,3); his ranking makes milk from sheep the most nutritive, followed by that from goats, while the order for the most purgative is mare's, asses', cow's, then goat's. His reasoning seems much affected by the supplementary profit to be made when raising sheep for their wool.

¹¹ 'Il Panunto' *op. cit.*, c. 234 puts it very well: 'the milk from goats has good colour, white and clear, as well as a nice odour, and it is in between thin and viscous' ('*fra il sottile & il grosso, che se una goccia di esso si mette sopra un'unghia...non si sparga, ma rimanca larga di basso, & di sopra aguzza*). Pantaleo *ms.cit.*, fol.10 cites the hierarchy after Averroës, noting how much fatter is the sheep's milk and apt to become cheesy and curdling in the stomach, though the cow's is '*grossius & cum hoc butirosius & pinguius*'. He says that he will not speak of sow's milk, but this is after having discussed – unconvincingly – that one Rabbi Moyses holds, possibly after Galen, that it is next best after that of a woman, a pig being so close in temperament to the human. Savonarola (*op. cit.*), p. 147, follows the common opinion, while noting that the human milk takes pride of place only 40 days after parturition and should be received direct from the breasts. – cf. below.

¹² Pliny, XXVIII, xxi.

¹³ *Op. cit.*, cap. 26: '*ut etiam qui phthoë contabuerant, ex ipse mulieris papilla lac sugare iuberent*'.

¹⁴ *Op. cit.*, c. 235.

¹⁵ The translation derives from the Clarke & Kaske edition, 1989. The work refashions and combines three other works by Ficino on ageing – see Maurice de Gandillac, 'Thèmes alimentaires chez Marsile Ficine', in *Pratiques et discours alimentaires à la Renaissance* (Colloque de Tours, 1979), ed. J.C. Margolin, R. Sauzet, Paris, 1982.

New York Milk Culture: Some History, Facts and Concerns

Una Bray

The United States' dairy industry is at a critical juncture. The cost of producing milk is rising steadily while the price paid to the farmer fluctuates, sometimes wildly, due to supply and demand, and price cannot keep up with cost. Some states enter regional dairy compacts to insure a fair price for milk. Other states, particularly those in the Midwest, perceive such compacts as threats to their control of milk prices. Farmland is taxed according to market value and urban sprawl is spilling over, making farmland close to cities attractive to developers and too expensive to maintain for grazing animals. Labour costs are mounting as children choose to leave the farm for more secure employment. Machinery costs are increasing over inflation. In general, farmers are under intensifying economic pressure to use new technology, particularly biotechnology, to produce more milk at minimal cost so that farms can stay solvent. While some technologies can bring cost down, there are serious questions as to whether the use of one technology, bovine growth hormone, can increase the quality of life on the farm, the quality of the milk produced, or the overall profitability of the farm. Every dairy state is grappling with these issues, none more so than the State of New York.

This paper will attempt to very briefly summarize the history and present situation of dairy farming in New York. Data from the United States Agricultural Statistics Service and Cornell University will be used to present some facts about the economic environment facing dairy farmers and discuss such trends in the dairy industry as increased herd size and frequency of milking. A small sample of 'upstate' dairy farmers describe the changes that have taken place over the last 30 years, and the pressure that they, and farmers like them, feel as they attempt to compete with large, industrial-sized farms in a market where the price of milk is controlled while the cost of producing milk increases yearly, and farmland is giving way to development and urbanization.

New York farms

New York State has a land area of approximately 30.6 million acres. Of these, farmland accounts for about 7.3 million acres and there are farms in all but six of the 62 state legislative subdivisions, called counties.[1] The percentage of state land devoted to farming has decreased continuously since the peak year of 1900, when almost 23 million acres of New York was farmland. One must be careful when making comparative statements, as the definition of a farm has changed nine times since 1850. Currently, the State of New York defines a farm to be, 'Any place for which $1,000 or more of agricultural products were sold, or normally would have been sold, during the census year'.[2] Using this definition, there are about 32,000 farms in New York.[3]

Dairy farming in New York

According to the 1977 *New York State Farm Census*,[4] some 65 per cent, or 4.7 million acres, of New York's total farmland is under crops, much of which is best suited for growing forage crops.[5] For more than 150 years, dairy cows have converted these crops into the liquid gold of dairy products, with current annual state sales amounting to 1.5 billion dollars.[6] Most visitors to New York would be surprised to learn that, outside of the major metropolitan areas, New York is considered a dairy state and dairying is *the* dominant 'upstate' industry. In fact, New York ranks third nationally in number of dairy cows and amount of milk produced, behind California and Wisconsin. Here are some comparative data, from 1996:[7]

STATE	Number of cows in thousands	Pounds of milk per cow per year	Yearly milk production (lbs)
Wisconsin	1,449	15,442	23,147,558
California	1,276	20,267	25,860,692
New York	702	16,396	11,509,992

Of New York's 30,000 farms, 8,162 (about 26 per cent) classify themselves as dairy farms and such farms can be found in all but seven counties.[8] These farms are home to more than 700,000 milking cows, most from one of the six major North American dairy breeds: the Ayrshire, Guernsey, Jersey, Milking Shorthorn, Brown Swiss, and Holstein. (These are but a small percentage of the more than 1,000 different breeds of domestic cattle worldwide.)[9]

The cow most responsible for the success of the New York Dairy industry, due to its prodigious milk output, is the Holstein. She is America's most popular dairy breed.

The Holstein – a genetically-enhanced cash cow

The Holstein cow, often referred to as the Friesian, accounts for 90 per cent of the national dairy herd and about the same percentage of the New York State herd. This is the massive black and white cow of advertising fame, representing organizations as diverse as Agway, an international supplier of farm feed and equipment, and Ben and Jerry's ice-cream. She is the 'poster cow' for the entire state of Vermont, most milk companies, farmland preservation movements, organic dairy farms, some divisions of the Monsanto company, veterinary offices, and many more purveyors of products and services. In the United States, she is the cow depicted on postcards, tee-shirts, coffee mugs, shopping bags, garden statues, television commercials and children's toys.

Holsteins have been in North America since the early seventeenth century, but in very small numbers. A few were brought by Dutch settlers as early as 1613, a few more by the West India Company in 1625. The animals were described as being black and white, mild in nature, the size of oxen, with very straight, handsome bodies.[10] This description is still used to describe the Holstein today.

The Holland Land Company sent eight cows to New York in the late 1700s. Records indicate that a bull and two cows were imported by a sheep breeder in Vermont in the early nineteenth century, and, during the same period, Dutch cattle were imported by the Le Roy family of upstate New York. According to the Holstein Breeders Association, all traces of these early Holsteins were lost by cross-breeding.

The 'real beginning' of the breed in the United States is traced to one Winthrop Chenery of Belmont, Massachusetts. In 1852, Chenery purchased a cow from the master of a Dutch rum ship. He was so pleased with the animal's milk production that, in 1857, he purchased two more cows and a bull. His animals became ill with the disease rinderpest, and were put to death by the local authorities. One young bull was spared. The intrepid Chenery then imported four cows and an additional bull.[11]

By 1871, the Holstein captured enough interest to warrant the formation of a breeders' group, complete with pedigree records and ideas about improvement of the breed. The group called itself The Association of Breeders of Thoroughbred Holstein Cattle, with Chenery as the first president. Rival associations formed, using the alternative name Friesian but, by 1885, the rivals more or less merged into the Holstein-Friesian Association of America, and the breed became officially known as Holstein-Friesians. This ended in 1977 with the readoption of the name Holstein to define and designate the breed that was developed by more than 75 years of 'selected, single-purpose, closed herd book breeding'.[12]

The Holstein breeders used careful genetic practices to produce cows that made very large quantities of milk with high fat content, and now these improved 'nineteenth century Friesian imports from the European continent, bred and highly selected for upwards of 100 years in the United States',[13] provide the genetic pool for the Holstein stock returning to Europe and elsewhere in the world.

Today, the dairy farmers I've met are experienced geneticists, who pride themselves on the cows they've bred, as well as on the quantity and quality of the bull sperm in their holding tanks. Some of the sperm is as old as 45 years. And, yes, a cow can be bred with her ancestor of ten, or more, generations before.

Years ago, E.B. White said:
> 'I'm sorry for cows who have to boast
> of affairs they've had by parcel post.'

Today, we might say:
> 'I pity the cow as she tries to cozen
> with the sperm of a relative, forty years frozen.'

US milk production

The effect of careful breeding by owners of Holstein and other breeds, is obvious when one looks at production data.[14] In 1944, there were 25,597,000 dairy cows in the United States, producing an average per cow of 4,572 lb of milk. This amounted to a total of 117 billion lb.

In 1996, the last confirmed data report 9,372,000 cows producing an average 16,433 lb of milk per cow, giving a total of 154 billion lb. The preliminary figures for 1998 indicate the trend towards higher yield continues, with the average per cow rising to 17,192 lb for 9,158,000 cows, with a total production of well over 157 billion lb. But, as will be shown below, the trend seems to be slowing.

In order to understand how much the genetic improvement of dairy cows has affected milk production, compare US milk production figures with those from some of the other developed countries. The latest available data come from 1997.[15]

Country	Average Milk/Cow	Country	Average Milk/Cow
Japan	18,423 lb	Netherlands	14,476 lb
US	16,880	Canada	14,221
UK	15,611	Belgium	12,256
Sweden	15,363	France	12,036
Denmark	14,623	Australia	10,320

The US is second only to Japan. American breeders are now importing dairy cattle and sperm from Japan, hoping to use some of the stock produced in Japan from the 'highly selected' US breeding pool they once exported.

Bovine growth hormone

Bovine Somatatropin (bST) is a natural growth hormone, originally taken from the pituitaries of cows. Recombinant bovine growth hormone (called either rbGH or rbST), is a genetically-engineered form of the hormone that was developed in the 1980s by several companies at once. Monsanto won the patent and rbGH has been commercially available in the United States since 1994, and has been used by up to 30 per cent of US dairy farmers on some or all of their herd. Real data about rbGH use is difficult to come by. It is not published as part of the USDA's *Dairy Statistics*, and a search for information about growth hormone use proved unsuccessful on the USDA web site. How widely is rbGH being used? Has use of the hormone significantly increased the quantity of milk produced per cow in the United States? Clearly, with the proper use of the hormone, milk production is increased in some cows, but it is too early to evaluate the long term use of the hormone, or how effective the hormone will be when used on large numbers of cows with varying milk productivity.

A look at the available data[16] gives the following pattern of increased milk production over ten-year intervals:

	Increased milk production per cow
1948-1958	31%
1958-1968	39%
1968-1978	23%
1978-1988	26%
1988-1998	21%

These data seem to indicate a slowdown in the growth trend, and after years of genetic manipulation, this might be expected. In fact, one may speculate that growth hormone is inflating a growth curve that was approaching a natural limit.

Cornell University does publish an annual report containing some information about the New York experience with the use of rbGH.[17] The authors report on 77 farms that have been followed from 1993 to 1997. Forty of the farms did not use rbGH, and 37 used the hormone over the five-year period. Here is an excerpt from the report:

	Non-user Farms		Farms using rbGH	
		% change from 1993		*% change from 1993*
Farms net worth	$468,041	13%	$1,488,045	26%
Debt to asset ratio	.21	-9%	.43	13%
Avg. number of cows	89	11%	451	38%
Cows per worker	32	6.6%	43	13%
Operating cost of producing milk (per cwt)	$11.01	11%	$11.66	6.5%

Other data from Cornell's *Dairy Farm Management Summary* (1997) indicate that farms using rbGH experienced an increase in milk sold per cow, from 20,256 pounds in 1993, to 23,081 pounds in 1997. This contrasts with a decrease in milk sold per cow in the non-rbGH users, from 17,428 pounds per cow in 1993 to 17,147 in 1997. The rbGH farms were larger and grew more rapidly in net worth, operating costs and number of cows, than the non-hormone group. Debt-to-asset ratio also increased, while it dropped in the non-rbGH group.

The rbGh farms need to be larger because they require a greater number of cows, producing more milk, to be cost effective. All New York dairy farmers work under the pressure of a very small profit margin. The larger farms (more than 200 cows) need more investment in buildings, machinery and supplies. Debt supports those investments and creates enormous pressure. If the milk price drops, that large debt cannot be repaid.

The Cornell data agrees with the anecdotal information I have obtained from my small (non-scientific) survey of area farmers. They interpret the statistics a little differently to Cornell. The farmers say that those using rbGH spend more money per cow because rbGh is expensive, requires very careful management and, therefore, more workers to maintain cow health. New facilities are needed to milk larger numbers of cows more often (four times per day is considered common and five times is not unheard of). There are increased veterinary and medical costs to care for cows with swollen joints from standing on concrete, as well as cows with mastitis and other disorders.

The most troubling concern raised by the farmers has to do with cow age. Cows are bred after they 'freshen', and begin milking after they calve. Traditionally, they hit their milking peak at about five to six years. Indeed, I was introduced to cows, aged five or six, whose milk capacity was, astonishingly, 100 lb or more per day on two milkings!

With rbGH, many cows are lasting only two years as milkers, then they are sold at the meat market. This requires new cows every two years to replace those too ill to keep on. Replacements cost money, so the pressure is on the cow and the farmer to make up for the eroding profit caused by the recycling of cows before their time. This means more milking and more stress on cows.

One farmer, who prides himself on his herd of 100 cows as well as on his legendary talent with genetics, said that he would not attend the cow sales any more because he could not bear to see the condition of the cows discarded after working to make a farmer's bottom line.

Concern about Bovine Growth Hormone extends beyond animal welfare consideration. There is some evidence of a link between rbGH and increased levels of insulin growth factor (IGF-1). This occurs naturally in milk, but elevated levels may be associated with high levels of prostate and breast cancers.

The US Food and Drug Administration assures the public that the small amount of IGF-1 that may be retained after drinking rbGH-enhanced milk, would probably not travel beyond the stomach of the drinker, and that the amount in question is only a small fraction of a per cent of what the body makes itself. But, hormones are substances that are present in very small amounts and exert very strong and varying effects on various types of tissue. Since there is tentative experimental evidence that this hormone can pass from the gut to tissue in some studies involving rodents,[18] there is mounting concern about the safety of the rbGH use.

Furthermore, it is argued that rbGH use will make it impossible to choose good bull stock because it will not be clear whose daughters are better milkers, which is how a bull is judged in dairy farming. This is also true of cows that are bred to increase milk in their daughters. How will the effect of rbGH be distinguished from genetic selection? Many farmers are concerned that this may be the end of the scientific breeding tradition that made the US Holstein the most important dairy cow.

Another concern raised by those who worry about rbGH in the milk supply is, that 'if BST results in a large increase in milk supply and federal prices drop, then farm milk price could fall drastically'.[19] The conclusion may be, that 'in the long run, dairy farmers who remain in business may not be any better off than they were before'.[20]

In fact, they may be worse off. Many countries that traditionally purchase US meat and milk products have refused to purchase products from animals injected with rbGH. A ban on hormone-treated US products will eventually have a serious effect on the US dairy farmer.

The pressure is mounting on the farm and in the marketplace. On the one hand, the New York 'family farm' may be fast disappearing because cost is soaring, while profit is growing more difficult to achieve. Purity of the milk supply must be weighed against survival of the farm. In the marketplace, rbGH is encouraged by those that sell it. Monsanto, the company that owns the patent, will not retain it forever and money must be made while they retain the rights. This means that farmers must be persuaded to buy the product, researchers must be persuaded to support use of the product, and countries importing US meat and milk must be persuaded to allow the product in. Otherwise, Monsanto might not make that profit.

BIBLIOGRAPHY

1999 Dairy Statistics, US Dept. Agriculture, National Agriculture Statistics Service, 1999 Statistics.
Castle, M., & P. Watkins, *Modern Milk Production*, Faber & Faber Pub. Co., 1979.
Dairy Farm Management, Business Summary, New York State, 1997, W. Knobloch & L. Putnam.
The Economist, 7/3/99.
Giblin, J., *Milk, The Fight for Purity*, T.Y. Crowell Pub.Co., 1986.
Jarus, L., *Potential Effect of Two New Biotechnologies on The World Dairy Industry*, Westview Press, 1996.
Knobloch, W., L. Putnam, B. Stanton, *Census of Agricultural Highlights, New York State, 1977* Cornell University Press.
Mansfield, R.S., *Progress of the Breed: The History of US Holsteins*, Holstein- Friesian World Press, 1985.
Novakovic, A., & M. Kenniston, 'Possible Implications of BST on Dairy Markets and Policy', Dept. Agricultural Economics, Cornell University, USDA Policy Document.
Rath, S., *The Complete Cow*, Voyageur Press, 1998

REFERENCES

[1] *Census of Agricultural Highlights*, Figure 2.
[2] *Census of Agricultural Highlights*, page 2.
[3] *Census of Agricultural Highlights*, Figure 1.
[4] 1977 is the latest year for which data is available, due to the 5 year interval between census collection.
[5] *Census of Agricultural Highlights*, Tables 1 and 2.
[6] One billion is used as 1,000 million.
[7] From *1999 Dairy Statistics*, Table 8-9—Milk.
[8] *Census of Agricultural Highlights*, Figure 26.
[9] For an entertaining listing of cow varieties see Rath.
[10] Mansfield, page 7.
[11] Mansfield, page 7.
[12] Mansfield, page 10.
[13] Mansfield, p. XVII.
[14] From *1999 Dairy Statistics*.
[15] From *1999 Dairy Statistics*.
[16] 17 From *1999 Dairy Statistics*.
[17] *Dairy Farm Management*, Table 58.
[18] 'Turning Dairy Cows into Milk Machines', Humane farming association (www.hfa.org). See also *The Economist* 7/3/99.
[19] 'Possible Implications of BST', p. 2.
[20] 'Possible Implications of BST', p. 2.

E. van Hove, Making Farmhouse Gouda.

Farmhouse Gouda: A Dutch Family Business

Janny de Moor

Gouda,[1] one of the most important traditional Dutch cheeses,[2] derives its name from the city of Gouda, situated centrally in the grazing-land of Zuid Holland (South Holland). From the viewpoint of agriculture the river-clay of this part of the Netherlands suited nothing so well as for breeding cattle. The presence of prosperous cities guaranteed brisk sales of dairy products. The city of Gouda was the main market for the cheese made at these farms.

The picture accompanying this paper is very dear to me. It hung on the wall of my parental home when I was young. My father stemmed from a family of cheese-makers in central Holland and although he himself was one of the many young men who bade stock farming a farewell in the 1930s, I was instructed in the secrets of Dutch cheese-making early on. My father used to explain to me the various phases of the process depicted in this reproduction of a painting by E. van Hove. When I went to stay with my grandmother and aunts during the summer holidays I marvelled often at the accurateness of this painting which represents all major stages of the making of farmhouse Gouda.[3]

The queen of Gouda-making is the farmer's wife. On her wooden shoes she stands on the well-scrubbed red flags of her sanctuary, the special cheese-making room. She is cutting the curd with a kind of rack. This required a lot of strength, but it was especially tiring because she had to stoop to reach the middle of the enormous wooden tub in which she was working.

Farmhouse Gouda cheese is made from raw milk which – still lukewarm – was brought twice a day to the farm directly from the cows in the meadows. As you may discern in the oval picture on the left wall, this was supposed to be a man's task, but in other pictures I have seen my mother and other young women toil under the wooden yoke as well.

Directly below the oval picture is a little shelf with an alarm, measuring glass and bottles of a starter – a culture of lactic acid bacteria to sour the milk, which helps the clotting and improves the taste – and calf rennet, an enzyme from calf stomach to clot the milk.[4] Until very recently the farmer's wife made her own culture of lactic acid. This was a tricky business which often took her sleep away as she lay thinking 'Did I do everything right?'

The fresh milk was sieved from the bucket into the galvanized churns on the left.[5] From the churns the lukewarm milk is poured into the tub and first the *zuursel* (starter) and half an hour later the *stremsel* (rennet) is added. After about 30 minutes the milk becomes a soft clot which is cut with the rack into a mass of small pieces of soft curd floating in the whey. The cutting takes approximately twenty minutes by hand (mechanically fifteen minutes).

Some of the whey is drained into the bucket to the left of the tub. The curd settles to the bottom. In order to force it to release still more whey, hot water (80°C) is added twice. Again the mixture is stirred for about fifteen minutes. The shrinking of the curd to exactly the right grain is important to the flavour and texture of the Gouda, and is the cheese-maker's most valued expertise. More moisture means that the cheese will be softer and must be consumed young. Less moisture renders the cheese more suitable for ageing.

The whey is collected in the tall vats you see standing around. The cream slowly rises to the surface and formerly was often churned into so-called whey-butter. Nowadays it is often fed to the calves because it is an excellent nutrient for them.[6]

When most of the whey has been drained, the curd is collected in relatively flat wooden moulds lined with special cheese-cloths with wide pores through which remaining whey may ooze out. Heaving the curd from the bottom of the tub required great strength and I have often seen my grandfather and uncles help their wives with this particularly heavy chore. Typical for the Gouda is the lid (*volger*) put on the curd in the mould before it goes into the press where it remains for about five to six hours. After about 45 minutes the cheese has to be turned.[7]

Subsequently the fresh cheese is put into a brine bath for four to six days and has to be turned daily)this phase does not seem to be depicted in van Hove's painting). Afterwards, they were laid out on long tables and had to be turned daily for several weeks. If I tell you that normally each Gouda weighs about twenty kilograms, you can imagine that my aunts were fairly muscular women! To prevent the cheeses from getting mouldy, they had to be scrubbed several times. Since 1957 the cheese is covered in a breathing kind of plastic. But fungi are still a cheese-maker's nightmare and the Dutch government permits the addition of a tiny amount of Natamycine (a mould-killer) to the plastic. Of course the wisdom of this is questioned by some nowadays.[8]

And finally the farmer himself, leaning in over the hatch, smoking his pipe and smiling approvingly at his hard-working wife. From a modern point of view one might be inclined to condemn the idling fellow. But then we might forget that at this hour in the morning he already has a busy working day behind him, milking the cows at five in the morning (and again around five in the afternoon). And as I said, the couple regarded their cheese-making as a family business in which both took pride and in which the one stood for the other if need be. I interviewed many farmer's wives who have now retired from making cheese themselves, but none of them complained about the hardship. I should add that the older children were expected to shoulder about a third of the workload of a cheese-farm. Nevertheless, it was the man who went to town to deliver the ready product. Until 1957 it was even forbidden by law for a woman to go to the market to sell the cheese she had made herself.[9]

As I said, formerly the city of Gouda itself was the main market for farmhouse Gouda. A beautiful *waag* – a building where the cheese was weighed according to fixed standards – recalls this glorious past. Nowadays there are still two markets for farmers' Gouda, in Bodegraven on Tuesday and in Woerden on Wednesday. Most Gouda cheese, however, is sold not on the market, but directly to the customer and to wholesale merchants. Since the fourteenth century it has become customary in large areas of Holland to sell the bulk of Gouda to merchants who visit the farms beforehand to inspect the quality of the cheese and to place orders if they find it satisfactory.[10] Every four to five weeks the farmer delivers some of the cheese in town. I vividly recall these festive occasions. Uncle Jos even had a beautiful white horse to draw the wagon especially for these trips. One of the children was allowed to accompany him on the driving-box and was envied by all the others who had to stay home.

Gouda's past

Cheese-making was an invention of the Stone Age. Cheese was already a common dairy product with the Egyptians and Sumerians from *ca.* 3000 BC.[11] Probably it was discovered accidentally when a cattle breeder tried to store some milk in a rennet-sack. It proved to be a much better way to preserve milk than anything else. It seems likely that in Holland too cheese-making started in prehistory. In archaeological excavations of cattle farms dating from the Iron Age, pottery cheese-moulds were found.[12]

When the Romans invaded our country in 57 BC Julius Caesar describes the fare of our forefathers as milk, cheese and meat.[13] The Roman soldiers themselves were accustomed to rations which included hard cheese made from raw milk.[14] It is possible that in addition to the word *kaas* (cheese), which the Dutch borrowed from the Latin *caseus*, the inhabitants of the Low Lands learned to make hard cheese from the Romans.

About the history of cheese in Holland up to the late Middle Ages little is known. However, in the thirteenth and fourteenth centuries the dykes and ditches were improved which resulted in increased possibilities for breeding large cattle.[15] Cow's cheese seems to have been of the Gouda type, compressed with a hard rind in contrast to the more common soft cream cheese.[16] Among many other varieties of cheese, documents dating from the middle of the fourteenth century mention *perskaas* (pressed cheese) which is probably Gouda.[17] The *Teutsche Speiskammer* of 1550 observes 'im Niderland haben die Holendische kaesz das erst lob' (in the Netherlands the cheeses of Holland get the highest praise).[18] Dutch cheese was exported in large quantities, especially the well-known Edammer.[19]

In the sixteenth century, farmers started to devote greater care to Gouda's rind. They began to rub it with salt to protect the cheese from insects. To improve the taste they let it ripen somewhat longer. Furthermore they regularly scraped the mould off.[20] Economically, cheese became a most important national product. In 1567 the Italian traveller Ludovico Guiccigardini gave a description of the Netherlands which the Dutch found so flattering that they translated it: '[A country] where so much cheese and butter is made that a person who did not see it for himself on the spot would not believe it.'[21]

Dutch cheese was now exported all over Europe. Boekel cites evidence of a transport from Hoorn to Spain and Portugal in 1586.[22] Far more common were cargoes of cheese to France (Paris, Rouen, Nantes, Bordeaux), Northern Germany and Scandinavia which were shipped from Amsterdam whereas Rotterdam provided the countries along the Rhine and Maas.[23] Both cities solicited Britain's custom.

Still-lifes from the seventeenth century by painters like Clara Peeters (*ca.* 1612) and Floris van Dijck (1613) show a great variety of cheeses, tastily described by Simon Schama, 'venerable golden *overjaarige* Gouda sat beside green Texel or younger Edam'.[24]

From this period too, date the first more-or-less accurate descriptions of cheese-making. Schoockius tells about cheeses made from whole raw milk, for example Gouda, and from skimmed milk, for example Edam and Leyden. Farmers experimented with various recipes. Cheeses from skimmed milk were scented with cumin seed or cloves, even coriander seed and caraway seed, and often got an infusion of sheep dung.[25] Fortunately this idea was later abandoned in favour of parsley juice.

Several Dutch poets of the seventeenth century praise the qualities of cheese. Jacob Westerbaen, a poetaster who specialized in verses describing food, sets off the honest and frugal character Kees against the flabby rich Tijs:

> Kees eats only cheese and bread,
> and says it has strong flavour.
> Tijs eats partridges,
> but they give him no savour.[26]

Doctor Stephanus Blankaart, a seventeenth-century general practitioner in Amsterdam, makes cheese say:

> End up your meal with me
> and accompany me with your best beverages.
> You will thank me for it more
> than for a dish of rice-pudding.[27]

Despite the somewhat unfair comparison the good doctor makes, it is interesting that he advises to conclude a fine dinner in much the same way as we still do. And as they did already in the early Middle Ages.[28]

In the Dutch Golden Age the joys of life were seen not as the well-earned reward for hard work, but as the undeserved blessings of God.[29] Melchior Fokkens describes Holland in biblical metaphors as the promised land and sees cheese as one of its main assets:

> As it was said by men in the Olden Times that there was a land that flowed with milk and honey, truly it is in our Holland and here in Amsterdam that the town overflows with milk and cheese and butter brought to market every week. So that truly may it be said that our Holland overflows with butter, cheese and milk and that these blessings we receive from the hand of the Almighty.[30]

In the eighteenth century, Dutch export of cheese took on such proportions that countries like France and Britain decided to put a ban on its import. Not for long, however, because already in 1793 Rotterdam alone exported 2,038,727 pounds of cheese to England.[31] At the beginning of the nineteenth century J. le Francq van Berkhey gives precise descriptions of the major Dutch types of cheese: Leyden, Gouda, Edam and green sheep cheese from the island of Texel. The Gouda which he calls *zoetemelksche, goudsche kaas* (cream Gouda cheese) was not easy to make:

> Its manufacture requires not so much a lot of work, but rather good care and attention, because it is made from milk which comes directly from the cow and therefore contains not only lactic substances but also cream or butter.[32]

He goes on to describe the production process in great detail which allows us to note some differences with the present day making of farmer's Gouda. The curd was not cut with a rack but was stirred with a wooden porringer or with the hands. The adding of hot water was already done and the shape of the cheese moulds was identical. But a brine bath was still unknown; one continued rubbing the rind with salt. The cheese was stored on shelves in the cheese room which remained closed by day to keep the flies out. Only in the early morning or evening one let in some cool air.

> During the daily turning of the cheeses one notices if there are holes or spots where cheese-maggots come out, and discovering them, one takes them out with a knife, leaving the hole open till the next day to see if more maggots will come forth. When all are removed, the hole is stopped with some other cheese. But if the maggots are too numerous, pepper should be poured in the holes.[33]

People were not squeamish about their food in those days! Le Francq van Berkhey plays down the amount of work involved in cheese-making, but the fairly popular writer Hildebrand had a keen eye for the workload of the farmer's wife:

> There she stands, she who, after his beasts, rates highest in the esteem of her dearly loved husband Dries Riek. I say: after his beasts. For if his beasts die, buying replacements costs him money. But a new wife is found for nothing and might even bring in a penny herself. …The true destiny of the North Holland farmer's wife is cheesing, cheesing, always cheesing. Continually taking good care that the milk which is brought in morning and evening does not leave the door but in the shape of healthy cheese without a crack. And this creates so much work for her that one wonders how she finds the time to bear children. Yet she gets them in large numbers. But as soon as the 'puppy' (new-born) has been admired by the neighbours for some three days … she promptly leaves the delivery room to go back to the cheese tub.[34]

Understandably the move from manual cheese-making at the farm to machine production in factories[35] was defended by a sudden concern for the fate of the cheese-making wife. Liberated from her hard labour at the cheese tub, she would be able 'to devote herself entirely to her vocation as wife and mother'.[36] Was he dreaming of an even fuller quiver? However, as I said earlier, most couples took considerable pride in their cheese-making which they rightly regarded as skilled business in which both husband and wife had their own responsibility and territory. Dutch factory cheese is of an internationally-recognized quality. But farmhouse cheese undoubtedly has more character. Despite ever more regulation to achieve a standard quality, every farmhouse cheese has its own unique taste and texture. A farmer's wife who regularly enters her cheese in competitions and has won several prizes told me she is able to recognize her own Gouda among hundreds of others. Even though many farmers succumbed to the temptation to stop making cheese at home and sold their milk to the factory, some did not want to give up the old tradition.

So a movement to preserve the artisanship of making cheese at the farm was started which in 1914 resulted in the founding of an Association of Cheese Producers (*Bond van Kaasproducenten*), though it is nowadays called *Bond van Boerderij-Zuivelbereiders* (Association of Farmhouse Dairy Producers) – with a woman as president, naturally. This organization has done a lot to maintain and improve quality. In 1938 Southern Holland and Utrecht still produced 30 million kilograms of farmhouse cheese, mainly Gouda. But the number of active producers continued to fall. According to C. Schiere the main reason why many farmers stopped making their own cheese was social changes. Such a family business could only be run successfully if father, mother and a sufficient number of children devoted their lives to it. When it became unacceptable to burden the wife with two tasks (housekeeping and cheese-making), when the size of families decreased and children opted for a career outside the village, it was simply not feasible any more to continue.[37]

Yet the enthusiasm remained. In 1947 the first issue of *De Zelfkazer* (The Independent Cheese-maker) appeared. Despite gloomy predictions that it would be impossible to fill its pages every fortnight, the magazine still exists. It informs its readers about new developments like the *dagkaasinstallatie* (machinery making it possible to make cheese only once per day, not twice after every milking) and deep-freeze *zuursel* which took away one of mother's major worries. Since her husband too has got all kinds of new appliances, especially the milking-machine, the burden of making your own cheese has been alleviated considerably. But obviously nothing can stop the children from choosing a different profession and therefore the question must be raised: does farmer's Gouda still have a future?

Gouda's future

If you travel through Holland you will notice signs pointing in the direction of farms where *boerenkaas* (farmer's cheese) is still made and sold directly to the public. Because this eliminates intermediate trade and because a guided tour through the farm is appreciated by the visitors, it is profitable for the producers. Moreover, the growing number of people who – rightly or wrongly – are suspicious of industrial food processing has created a whole new market for biologically 'safe' products directly from the farm. Many leaders in the Association advocate a policy further developing this new potential.

I must confess that I am rather sceptical about the future of farmer's Gouda. The total farmhouse cheese production in the Netherlands has shrunk to about 9,000 tons a year, made by about 500 farmers. This is a mere 1.5 per cent of the total national production. Receiving visitors at any time of the day puts an extra burden on the family. Everything has to be nice and clean, tea and coffee must be served, and cheese-making people are not by chance rather taciturn. Of course it is always inspiring to hear from your own customers how much they appreciate the work of your hands. But advanced research has led to spectacular improvements in factory-made Gouda too. And further mechanization at the farm will tend to bring farm and factory still closer together.

However, I am sure that there will always be a niche for exclusive products. And exclusive it is, the taste of real Gouda farmhouse cheese: a kind of tickling, electrical sensation in the mouth. A rich flavour created by enzymes, which break down the fat in raw milk, making farmhouse cheese tastier than factory Gouda, made of pasteurized milk. Moreover, farmhouse Gouda is fatter: 55 per cent of dry matter whereas cheese from the factories has only 48 per cent. That accounts for its extra smoothness. And finally, who would overlook the love put into this cheese? Makers of farmhouse Gouda never get rich; they do it out of a kind of vocation.

At the moment every cheese has a Dutch State Seal of Approval guaranteeing that it conforms to the strict government rules for Gouda. The farmers themselves have founded a new organization for quality control which issues licenses which are comparable to the French *appellation contrôlée*. All very laudable initiatives. But new European guidelines will make the farmers themselves responsible for the safety of their products by 2004. That will make producing your own cheese even more burdensome than it already is because not only every step in the process has to be kept under strictest control – responsible cheese-makers are accustomed to that – but every farmer will have to take expensive insurance against liability claims. Twice a year an overall inspection of the farm will take place which may result in fines and more restrictive regulation. Some Gouda farmers expect that this will be the end of their family business. But customers fortunate enough to have tasted the real farmhouse Gouda, preferably the two-year-old variety, will no doubt agree that it is a speciality deserving to survive.

REFERENCES

[1] For kindly providing information I want to thank Ir Tineke van der Haven (Quality Expert of the National Reference Centre for Agriculture), Ing. Th.S. Zwart (Director of the Foundation for Quality Supervision of Farmers' Dairy Products and Secretary of the Association of Farmhouse Dairy Producers), Aad Vernooij (Dutch Dairy Bureau) and last but not least Bep and Adri van Dijk, devoted cheese-makers in Wilnis, also for their hospitality and the demonstration of their skill.

[2] Other traditional types include Edam, the spherical cheese of Noord Holland (because of which Belgians call us 'Cheeseheads'), Leyden (a meagre cheese with cumin seeds and a red rind) and Frisian (with cloves and cumin seeds), all made from skimmed milk.

[3] For a more technical description than the one that follows, see Tineke van der Haven, Henk Oosterhuis, *Rondom de boerenkaas*, Lelystad: PR, 1999, pp. 6ff.

[4] Nowadays also some other substances are added, like potassium nitrate to prevent undesirable formation of gas in the cheese, and calcium chloride to stimulate clotting in wintertime.

[5] Upon seeing the picture, my knowledgeable friend Tineke van der Haven pointed out to me that here the painter has permitted himself an artistic liberty: in actual practice the sieve and the churns would long have been removed and cleaned. Cheese-making requires utmost cleanliness.

[6] This is the reason why the potassium nitrate is only added at the second 'heating'. Nowadays Turkish immigrants in the Netherlands are eager to buy the whey-butter. They even take it with them to Turkey as a special treat, better than their own *kaymak*.

[7] Wooden moulds are warmed beforehand and the modern lighter plastic moulds are covered with plastic to keep them warm: the rind will be better if the cheese does not cool too soon.

[8] T. van der Haven, 'Ook de kaaskorst heeft zijn problemen ...', *De Zelfkazer*, February 1997, No. 2, pp. 7–8.

[9] Sabine de Rooij, *Werk van de tweede soort: Boerinnen in de melkveehouderij*, diss. Wageningen 1992, p. 171.

[10] P.N. Boekel, *De zuivelexport van Nederland tot 1813*, Diss. Wageningen. Utrecht: Drukkerij Fa. Schotanus & Jens, 1929, p. 23.

[11] A. Zaky, Z. Iskander, 'Ancient Egyptian Cheese', *Annales du service des Antiquités d'Égypte* 41 (1942), pp. 295–313; A. Lucas, *Ancient Egyptian Materials and Industries*, London 1959, p. 383; H. Limet, 'The Cuisine of Ancient Sumer', *Biblical Archaeologist* 50 (1987), pp. 132–147. Cheese also occurs in documents of the ancient Babylonians, Hittites, Ugaritians, Phoenicians and Israelites.

[12] See for the most recent finds http://archweb.leidenuniv.nl/fa/recentoz/md1007bn.html.

[13] C. Iulius Caesar, *De Bello Gallico*, Libri VII, Liber VI, cap. 22: 'Agri culturae non student, maiorque pars eorum victûs in lacte, caseo, carne consistit'. For an English translation see http://classics.mit.edu/Caesar/gallic.6.6.html.

[14] J. André, *L'alimentation et la cuisine de Rome*, Paris: Les Belles Lettres, 1981, p. 154.

[15] L. Burema, *De voeding in Nederland van de Middeleeuwen tot de twintigste eeuw*, Assen: Van Gorcum, 1953, p. 14.

[16] J.M. van Winter, in A. Riddervold, A. Ropeid (eds.), *Food Conservation: Ethnological Studies*, London: Prospect Books, 1988, p. 60.

[17] F. E. J. M. Baudet, *De maaltijd en de keuken in de Middeleeuwen*, Leiden: A.W.Sijthoff [1904], p. 101. Because it is distinguished from 'Haarlemsche' *kaas* it can hardly be 'Edammer'. J. Bieleman, *Geschiedenis van de landbouw in Nederland 1500–1950*, Meppel/Amsterdam: Boom, 1992, p. 61, surmises that hard cheeses have been made since the 12th century or earlier.

[18] Baudet, *De maaltijd en de keuken*, p. 101.

[19] Vernooij, *Hard van binnen, rond van Fatsoen: Geschiedenis van de Nederlandse kaascultuur*, Het Nederlands Zuivelbureau, 1994, p. 71, on the basis of a toll ledger for the years 1439–1441 from the city of Kampen.

[20] Bieleman, *Geschiedenis van de landbouw*, p. 61.
[21] '(Een land) waar soo groote menichte van kase ende botter ghemaeckt wordt, dat de gehene die het terselver plaetsen niet teghen woordichlijck gesien en hadde, nimmermeer gheloovenen soude'. Quoted by P.N. Boekel, *De zuivelexport van Nederland tot 1813*, Utrecht: Schotanus & Jens, 1929, p. 12.
[22] Boekel, *De zuivelexport*, p. 161.
[23] Boekel, *De zuivelexport*, pp. 123, 137.
[24] Simon Schama, *The Embarrassment of Riches: An interpretation of Dutch Culture in the Golden Age*, London: Fontana Press, p. 161.
[25] Boekel, *De zuivelexport*, pp. 43–44.
[26] 'Kees eet maer Kaesenbrood, en zeyt 't heeft roock en smaek. Tijs eet patrysen, en't en geeft hem geen vermaeck.' Quoted by Schama, *The Embarrassment of Riches*, p. 162.
[27] 'Besluit uw maaltijd maar met mij, / En mengt mij met uw beste dranken. / Gy zultmij daarvoor meer bedanken / Als voor een schootel rystenbry.' S. Blankaart, *De Borgerlijke Tafel*, Amsterdam: Jan ten Hoorn, 1683.
[28] 'Na den eten een luttelkijn ghegeten verduwet sy die spise' (After dinner a little bit (of cheese) digests the food.), Baudet, *De maaltijd en de keuken in de Middeleeuwen*, p. 101.
[29] See for this typical Calvinistic attitude, Janny de Moor, 'Dutch Cookery and Calvin', in: Harlan Walker (ed.), *Cooks & Other People: Proceedings of the Oxford Symposium on Food and Cookery 1995*, Totnes: Prospect Books, 1996, pp. 94–105.
[30] Translation Schama, *The Embarrassment of Riches*, p. 130. Dutch text: 'En land was toegezeyt, dat vloeyde van honig en melk, waarlijk het is ons Hollandt, en hier in Amsterdam, waar is een landt en stat daar de melk en kaas zoo goet en overvloedig is als hier? Daar zulk een markt vol kaas en boter alle weeklijke maandagen te koop wordt gebracht? Met waarheyt mag men zeggen dat dit ons Hollandt als van boter, kaas en melk overvloeyt, en zulk een zegen heeft deze plaats zoo mildelijk van Godts hand ontvangen'.
[31] Boekel, *De zuivelexport*, p. 136.
[32] J. le Francq van Berkhey, *Natuurlyke historie van Holland*, vol. 9, Leiden:P.H. Trap, 1811, p. 447.
[33] Le Francq van Berkhey, *Natuurlyke historie*, vol. 9, pp. 452f.
[34] 'Daar staat zij nu, die na zijn beesten, het hoogst staat aangeschreven in de schatting van Dries Riek, haar welbeminden echtgenoot. Ik zeg na zijn beesten. Want als zijn beesten sterven, kost de inkoop van andere geld; een vrouw is om niet terug te vinden en brengt mogelijk nog wel een stuivertje mee ... De bestemming der Noordhollandsche boerin, als zodanig is keezen, keezen, altijd keezen; is bestendig zorgen dat de melk die 'ochtends en 's avonds "na melkerstaid" wordt binnengebracht, de deur niet uitga dan in de gedaante van gezonde nietbarstende kaas. En dat geeft haar dagelijks zoveel werk, dat men niet weet hoe zij den tijd vindt om kinderen te krijgen. Nochtans krijgt zij die in grote menigte. Maar ook als het "puppie" (de pasgeborene) een dag of drie door de buren is "gekeken" ... verlaat zij de kraamkamer alweer,en begeeft zich oogenblikkelijk aan de kaastobbe.' Hildebrand (N. Beets), *De Nederlanden: Karakterschetsen, kleederdrachten, houding en voorkomen van verschillende standen*, 's Gravenhage: Nederl. Maatsch. Van Schoone Kunsten 1841, pp. 161–164.
[35] This took place between 1860–70, Burema, *De voeding in Nederland*, p. 218.
[36] 'om zich ten volle te kunnen wijden aan hare roeping als vrouw en moeder', H.B. Hylkema, *Leerboek voor zuivelbereiding*, Leeuwarden: R. van der Velde, eerste druk 1896, 2e herziene en vermeerderde druk, z.j., p. 9.
[37] C. Schiere, *Het bedrijf van de zelfkazers in Utrecht en Zuidholland*, Utrecht: Da Costa, 1938.

Milk and Dairy Products in the Roman Period

Carol A. Déry

Milk as a beverage

In its liquid state milk did not feature as an important component in the diet of the typical Roman urban dweller of the Late Republic and Imperial periods. This was due, in the first instance, to reasons of practicality: milk was produced on farmsteads outside the towns and cities, and in the hot Mediterranean climate it was often difficult to keep milk fresh during transport to urban centres. In the second instance, intellectual Romans associated the drinking of milk with barbarians (non-Romans) and nomads, whom they considered to be unrefined and uncultured because they were pastoralists as opposed to settled agriculturalists, living off their animals instead of farming the land. Such peoples were classified as 'eaters of flesh and drinkers of milk' in ancient ethnography and historiography, a designation which signified their place on the periphery of the civilized world, and was applied to the Celts, Britons, Germans, and Scythians for example.[1] Although it was acceptable for the Roman ancestors to be portrayed as consuming milk, in later times milk-drinking became unfashionable in Rome among the adult population because it was regarded as an uncivilized activity on account of the mental connection with barbarians. It was, seemingly, still acceptable for children to drink milk.

Of course in the countryside, where practicality necessarily superseded ideology, milk was regularly consumed by the peasantry, who drew a steady supply of fresh milk from the animals they kept. For these people milk and dairy products constituted a valuable source of nutrition, especially protein and calcium. Sheep's and goat's milk were the most common types, partly because these animals were more abundant in Italy than cows, but goats also yielded prolific quantities of milk, more than any other animal in proportion to their size, so it has been claimed.[2] Mare's milk and camel's milk are also mentioned in the literary sources, though camel's milk was uncommon in Rome and was restricted in the main to the provinces of Africa and Asia where these animals lived. Pliny noted that camel's milk was agreeable if diluted with three parts water, though he omits to say whether his recommendation derived from personal experience or not (*Natural History*, 11.96.236). Cow's milk however is little mentioned as a drink by the ancient authorities, and White even refers to it as something of 'a rarity in Roman Italy' (1970: 277). Country people sometimes flavoured their milk with a sprig of parsley according to Pliny (*Natural History*, 20.44.112).

Opinion differed as to which type of milk was the best. Varro thought sheep's milk the most nourishing, followed by goat's milk (*On Agriculture*, 2.11.1–2). Pliny awarded first place to goat's milk, holding cow's milk to be more medicinal and soothing, but sheep's milk to taste sweeter (*Natural History*, 28.33.123). Galen however thought cow's milk was the richest and thickest, followed by mare's, then goat's and sheep's (*On the Properties of Foods*, 3.14.7 = 6.684K). Mare's milk was considered greatly purgative, followed in this capacity by ass's, cow's and goat's milk. Pasturage and the seasons were held to affect the flavour and properties of milk. Milk that came from animals that were fed on dry solid food, like straw and barley, was adjudged most nourishing for humans, while the milk from animals that fed on grass was laxative in humans. Milk was of a more watery consistency in the spring when the grass was lush.

Colostrum (beestings – the first milk produced by the female, which is rich in nutrients) was also appreciated, and is included by Martial in his list of guest-gifts (*Epigrams*, 13.38).[3] It was probably the nearest thing to cream that the Romans knew.

Cheese

It is in the form of cheese that most of the milk produced in Roman Italy was consumed. Cheese-making presented a practical solution to the problem of trying to prevent milk from deteriorating too rapidly in a hot climate, and was a necessary task in those parts of the country too far removed from the town to make the transportation of fresh milk a viable option. On the farm, milk in the form of cheese could be stored to provide vital sustenance throughout the winter months when milk production was comparatively low.

Columella describes a method of making hard cheese (*caseus*) in his agricultural manual (*On Agriculture*, 7.8.2–7). Full cream milk, as fresh as possible, is curdled with animal or vegetable rennet and warmed gently. When the liquid becomes thickened, it is transferred into wicker baskets and the curds allowed to gradually separate from the whey. This part of the process could be speeded up by placing weights on top to hasten the draining of the whey. The curds are put afterwards into a cool place and sprinkled with salt to exude any remaining liquid. They are then pressed with weights, and sprinkled with more salt for nine days, after which the cheese is washed in clean water. Finally the moulded cheeses are placed on racks in a sheltered spot to dry. Cheese of this type could readily withstand transport over long distances, even overseas, and thus provided suitable fare for travellers and soldiers on the march.[4]

Soft cheese (*caseus mollis* or *caseus recens*) could be prepared more quickly, but needed to be eaten within a few days. The curds were simply removed from the wicker baskets, dipped into salt and brine and dried a little in the sun. Crushed pine nuts or thyme were sometimes placed in the pail before milking commenced to impart additional flavour to the milk.[5] 'Hand-pressed' cheese was also common: when the milk was thickened in the pail and still slightly warm, it was broken up and hot water poured over it. The curds were then shaped into rounds by hand or pressed into wooden moulds. Curd cheese was thus relatively quick to manufacture, but it did not stay fresh for long. Cheese made in the evening therefore had to be sold at the market the next morning.[6] Columella regarded cheese-making as a particular duty of the shepherd. Varro thought May to July the best time of year for cheese-making (*On Agriculture*, 2.11.4), and Columella similarly advised July (*On Agriculture*, 12.13).

Various types of rennet were used by the Romans to curdle milk, such as rennet from a hare, kid or lamb, or ass's milk. Rennets of a non-animal origin included the sap from a fig tree, the juice of the fig itself, the flower of the wild thistle or artichoke, safflower seeds, or vinegar.[7]

Cheese could be pickled in brine or vinegar and thyme, but smoking also produced an appealing flavour (Columella, *On Agriculture*, 7.8.7; Pliny, *Natural History*, 11.97.242). The Romans were fond of smoked foods like smoked sausages, but Velabran smoked goat's cheese, which was made near Rome, was particularly esteemed.

Cheese might be eaten at breakfast (*ientaculum*) along with bread and the previous day's left-overs, at lunch (*prandium*), or as part of the early evening meal (*cena*), where it was normally served as a starter (*gustatio*) or occasionally as part of the dessert course (*secunda mensa*).[8] Cheeses were also given as gifts at the Saturnalia, and four different varieties are mentioned in this connection in Martial's list (*Epigrams*, 13.30–33).

The *Moretum*, an epyllion formerly attributed to Vergil, affords a brief glimpse into the everyday life of the rural peasantry in Roman Italy, and into the importance of cheese in their diet. The meat-rack with its customary flitch of bacon that is so resonant of rustic frugality in Latin literature, is absent from Simulus' cottage, the protagonist of the poem. Here instead a round cheese hangs, suspended from the rafters by a string through the middle, along with an old bunch of dill (55–8) – evidently a hard cheese stored up against times of want. Cheese makes a second appearance in the poem in the eponymous dish, which is made in a mortar.[9] First garlic is crushed with some salt, then pounded together with some cheese and a handful of fresh mixed herbs gathered from the garden (parsley, rue, and coriander). A few drops of olive oil and vinegar moisten the mixture, but the end result is not so much a loose paste as a substance thick enough to be shaped into a ball (92–116).[10]

Ancient Italy produced many fine cheeses. There was Coebanum cheese, made from sheep's milk, that came from Liguria, and Sarsina cheese from Umbria. Vestine cheese was a smoked cheese made from goat's milk which was made by the Vestini in central Italy between the Apennines and the Adriatic. Velabrum near Rome also produced a good smoked cheese. Trebullan cheese from Sabine country was thought especially good for toasting. Luniensine cheese, manufactured in Luna on the Tuscany-Liguria border, was remarkable for its huge size, for each individual cheese was said to weigh up to 1,000 lb. A number of excellent varieties of cheese were also imported into Rome from the provinces. Cheese brought from the district of Nîmes in France, from the villages of Lozère and Gévaudan, was thought best when eaten fresh. Bithynian cheese had a characteristically salty taste, while the cheese produced from Gallic goats apparently had a somewhat medicinal taste. There was also Docleate cheese that came from the Dalmatian mountains, and Vatusic from the Tarenaise.[11]

In the Late Republic and Early Imperial periods goat's or sheep's milk and cheese were evidently the preferred types, but during the later empire, the milk and cheese of cows began to acquire greater acceptance than previously. Cow's milk cheese (*caseus bubulus*) is specified as an ingredient in two of the three recipes for *sala cattabia* in Apicius – Vestine cheese in the other (*Roman Cookery*, 4.1.1–3).

Varro thought that cheese made from cow's milk was the most nourishing, but was difficult to digest. Cheese made from sheep's milk was a little less nourishing, but was easier on the stomach, with goat's cheese having the least nutriment, but being easily evacuated. Fresh, soft cheese was considered more nutritious and less constipating than old, dry cheese (*On Agriculture*, 2.11.3; cf. Pliny, *Natural History*, 28.131). Plutarch advised that cheese be used cautiously, because like meat, it had a tendency to lie heavily in the stomach (we would say that it is high in protein). He also cautioned against drinking milk because of its richness (*Rules for the Preservation of Health*, 131E, 132B).

Curdled milk

Acidified or curdled milk (often called by its Greek name *oxygala*, but also known as *melca(e)*) was familiar to the Romans, and was made simply by adding sour milk to fresh (Pliny, *Natural History*, 28.36.135). It seems to have been a similar product to yoghurt, and was eaten plain, or mixed with honey, or with oil from unripe olives (Galen, *On the Properties of Food*, 3.15; Anthimus, *On the Observance of Foods*, 78). Curdled milk also appears in Apicius, where it is flavoured with pepper, *liquamen* (fish sauce) or salt, olive oil, and coriander

(*Roman Cookery*, 7.13.9). A recipe in the *Geoponica* says that it may be made by pouring vinegar into new earthenware vessels and putting them on the hot embers or a slow fire. When the vinegar begins to boil, the vessels are removed from the heat and milk poured into them. After being allowed to cool undisturbed, *melcae* is formed by the following day (18.21). Columella gives instructions for making *oxygala*, but the process he describes would seemingly result in something more like a cheese than yoghurt. A vessel with a stoppered hole near the base is filled with fresh sheep's milk, and a bouquet garni of marjoram, mint, onion, and coriander added. The vessel is then covered for five days, after which the whey is drained through the hole in the bottom of the container. The hole is once again stopped and the same procedure repeated for a further three days. The bouquet garni is removed, and replaced with some dried thyme, oregano and chopped chives. After another two days, the whey is drained out, and the remaining substance salted (*On Agriculture*, 12.8.1–2).

As for curds Pliny mentions that those of the goat were praised, and those of the roedeer, hare, and rabbit appreciated, with the curds of the latter apparently being a cure for diarrhoea as well (Pliny, *Natural History*, 11.96.239). They were sometimes served with honey (Athenaeus, *Dinner of the Sophists*, 1.147e). Whey also had its uses, being fed to pigs (Cato, *On Agriculture*, 150).

Butter

Butter (*butyrum*), from the Greek *boutyros* meaning 'cow curds', was little used for cooking by the Romans, who preferred olive oil for this purpose. However the Romans were well aware of the popularity of butter among barbarian nations. Pliny mentions that it was usually made from cow's milk, though also from sheep's and goat's milk (*Natural History*, 28.35.133). He remarks that the barbarians used it for anointing themselves, and that the Romans sometimes used it for anointing children (*Natural History*, 11.96.239).

Recipes

The *Roman Cookery* of Apicius illustrates some of the ways in which milk and cheese were used in both sweet and savoury dishes. Recipes range from a simple directive to cook salted meat first in milk then in water to make it sweet (1.8), to a gourmet terrine (*patina ex lacte*): mallow, beets, leeks, celery sticks, vegetable puree, boiled greens, chicken pieces that have previously been cooked in stock, boiled brains, Lucanian sausages, halved hard-boiled eggs, chopped pork sausages stuffed with Terentian sauce, chicken livers, fried fillet of hake, jellyfish, oysters, and fresh cheese are arranged in layers in a dish and topped with pine-kernels and peppercorns. Then a mixture of milk and eggs flavoured with pepper, lovage, celery-seed and asafoetida is poured over and allowed to set.

Afterwards it is garnished with fresh corals of sea-urchins, and sprinkled with pepper (4.2.13).

In the recipes for *Pullus tractogalatus* (Chicken with a milk and pastry sauce – *Roman Cookery*, 6.9.13) and *Pullus Varianus* (Chicken à la Varius – *Roman Cookery*, 6.9.11) the milk forms the basis of a sauce to accompany the chicken. It is similarly employed in *Pultes tractogalatae* (Pottage with milk and pastry – *Roman Cookery*, 5.1.3) which is for a sauce to serve with lamb.

Dulcia (sweetmeats) may be made by first soaking must-cakes or wheat bread in milk, and then either baking or frying them, before coating them with honey (*Roman Cookery*, 7.13.2–3). Pliny mentions a similar sweet in which Picene bread (made from groats and raisin juice), was soaked in milk or honeyed wine before eating (*Natural History*, 18.27.106). Another sweet was made by pounding pepper, pine-kernels, honey, rue, and *passum* (raisin wine), and cooking these in milk and pastry. The resulting mixture was thickened with a few eggs, then topped with honey and pepper (*Roman Cookery*, 7.13.5). *Tyropatinam* is something of a misnomer since the word suggests a dish made with cheese, but cheese is not included among the ingredients – it is actually a set milk custard flavoured with honey (*Roman Cookery*, 7.13.7). *Ova spongia ex lacte* meanwhile was a type of omelette made with milk and served with honey and pepper (*Roman Cookery*, 7.13.8). *Patina versatilis* was a sweet nut omelette in which chopped nuts and pine-kernels were mixed with honey, pepper, *liquamen*, milk, eggs, wine, and then cooked in a little oil (*Roman Cookery*, 4.2.2 = 4.2.16).

Patina cotidiana (everyday patina) was a dish of boiled brains, pepper, cumin, asafoetida, *liquamen, caroenum* (reduced wine-must), milk and eggs. The instructions say simply to cook over a low fire or in a bain-marie. Depending on the method used, this might result in a type of omelette, scrambled eggs or a set custard (*Roman Cookery*, 4.2.1).

For the intrepid cook there was a recipe for *cocleas lacte pastas* (snails fed on milk): 'Take the snails, clean with a sponge, and remove the membrane so that they can come out [of their shells]. Put into a vessel [with the snails] both milk and salt on one day, and milk only on the remaining days, and clean out the excrement every hour. When the snails are fattened to the extent that they cannot get back in their shells, fry them in olive oil. Serve with *oenogarum* (wine-flavoured fish sauce).' (*Roman Cookery*, 7.18.1).

Cheese was included as an ingredient in all the cake recipes in Cato's agricultural manual (75–82, 84). *Libum* was a simple cheese loaf, baked on top of leaves. The instructions for making *placenta* are complicated, but would seem to result in a layered cheesecake, with sweet, fresh sheep's cheese mixed with honey as the filling. *Libum* and *placenta* were both sacrificial cakes, for use in religious ceremonies. *Savillum* was similar to *libum*, but was spread with honey after baking and sprinkled with poppy-seeds. *Scriblita* was a honey-less version of *placenta*, while *spira* was a spiral-shaped version of *placenta* retaining the honey. *Spaerita* was similar to *spira*, but in this case the cheese and honey mixture was formed into large spheres, rather than spirals, which were baked on top of a crisp base. *Erneum* used the same ingredients as *placenta* but was cooked in a bain-marie rather than being baked. *Globus*, made from cheese and spelt grits, was a globe-shaped, deep-fried fritter which was served dipped in honey and sprinkled with poppy-seeds. *Encytum* was made in the same way as the *globus*, but was shaped like a ring-doughnut. Although it is only specified in the recipe for *placenta*, it is likely that sheep's cheese would have been used in all these recipes.

Tyrotarichum was a cheese and salted fish dish that was often considered poor fare (Cicero, *Letters to his Friends*, 9.16; *Letters to Atticus*, 14.16). A more fancy version appears in Apicius: 'Cook any salt fish in oil; bone. Take cooked brains, boned fish, chicken liver, hard-boiled eggs, and soft cheese washed in warm water, and heat them in a shallow pan. Then pound pepper, lovage, marjoram, rue berries, wine, honeyed-wine, oil, put in the pan, and allow to cook gently over a low fire. Bind with raw eggs, garnish, and sprinkle with finely ground cumin' (*Roman Cookery*, 4.2.17).

Cosmetic and medicinal uses of dairy products

Ass's milk was frequently used as a beauty treatment by wealthy women as it was thought to impart whiteness to their skin and smooth out wrinkles. Poppaea, wife of the emperor Nero, was notorious for bathing in ass's milk, and reputedly took 500 female asses with her everywhere she went in order to have a ready supply of milk for her bath (Juvenal, *Satires*, 6.468–70; Pliny, *Natural History*, 11.96.238; 28.50.183). In Egypt, Cleopatra had entertained much the same reputation for her use of ass's milk as part of her beauty regimen.

Olive oil was traditionally used by the Romans as a moisturizer applied to the skin after bathing, but in cold climates where the supply of olive oil was frequently limited, butter was used instead (Pliny, *Natural History*, 11.96.239; Galen, *On the Properties of Foods*, 3.14.6=6.683K). Butter also found a use as a spot remover, as did sour cheese mixed with *oxymel* (honey vinegar).

In his *Natural History* Pliny dilates at some length upon the medicinal uses of milk, butter, and cheese (28.33.125–130). For example, fresh cheese mixed with honey was thought good for bruises. A cheese called *saprum* (from the Greek meaning 'rotten') could cure coeliac afflictions if taken in a drink after being pounded with salt and dried sorb apples. Carbuncles of the genitals could be treated by the application of pounded goat's cheese. Ass's milk could cure gout. Butter mixed with honey and rubbed onto the gums was believed to alleviate toothache and mouth ulcers. A junket made from goat's milk and honeyed wine, and stirred with a branch from a fig tree, could be used in the treatment of a range of ailments from epilepsy, depression, paralysis and leprosy, to joint pain.

Milk in Roman ritual

Milk was offered as a libation from the earliest times in primitive Roman religion, but when vines began to be cultivated in Italy, its function was to a large extent overtaken by wine. Pliny attests to the antiquity of milk in ritual when he says that Romulus, the mythical founder of Rome, made libations of milk, not wine (*Natural History*, 14.14.88; cf. 14.14.91). Pseudo-Tibullus mentions that milk was used to douse bones in funeral rites, and was also mixed with wine as a libation (*Elegies*, 3.5.34, 3.2.20). Details are frequently lacking on the precise nature of the offerings made to various deities, but we do know that milk played a part in the cult of Mercury (Festus, 423.9L), and was offered to the Egyptian goddess Isis, and the Persian god Mithras, along with honey and oil.

Some of the cakes mentioned by Cato are sacrificial cakes that would have been offered to the gods, and these included cheese among their ingredients. We may wonder if the presence of cheese had less of a culinary significance and more of a symbolic one – precedent was important in religion, and if wine had largely replaced milk as a liquid offering, perhaps the presence of cheese in the cakes was one way in which milk was reinstated to its former place in ritual. Alternatively, cheese was considered a primitive food in itself, and this fact alone would sanction its presence in offerings to the gods.

BIBLIOGRAPHY
André, J., (1981) *L'Alimentation et la Cuisine à Rome* (Paris).
Frayn. J., (1979) *Subsistence Farming in Roman Italy* (London).
White, K. D., (1970) *Roman Farming* (London).

REFERENCES
[1] See Shaw, B. D., (1982/3) 'Eaters of Flesh and Drinkers of Milk: The Ancient Mediterranean Ideology of the Pastoral Nomad', *Ancient Society*, 13/14, 5–31. Also Caesar, *Gallic War*, 4.1, 5.14, 6.22; Tacitus, *Germania*, 23; Columella, *On Farming*, 7.2.2; Pliny, *Natural History*, 11.96.239.

[2] White (1970), 315; Vergil, *Georgics*, 3.308–310; Columella, *On Agriculture*, 7.6.4.

[3] See also Varro, *On Agriculture*, 2.11.2.; Pliny, *Natural History*, 28.33.123. 'Colostrum' and 'soft cheese' are employed as terms of endearment by the playwright Plautus (*The Little Carthaginian*, 367).

[4] Cheese-making equipment has been found on several Roman military sites in Britain such as Corbridge on Hadrian's Wall. This indicates that at least some of the cheese consumed by the soldiers was made on site.

[5] The Roman milking-pail was called a *mulctra* or *mulctrum*, and was actually more of a shallow bowl than the high-sided bucket with which we are familiar, as may be seen from mosaics and wall-paintings. In Ps. Vergil, *Copa* 17, cheese is placed in reed baskets to dry.

[6] See Vergil, *Georgics*, 3.400–2.

[7] Varro, *On Agriculture*, 2.11.4–5; Columella, *On Agriculture*, 7.8.1; Pliny, *Natural History*, 11.96.237; Tibullus, *Elegies*, 2.3.14.

[8] For example, Martial serves lettuce, leeks, tunny fish, roasted eggs, cheese and olives as hors d'oeuvres (*Epigrams*, 11.52). When Jupiter and Mercury in disguise visited Philemon and Baucis in one of the most celebrated scenes of ancient hospitality, they were served olives, Cornelian cherries, endive, radishes, cheese and roasted eggs (Ovid, *Metamorphoses*, 616ff).

[9] It has been suggested that the name of the dish 'moretum' derives from the mortar 'mortarium' in which it was prepared, but the etymology is by no means certain.

[10] Margaret Visser (1993) has argued for *moretum* being an early version of pesto in '*Moretum*: Ancient Roman Pesto' in H. Walker (ed.) *Spicing up the Palate: Studies of Flavourings – Ancient and Modern*, (*Proceedings of the Oxford Symposium on Food and Cookery, 1992*), (Totnes), 263–274, though I am inclined to disagree with her on this point. Other versions of *moretum* appear in Apicius (*Roman Cookery*, 1.21) and Columella (*On Agriculture*, 12.59.1–4).

[11] Various types of cheeses are mentioned by Pliny, *Natural History*, 11.97.240–2; Martial, *Epigrams*, 1.43; 3.58; 11.52; 13.30; 13.31; 13.32; 13.33.

Carabao Milk in Philippine Life

Doreen G. Fernandez

To the Filipino who grows up in the provinces, the carabao (*Bubalus bubalis* or *Bos bubalis,* the water buffalo) is a dear and familiar part of the landscape. Generally black or dark grey, with occasional white ones (albinos), they are found in every province, and named in many native languages: *kalabaw, karabaw, karbaw, kabaw, kalakyan, damulag, nuwang, pagad, dueg.*

I remember carabaos grazing in the fields, often with white herons standing on their backs. As children we used to envy the boys, called *bakeros* (from the Spanish *vaqueros),* who had charge of them, sitting easily on their backs even as they led them to the river to wallow, and out again, to be tied to a tree or fence. We were allowed to ride them, with a *bakero* in attendance, and were fascinated by the way their skins slid from side to side as we rode.

Often they pulled carts, sometimes with us in them, and the mischievous *bakeros* would show us how their tails would lift to signal a bodily function, and how they could stop this by making a cross sign with their fingers in front of the lifting tail. Even carabao dung, in large round cakes, we were familiar with. When dry, they could be lifted whole, and used as fuel. The most imaginative use I have heard of was in Malolos, Bulacan, where, fully dried, soaked with kerosene and hung up on trees, they were lit, to supply light for the *sinakulo,* the Holy Week Passion Play. Their light is said to have lasted exactly as long as the play did.

The caraboa's primary function is, of course, to pull ploughs and loads (such as tree trunks out of a forest), with patience and great strength. Both male and female carabaos are beasts of burden and work, thus figuring highly in the agricultural production of the country. Currently, the carabao population stands at about two million. The Bureau of Animal Statistics (BAS) and the National Statistics Office (NSO) have announced that in 1998 627 head were butchered daily – a total of 225,790, the highest in six years. Although Executive Order No. 626 bans the slaughter of male carabaos below seven years old, and of female carabaos below eleven, it is often violated.

Carabaos are butchered because carabeef can compete with beef in wet markets, and is even preferred in some areas, such as the province of Leyte. Carabao horns are made into ornaments and handles, and uniquely into the instrument with which coconut vendors carve the flesh out of the nut – unbroken, still containing the coconut water.

Carabao's Milk

The females supply a bluish-white milk, which to Filipinos is special, and superior to fresh cow's milk (largely unavailable to most) or the canned (evaporated, condensed, filled) milk that the American influence imprinted into their diets. The mother carabao is milked when its calf is about three to four months old, and is gradually being weaned. By the time it is a month old, the calf starts eating grass. The mother carabao is thus milked six or ten months after having given birth, and can give about 3,000 cc of milk every day (Santos, 1989, 6).

Carabao's milk is more dense (less water) and richer (more fat, more protein) than cow's milk.

Composition of Milk

	Water (%)	Protein (%)	Fat (%)	Milk Sugar (%)
Cow	87	3.5	3.9	4.9
Carabao	78.5	5.9	10.4	4.3
Goat	87.9	3.5	4.3	4.3
Sheep	82	5.8	6.5	4.8
Reindeer	63.3	10.3	22.5	2.5

(Agriscope, 1995 in Matias and Roxas, 1998)

It is home-pasteurized by heating for ten minutes in a covered double boiler containing one quart of vigorously boiling water; or in bottles set in a pail filled with cold water that is heated to 145°F for an hour, then cooled gradually; or in a pressure cooker. The Philippines, in 1989, had about three million carabaos, with perhaps a fourth of them nursing. Paterno Santos calculates that if a 'milking carabao produces half to one liter of milk daily, that would mean about 144 million liters a year', a potential full-blown industry. However, carabao's milk never became an industry, nor did it compete with the 99 per cent import-dependent milk industry (Magboo, 1986, 10), of which Senator and former Secretary of Health Juan Flavier has ruefully said: 'Local dairy production contributes a pitiful 3 drops of milk for every Filipino' (quoted in Senate Bill No. 746, the National Dairy Development Bill). Carabao's milk was, and still remains, a cottage enterprise.

The number of carabaos in the country has since diminished because of mechanization (tractors instead of ploughs), butchering (once illegal) and other reasons, and so the milk supply has diminished as well. Even when the supply of carabao's milk was plentiful, however, in the Philippines its predominant use was, and still is, in the making of cheese, sweets and desserts.

It is of course drunk in the glass, and I remember that it tasted richer and fatter than the evaporated canned milk (diluted) that we had got used to. Milk, by the way, is not given to Philippine children as much as it is to American children, not only because of cost and availability, but also because many Filipinos are lactose-intolerant, lacking an enzyme for the digestion of milk.

My own close encounters with carabao's milk happened on the farm, during the Japanese occupation (the Pacific war). Like most families, ours left Silay, Negros Occidental, the town where we lived and went to school, for what was called 'evacuation', when the Japanese came. We moved to Hacienda Buenbano, in Saravia, a farm surrounded on three sides by the Malugo river. There my father kept a small herd (*manara*) of carabaos. From them animals were culled to be trained to draw ploughs and other loads. While within the herd, they were still 'wild', not yet work animals.

Every morning, carabao's milk would be brought to our house in tall blue or green bottles. My mother would pour this into containers in a cool, bamboo-and-nipa house. She would allow the cream to rise to the top, where it would form a thick layer, which was scooped off, and from which she would make butter. This was not usual, but my mother, a physician, experimented with many things (making soap, tooth powder, etc.) and usually succeeded. In an improvised wooden churn (like a small mortar), she churned the cream, and produced a rich butter – white, because of the carabao's diet – but delicious. She fried bananas and other food in it, and was amused that we children sometimes complained, 'Oh no! Butter again?' This in wartime.

Cheese: kesong puti

A more conventional, native way with the milk was to make it into cheese, what we call *kesong puti,* or white cheese. This was done simply by adding vinegar, and then shaping it into oval patties with the hands. This simple cottage-type cheese is made in many places in the Philippines, and my mother had learned how to do it as a child in Nueva Ecija, in central Luzon.

I have since tried cheeses from other provinces: in Cavite it is made soft and spreadable, and called *kasilyo* (probably from *quesillo*). On the sidewalks of Manila it is often available, square-shaped and moist, wrapped in banana leaves then packaged in twos in a length of banana stalk. Supermarkets often stock it. In Pagsangjan, Laguna, it was served to us in a bowl, to be eaten with rice or with hot *pan de sal* (brown, crumb-sprinkled rolls). This was so good that we asked about the maker, and were directed to Aling Nora in Lumbang, Laguna.

Nora del Valle makes her cheese in her kitchen. She processes daily the milk delivered to her in bottles. She adds vinegar and salt, then cools the product with an electric fan, and when it is the right consistency, stores it in her refrigerator or wraps each one in banana leaves (at P5 and P10 sizes). Her daily output fills advance orders, or is bought by regular or new patrons, who knock on her door. If they wish to take the cheese to relatives in the US, she packs it in plastic boxes and freezes it. Her cheese is soft, very slightly salty, and one of the best I have tasted.

Kesong puti, as already mentioned, we have for breakfast with hot *pan de sal* and perhaps, especially at Christmas time, with slices of *jamon China* (Chinese ham cooked and glazed the Spanish way) as well. It can also be fried and served with rice, or served on crackers, or (fried or fresh) eaten with fruit, or used in salads as Feta cheese is. Young chefs have experimented on serving it as appetizers – rolled in *lumpia* (egg roll) wrappers, or wrapped in squares of filo pastry. It has thus entered contemporary cuisine as a soft cheese congenial to Filipino and foreign dishes.

Sweets: dulce gatas/mazapan de leche; pastillas de leche

Some of the milk we would cook with sugar, stirring the mixture continuously until it thickened and changed colour to a light beige. The resulting sweet we called *dulce gatas* (milk candy). It was transferred while hot into cans or wide-mouthed bottles, and allowed to cool. It was delicious, grainy and chewy. We spooned it out, and walked around eating from the spoon. When we grew up and went away to school in the city, we longed for it, asked for it on vacations, had it sent to us in the city. Now a cousin markets it in neat blocks covered in coloured foil, and calls it *mazapan de leche* (milk marzipan).

This light-brown fudge-textured candy is the Visayan version of what is nationally called *pastillas de leche* (milk pastilles). Although found in many provinces, the epitome of the art is considered to be that of Bulacan, especially that from the town of San Miguel de Mayumo (literally, Saint Michael of Sweetness). These are sold in the small souvenir stores in San Miguel and Malolos, and in the many outlets all over Manila of the enterprise called Bulacan Sweets, which also exports these delicacies.

Bulacan *pastillas* are usually pristine white (although the many versions, include a light-brown one sometimes called *tostado*), a very much refined sweet produced by hours of patient stirring. A friend from Malolos cites the number of unmarried ladies in Bulacan, and says that it was their leisure, their patience, and their dedication to the care and feeding of their families that produced such a lovely thing.

The mixture, often flavoured with *dayap* (native lime) rind, is spread thinly on a board, cut into *pastillas* rectangles, and wrapped in white paper. For special occasions such as fiestas, however, these are further wrapped in coloured tissue paper – ornamentally hand-cut with scissors in the way of Chinese paper cutting. The designs range widely: flowers of many kinds, carabaos and nipa (palm-thatched, bamboo) huts, messages like *Recuerdo* (souvenir), *Maligayang Pasko* (Merry Christmas), *Manigong Bagong Taon* (Happy New Year), *Maligayang Kaarawan* (Happy Birthday), or one's name. Traditionally were these cut freehand, without using patterns, with little manicure scissors. Now some makers of *pastillas* wrappers draw designs on folded paper, cutting many at a time. They have also been known to use the same art to make decorative hangings.

It is in the same town, by the way, that pickles used to be arranged in designs, even landscapes in the bottles, and fruits preserved in syrup were carved in floral patterns. The underlying philosophy is, food can be beautiful, so why serve it plain? Thus the graceful *pastillas* wrappers that are meant to hang from the edge of a footed dish, and to be discarded when the candy is eaten – an ephemeral art that is worth its moment of visual pleasure.

Milk as viand

It is the people of Pampanga, however, who have the most ways with carabao's milk. Claude Tayag, an artist and chef from Angeles, Pampanga, lists ten ways in which Pampangos savour the milk.

The first is to pour it over rice, sprinkle both with some rock salt, and have it for breakfast, with any of the following: dried fish, preferably *tunsoy* (*Sardinella fimbriata;* fimbriated sardine), *pindang damulag* (carabeef *tapa/jerky* fried crisp), or ripe mango.

It could also be mixed with *duman* (the sweet and milky young green grains of *Milagrosa* rice, husked, pounded, fragrant) and sugar. This was served as a mid-morning (*segundo*) or afternoon snack. The tiny jade-like grains could also be soaked in hot chocolate made with carabao's milk – a rare and delicious treat.

For breakfast or for dessert, the milk is poured over buttered *camote* (sweet potato) balls. The latter, he swears, have a taste and texture very similar – and even better, he feels – than *marron glace*.

Desserts: tibok-tibok, leche flan, helado de mantecado

Pampangos are known for *tibok-tibok,* a blancmange or milk pudding unique to the province, sprinkled with toasted coconut, white, soft, quivering, and delicious. The name, according to the late E. Aguilar Cruz, Pampango writer and gourmet, probably originated from the quiver (like a heartbeat) seen when the pudding is shaken. It is made with pure carabao's milk and sugar thickened with cornstarch, its taste and texture very similar to *panacotta*.

Haleang gatas (milk candy/jam), similar to the Visayan *dulce gatas*, is sold by weight in jars. In Arayat town, it is used as a topping for *halo-halo*. Adoracion Tayag, Claude's mother, uses carabao's milk as a creaming agent in her *haleang ube* (*Dioscorea alata;* greater or purple yam).

A rich man's *brazo de Mercedes* (a meringue roll) has a thick roll of *haleang gatas* wrapped around meringue.

A hot chocolate drink is made when the milk is mixed with a generous amount of chocolate paste (with some ground peanuts in it) and beaten to a frothy consistency.

Finally, writes Tayag, the milk is poured over crumbled *puto seco* (a dry, slightly sweet biscuit).

From Apalit, Pampanga, comes news of a luxurious *leche flan* (*crème caramel* or baked custard) also made with carabao's milk. The recipe is given below – from Gene Gonzalez's book *Cocina Sulipena*.

The Arnedo-Gonzalez family of Apalit, Pampanga, lived right on the border of Pampanga and Bulacan provinces, in a place called Sulipan. Their large ancestral house was always filled with guests – friends, relatives, Spanish dignitaries including the Governor-General, foreign guests including European nobility, and later American dignitaries (including William Howard Taft).

Their guests, writes Andrew Gonzalez, FSC in *Camelot by the Rio Grande de Pampanga*:

> came by horse and carriage, but mostly by *banca* (boat) and *casco* (river boat) as the Rio Grande was the favorite way of transport to the town and to the Arnedo home which had a small pier for landing.
>
> Because the trip was long, guests stayed at least two nights and three days. On such days the hospitality and cordiality of the Capampangan, which had become sophisticated through many accounts of European hospitality, demonstrated itself most profligately.

The Sulipan chefs cooked dishes from Filipino, Spanish, Chinese and European cuisines, and designed and executed the French menu to celebrate the inauguration of the First Philippine Republic in 1898.

This is the heritage that Gene Gonzalez, chef, restaurateur and great-grandson of the original Arnedo-Gonzalezes, celebrates in his book *Cocina Sulipena* – after professional and rigid kitchen-testing by his staff.

Still another carabao's milk recipe in the book is one for *helado mantecado*, French vanilla ice-cream. It calls for 6 litres of fresh carabao's milk or 2 ½ litres cow's milk mixed with ½ litre cream, grated *dayap* rind, 12 egg yolks, powdered sugar, rum or brandy, and vanilla. It is blended in a hand-turned mixer till velvety, then frozen overnight. Today Gene still serves this home-made, luxurious ice-cream in his restaurants, in an old ice-cream maker (*garapinera*) which he has reconditioned.

The Sulipan of the Arnedo-Gonzalezes is gone now, having lost its name and identity to the two border towns of which it was part: Calumpit, Bulacan, and Apalit, Pampanga. The cuisine survives, however, in Gene's cookbook and restaurants.

Some contemporary uses

Traditional cooking has left some contemporary heirs. In the 1950s the tastiest, best commercial ice-cream available was that made by Selecta, from carabao's milk flavoured with *ube* (purple yam), or *macapuno* (sport coconut), or *mantecado* (French vanilla). Selecta eventually sold its plant and products to a large food-manufacturing company. The Arce family who owned it, however, retained a small ice-cream business that now supplies much-appreciated Arce Dairy ice-cream (3,000 gallons daily) through outlets and a few supermarkets. Arce ice-creams that still contain carabao's milk use whole fruit, not purees, and are labelled: Super Deluxe, Super Special, and Regular (Herrera, 1996, 30).

The Dairy Training and Research Institute (DTRI) of the College of Agriculture, University of the Philippines at Los Banos, Philippine Carabao Center, has suggested the potential of carabao's milk for the following products that could be made in the home for profit:

Pasteurized milk	toned milk; fresh whole milk.
Flavoured milk	flavoured milk: chocolate, vanilla, coffee.
Cheese	*kesong puti* (soft) cheese, hard cheese, Danish-style cheese.
Pastillas de Leche	
Curd	for cheese spread.
Whey	for vinegar, *nata,* syrup drink.

(Gomez and Lihat, n.d.)

Experiments at DTRI had shown their viability, and some of these were available for a while at the U.P. Los Banos dairy store – but unfortunately not for long. Those who sampled them and developed a taste for them (especially the hard cheeses) soon found that they were no longer being manufactured, for reasons unexplained.

Carabao's milk is no longer available in commercial quantities, and products in commercial quantity are hardly made with it. Still available, however, are the cheeses and desserts made in the small sources: households, where it is made just for the family; small businesses that make *tibok-tibok* and *leche flan* for customers; single-person or family entrepreneurships like Aling Nora's, or that of the *pastillas* makers, that accept orders for specialities made in the same ways they had been made through a long native tradition.

Carabao's milk is thus today even more of a special food, available mostly in the provinces, home-delivered if at all, and not in commercial quantities, savoured by the few who have access, treasured in memory by those who, dwelling in the cities, can seldom avail themselves of the milk and its products, and thus of the rich goodness flavoured by home and tradition.

BIBLIOGRAPHY

David-Perez, Enriqueta. *Recipes of the Philippines.* Manila: National Book Store, 1973 (19th printing).

Gomez, Iluminada V. and Thelma S. Lihat. 'Mga Produkto mula sa Gatas ng Kalabaw: Pasteurized Milk, Chocolate Milk, Kesong Puti, Pastillas', Dairy Training and Research Institute (DTRI), College of Agriculture, University of the Philippines Los Banos, Philippine Carabao Center, n.d.

Gonzalez, Gene R. *Cocina Sulipena: Culinary Gems from Old Pampanga.* Manila: Bookmark, Inc., 1993.

Herrera, Alicia A. 'The return of a classic.' *Business World,* May 23, 1996, 30.

Matias, Job M. and Rudyard R. Roxas. 'Getting to know more about milk', *Animal Husbandry & Agricultural Journal,* January 1998, 6.

Magboo, E.C. 'Hybrid carabaos give more meat and milk', *Weekly Agribusiness,* January 19, 1999, 10.

Maulana, Nash B. 'Carabao next century will look more like cow', *Philippine Daily Inquirer,* May 31, 1999, 1, 6.

Santos, Paterno R. 'Producing & pasteurizing CARABAO MILK milk for rural families', *Agribusiness Weekly,* June 23–29, 1989, 6.

Acknowledgements

Anna Asuncion Angeles, Manila; Alicia Lucero Gamboa, Makati, Metro Manila; Gene Gonzalez, San Juan, Metro Manila; Claude Tayag, Angeles, Pampanga; Nora del Valle, Lumban, Laguna.

Appendix: Recipes

Pastillas de leche

1 gallon fresh carabao's milk
1 ½ cups sugar
rind of 2 lemons

Place milk in a copper vat or pan over low charcoal heat. Stir until the milk has evaporated to one-fourth of its original quantity. Add sugar and rind and cook over low heat, stirring constantly, until the mixture forms a soft ball. Pour the paste on a sugared board, cut into 5 x 1.25 cm pieces, roll in sugar, and wrap in fringed white tissue paper.

(David-Perez, p. 118)

Leche flan (Baked Custard)

In old Sulipan, the traditional dessert table was always a wondrous sight: an assortment of cakes, flans and tortes surrounded by preserves in syrup from different regions, little bejewelled pastries in beautiful épergnes of French crystal, and *pièces montées* of pastry made by the Sulipan patissiers.

1 ½ cups white sugar
14 egg yolks
1 egg
pinch of chopped lemon rind
3 cups fresh carabao's or cow's milk.

1. Dissolve 1 cup white sugar over moderate heat to caramelize it. Pour evenly on bottom and sides of mould. Set aside.

2. Preheat oven to 375°F. In a mixing bowl, combine egg yolks and whole egg. Add ½ cup sugar, lemon rind and milk. Mix thoroughly but gently so as not to form air bubbles. Pour into prepared caramel-lined pan. Put the pan in a large, shallow baking pan containing water [*bain-marie*]. Bake in the oven for 45 minutes.

It is difficult to work with carabao's milk. Too much heat will coagulate it, and the result will be a mealy, gritty flan. Be sure the milk you buy is pure, as some adulterate it with coconut milk. Cow's milk is an acceptable substitute, but the consistency of the flan will be less rich.

(Gonzalez, p. 117)

Helado mantecado (French vanilla ice-cream)

It is a wonder how Sulipena hospitality could produce desserts such as *helado* (ice-cream), fruit ices and sorbets considering that ice plants were introduced here only in the early 1900s. As early as the mid-1800s people already had recipes for ice cream and knew about the existence of the *garapinera* (ice-cream maker). According to Dr Serafin Quiason, our foremost authority on early Philippine trade, ice was imported to the country in the late 1800s. With the opening of the Suez Canal after 1869, the Americans brought ice, insulated by sawdust and straw, on ships from New England crossing the Atlantic to India. From India, the ships took the Pacific route. The ice was rationed among the rich who could afford to have it delivered daily in wooden iceboxes. This explains the existence of ice-cream recipes in the Philippines long before the ice plant became a franchise (usually awarded by the government to the richest in town).

Helado mantecado is rich ice-cream made smooth and creamy with carabao's milk, known for its high butterfat content and tangy flavour. The milk is often simmered to reduce its water content. Unfortunately, one cannot be sure about the purity of milk bought from unfamiliar sources. If pure carabao's milk is not available, substitute with a mixture of one part fresh cow's milk and one part thick cream. Check that the *dayap* possesses a citrus fragrance. Do not use the lemon hybrids which try to pass off as *dayap* in the market.

6 litres fresh carabao's milk or 2 ½ litres cow's milk mixed with 1 litre cream
grated rind of 4 small *dayap* (lime)
12 egg yolks
1 kilo powdered sugar
1 tablespoon rum or brandy

1. Put the milk in a large pot. Bring to a boil and simmer until it is reduced to half its quantity. Remove from heat and cool.

2. Mix *dayap* rind with egg yolks. Add sugar and beat until mixture is lemon in colour. Add it to the cooled milk. Add rum or brandy.

3. In the ice cream maker, beat the mixture until the texture is velvety. Freeze overnight. Serve with *barquillos* [tube-like wafers].

(Gonzalez, pp. 120–121)

Tibok-tibok (Milk Pudding)

Tibok-tibok is perhaps one of Pampanga's finest *kakanin* (sweetmeats). It is simply a sweet-based pudding of carabao's milk, fragranced and flavoured by *dayap* (lime) rind. To the native Sulipena, *dayap* is a most essential cooking ingredient, to which the native *kalamansi* [small native lime] is inferior. *Tibok-tibok* is a dessert bound by starch and served with *latik* (fried, toasted coconut milk) which, to some Pamapanguenos, is not necessary because it cuts the tangy flavour of the carabao's milk. You will need to get real *dayap* and carabao's milk for this recipe. Substitutes would completely change the character of this dessert.

> 4–6 cups carabao's milk
> 500–600 g glutinous rice, ground smooth
> 1 cup sugar
> grated rind of 2 *dayap* (lime)
> *latik* (fried coconut milk)

1. In a *kawa* (wok), cook the carabao's milk, ground rice and sugar at low heat. Add the *dayap* rind and stir continuously with a wooden spatula until smooth and thick. When the mixture does not stick to a dipped finger, remove from heat.

2. Grease a platter or tray with oil or butter on top with a greased banana leaf. Pour milk mixture in and spread evenly. Let cool. Serve with *latik*. Serves 10–12

(Gonzalez, p. 122)

A Spring-house in Pennsylvania: Design and Use

Rebecca Fitzjohn and Harlan Walker

A spring-house is a small building covering a spring to protect the water from contamination and to reduce freezing in winter. There are several designs in different parts of the United States, but the style we will consider seems to be centred on Pennsylvania. Our grandfather, Scott Paul Harlan,* had a farm near Philadelphia with a perfect example of the type in question and it was in use in the traditional way in the 1930s and 1940s, when we separately spent quite a lot of time there. We will describe the design and explain how its use, in particular in connection with the dairy, fitted into the rather unusual way of life on this farm.

When it is planned to establish a farm the first requirement, the *sine qua non*, has always been the existence of a regular and uncontaminated water supply. Our farm, known as Idle Dell, is said to have been built in the 1780s as two cottages on a large estate of which the much grander house stands about half a mile away across the fields; that house is said to date from the 1740s. At some stage the cottages were combined to make the elegant house that is there today. The essential spring, now in its house made of stone, is located about 100 feet in front of the farmhouse; from it a little stream runs down to a slightly larger stream, which today forms the boundary of the front garden or yard. This garden consists of a gently sloping lawn with several fully grown trees. Behind the house was a large vegetable garden, at least an acre, of which more anon. A barn of the traditional local design and other outbuildings were constructed across the lane that went down past the house.

The basic spring-house is just a small building over a spring with of course a door which is opened to let water be dipped and carried to the house. Sometimes it was merely a cave cut into a hillside. Our type was much more elaborate and much larger – specifically designed for dairying. The farm is about twenty miles from the middle of Philadelphia, which in the mid-eighteenth century was the second largest English-speaking city in the world – with a strong demand for dairy products.

---

* Our mother Katharine Harlan married an Englishman, Jeffery Walker, and we have mostly lived in England.

Figure 1. The spring-house at Idle Dell. Drawing made from a photograph of 1993. It was being restored and the roof is new.

The water bubbles up near the door (A) and is then led in a fairly wide channel (B) round three interior walls of the building before flowing out through a square hole and making its way down through the garden to the larger stream. The area A is about eighteen inches deep so that a jug can be dipped for water without disturbing the sediment. The channel B is about six inches deep in which various vessels could sit in the water to keep cool. In it were placed a number of large stones perhaps three or four inches thick on which shallower vessels, in particular pans of milk, were placed. The central area consists of large flagstones about eight inches thick (C); one can walk about on this area to get to the vessels in the surrounding channel. The building is sixteen feet by twelve feet with thick stone walls and a shingled roof.

Above this room there is, as is normal in this type of spring-house, another room with separate outside access. The prime purpose of this room is to provide further heat insulation to the spring-house itself, though it was a useful room and in some cases, though not at Idle Dell, even had a small fireplace. The window of this room can be seen in Figure 1. The window below, slatted and fly-screened, leads into the spring-house proper.

In front of the door of the spring-house there is a paved area with a roof over it which is an extension of the roof of the room above. As well as sheltering a working area this roof prevents the sun from reaching the door of the spring-house. This porch area is about sixteen feet by eight feet.

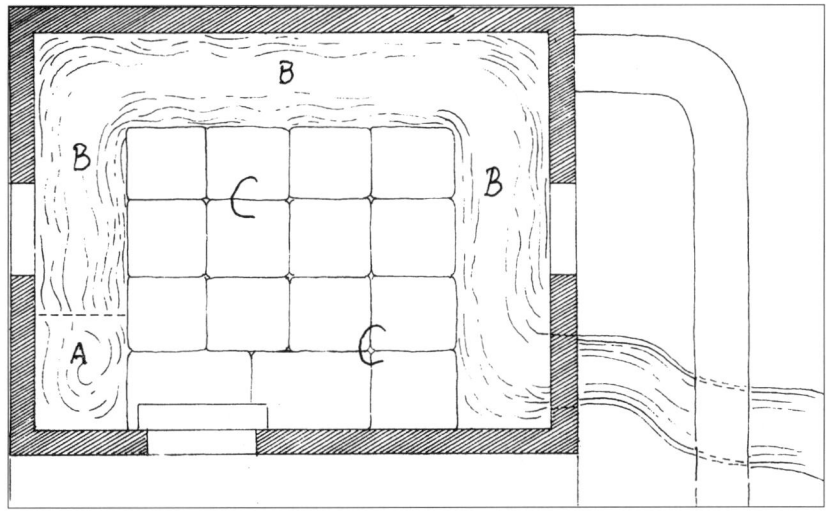

Figure 2. Plan of the spring-house.

When we lived at Idle Dell, the spring-house was fully used in just the same way as it had been in the nineteenth, and possibly eighteenth, centuries. The reason for this is curious and to explain it we will describe our way of life there. Our grandfather, a rather eccentric person, believed passionately that a gentleman living on his own farm should be self-sufficient. He should grow his own food. That was how he had been brought up and it was the proper way. Of course there were exceptions – coffee, sugar, flour, whisky – but that ought to be about the limit of bought items. We had pigs, chickens, eggs, muscovy ducks and a certain amount of game, mainly duck. The garden produced a huge variety of vegetables and fruit. And from the dairy there was rich milk, cream, butter and cottage cheese. It should not be thought that we are describing a backwoods farm just surviving in the years following the Depression. Not at all! The principal cash crop was the breeding and training of racehorses, at which our grandfather was extremely successful.

There were seven members of the family living on the farm at the time when Rebecca was there for four years, all women except for Grandfather. Also living in the house were Scott Tyndall and his wife Rosie. Rosie did all the cooking and an excellent cook she was. It was all done on an old-fashioned coal range that had to be lit every day. Scott's brother Allen lived in a cottage on the farm with his wife Bertha and their two little children. These two brothers were the only fully employed men on the farm. They were born on the Harlan family farm in Tennessee and had come north to work for Grandfather. Their father and grandfather had been with the family for many years. Their grandfather must have been born a slave, presumably the property of our great-grandfather. They could do everything – ride the racehorses, drive the tractor, work the fields, milk the cows, grow the many vegetables. Allen was an excellent cook and could don his white jacket and wait at table.

There was a large vegetable garden behind the house where tomatoes, sweetcorn, lima beans, snap beans, peppers, eggplant, beetroot, potatoes, onions, turnips, lettuce and cabbage were grown. Because there was plenty of space, the rows were much further apart than we are accustomed to in England. Also at the end of the rows there was a wide turning area; this was because all these cultivations were done with one of the work horses – first the ploughing and then the scuffling between the rows.

The spring-house was a vital part of the running of the household. We did have a small refrigerator in the house, but it was very limited in capacity with only small quantities of the most perishable foods stored there. Unpasteurized milk must be one of the most perishable of foodstuffs, so the spring-house was in constant use.

Allen usually did the milking, by hand of course, but from then on all the handling of the milk was done by Rosie. After straining it was brought to the spring-house where it was put in pans for the cream to rise. The pans were tinned steel about twenty inches wide at the top and six inches deep, but tapering to about ten inches at the base. They were carefully placed in the shallow water which continually flowed around the central area of the spring-house. The coolest area was nearest the deep water by the door and this is where the milk was placed first to cool it down as quickly as possible. Some milk was of course put into jugs for drinking, for we never drank skimmed or semi-skimmed as seems to be the fashion now. These jugs also sat with the cold water flowing round them. The pans of milk were moved along the water area after each milking to ensure that it was all used in rotation and not mixed. Each day the cream was skimmed off and collected in a crock, which also sat in the water, until there was sufficient quantity of mature cream to make butter.

Rosie had had no formal training but was, as was necessary, meticulous in the hygiene necessary in the dairy. All the utensils had to be scalded with boiling water every time they were used, including of course the wooden churn.

The by-product of making butter is of course buttermilk. This was a popular drink when nice and cold, but we English children never learned to enjoy it. It was used in cooking to make what the English call scones and the Americans biscuits. We had these everyday with our evening meal – hot and freshly baked. The normal drinks at meal times were milk or water for both adults and children.

As well as buttermilk, there was plenty of skimmed milk to be put to good use. This probably compares to the semi-skimmed milk of today because there was always a certain amount of cream left behind. No-one would dream of drinking it; it was fed to calves or made into lovely cottage cheese, left to drip in the spring-house. It was sweet tasting and fresh, not a bit sour or acid like the commercial varieties available in shops.

Not all the cream was made into butter. Occasionally we had ice-cream, made not by Rosie but by our Grandmother. It was delicious, nearly pure cream and flavoured with vanilla. Also a speciality of the house was Charlotte Russe, again made with nearly all cream. This was just for company.

Inevitably there was a certain amount of wastage – sometimes milk goes sour when it is not supposed to, or the butter won't come and no amount of turning the churn handle will make it. But we had pigs on the farm – Tamworths we think – and they seemed pleased to have it. The pigs were not for breeding or selling but for fattening and eating. We think we killed eight each year. When the pigs had been slaughtered – usually two at one time – the carcasses were hung in the porch area of the spring-house to be butchered. There was a good strong table in the porch for this and other purposes. After salting, the hams and flitches were taken to the smoke-house (near the back of the house) where they were smoked with hickory wood, cut from our own woods, and left in there for storage. The other meat was ground and made into sausages. These were not put into skins, so were not really sausages, but that is what we called them; the mixture, of which the recipe is given below, was just shaped into balls like rissoles. These were then smoked on open racks. Of course all this did not take place in a day, so the meat was hung from the beams of the spring-house to keep it cool and safe until it could all be used.

The porch of the spring-house was often in use during the summer. It was a good place to sit around the table and prepare the beans for dinner or fruit and tomatoes for bottling. We both particularly remember enjoying large water melons there. The huge slices made too much mess to eat indoors. The melons had been floating to cool in the cold water inside before we tackled them.

Idle Dell was a mixed farm growing various crops including wheat – and wheat has to be threshed. This was before the days of combine harvesters, so it was quite an event when threshing time came round. Seven or eight extra men had to be hired by the day and they had to be fed. We remember meat loaf with lots of vegetables and the pudding was certainly apple pie at which Rosie excelled. Some of these men were white and some black and, unbelievable as it sounds today, they could not eat together. The white men had their meal in the back garden by the house. The coloured men (as they were called at that time) had the table laid for them in the porch of the spring-house.

We must also mention the splendid supply of fresh drinking water. There was no need for ice. The last job for one of the children at two minutes before dinner was to run to the spring-house to fetch a jugful of water in a beautiful silver pitcher that had been in the family for many years. The cold water made a wonderful bloom on the outside and if you touched it you could write on it. We dipped water before every meal and the final job for the day was to dip a fresh jug for Grandfather when he went to bed. It had to be on his bedside table just before he retired at about nine o'clock. Sometimes we forgot and this was an unforgivable sin producing loud and terrifying verbal abuse.

Harlan Recipe for Sausage Meat

40 lb pork meat (proportion 10 pounds lean to 6 pounds fat)
2 cups rock salt • 2 cups dried sage
1 cup black pepper • 1 cup cayenne pepper
big pinch of saltpetre

Mix, make into balls and smoke. For best results the meat should be very tender from young hogs.

This sounds terribly peppery but the sausages were delicious and we agree that we have never tasted better.

Charlotte Russe

2 x 3 oz packages lady fingers
1 ½ envelopes unflavoured gelatin • boiling water
3 tablespoons sugar • 2 teaspoons vanilla • 2 cups heavy cream

Separate the lady fingers and line a glass bowl. Soak the gelatin in 6 tablespoons cold water for 5 minutes. Add 4 tablespoons boiling water, stir well and place over boiling water until the gelatin is dissolved. Cool, add the sugar and vanilla. Beat the cream in a large bowl until almost stiff, then beat in gradually the cooled gelatin mixture. Do not over beat. Spoon the mixture immediately into the lined bowl. Refrigerate until ready to serve.

******

Our original intention was to make a survey of the spring-house in general, but particularly in Pennsylvania. Harlan planned to visit the United States in the spring of this year to study this subject, but he was not able to do this, so rather than give information at best incomplete and very possibly inaccurate, we decided to limit this paper to our own experiences at Idle Dell.

Figure 3. Another view of our spring-house. The four wooden steps leading to the upper room had fallen down when the photographic basis of this drawing was made, but they would have been partly visible on the left of the building.

Nevertheless we have some details of other spring-houses and it may be useful to give some references even if some of them are rather vague. I give also web addresses if available. The Colonial Pennsylvania Plantation, Delaware County, a 'research and educational facility', has a spring-house (www.de.psu.edu/cpp/wc.htm). The Chadds Ford Historical Society has a spring-house (www.de.psu.edu/cfhs/spring.html). Charlestown Township has two spring-houses (www.charlestown.org). Rebecca visited and photographed spring-houses in Valley Forge and Chestnut Hill.

We have an article by Amos Long Jr. from an unknown publication, perhaps ten years ago, which shows several spring-houses in 'southeastern Pennsylvania' particularly in the neighbourhood of Oley. We made some very limited enquiries in New England but only traced one at Coggeshall Farm Museum, Bristol, RI. Telephone: 401-253-9062. We can't read the web address from which we got this: Visit RI History – Greater Providence Historic Sites. We were told that there were not many spring-houses in New England, but that might not be true.

We propose to continue our study of spring-houses and any information on the subject would be welcome. We would be particularly interested to hear of buildings of this type and function in Europe. So far we have not found any – hundreds of ice-houses, but no spring-houses. Was it invented in America?

The illustrations to this paper were made by Soun Vannithone based on sketches and photographs by Rebecca.

How Old is Old Cheese? *Gamalost* in Coffin-shaped Boxes and Eccentric Jars

Ove Fosså

Trying to find out what *gamalost* is may prove a difficult task if you don't read Norwegian. André L. Simon helpfully described 'Gammelöst' [sic] as 'a Norwegian cheese made all the year round'.[1] *The Simon and Schuster Pocket Guide to Cheese* offers a more detailed description, but of the cheese as it was made by the dairies many years ago, 'as intimidating in appearance as the Vikings who reputedly enjoyed it' and 'virtually inedible unless sliced very thinly'.[2]

The *gamalost* of today has a cylindrical shape, the weight around 1.7 kg, although a small size of 0.38 kg is also made. It is brown, lighter towards the centre, semi-soft with a grainy texture, and has no rind. Made from soured, skimmed milk, it has only one per cent fat (FDM). A soft, processed type is also available. It is usually eaten on buttered, dark bread, either spread with more butter, or with some dark syrup.

A single mould, *Mucor mucedo* var. *racemosus*, is responsible for the ripening of today's dairy-made cheeses. They used to have a variety of moulds, often including some of the *Penicillium* type, which gave the cheeses a greenish colour in the centre.[3]

The name *gamalost*[4] literally means old cheese. The dairy-made cheese of today is ready for consumption at an age of only two weeks, so the name may not seem quite appropriate. Now it is usually explained as referring to cheese of the old type. However, if we go back to the first detailed description of *gamalost*, published by Bishop Gunnerus in 1774, there is no indication that *gamalost* was considered to be a more ancient type than other cheeses. He said that it was called '*gammelost*' because it had to be somewhat old in order to be good. Gunnerus gave three alternative names for it: '*skiør-ost*' (curd cheese), '*suurost*' (sour cheese) and '*gammelost*'. It has also been described as '*ramost*' (pungent cheese).[5]

The imaginary descriptions

One of the most widespread misunderstandings about *gamalost* is the idea that it is mentioned in *Njal's saga*, one of the greatest Icelandic sagas, written c. 1280. This 'fact' appears in a very large percentage of all writings on *gamalost* this century. The first one to claim that he found the phrase '*forn ostr*', meaning old cheese, in *Njal's* saga was Olav Johan-Olsen in 1905. He said that he searched for information in the sagas, but it would seem more likely that he used the Old Norse dictionary published by Fritzner in 1883–1896. The entry on *forn ostr* correctly refers to the saga of the sworn brothers (*Fóstbrǽðra saga*, see below), but can easily be misread as referring to Njal's saga. Even the usually thorough Fredrik Grøn, in his work on Norwegian food traditions, did not detect this error, but quoted it and added that *forn ostr* was 'also' found in the saga of the sworn brothers.[6]

Cheese is not entirely absent from *Njal's saga* though. Melkolf the slave was asked to steal cheese and butter at Kirkby (*Kirkjubœ*). The theft was proven by Mord, by using the cheese-mould from Kirkby: 'He asked Otkel to fetch his wife's cheese-mould, and when it was brought he laid the slices into it. They fitted perfectly.'[7]

The Icelandic/Old Norse term used here was *ostkista*, apparently used only in this saga. *Kista* could also mean chest or box, and *ostekiste* more recently was the name for the boxes where *gamalost* were stored to mature. This episode has also been interpreted as 'proof' that the cheese was *gamalost*, but there is no such evidence in the saga. Grøn discussed this in more detail, but concluded that the type of cheese could not be established.[8]

In another chapter, a fire was put out by women throwing sour whey on the flames. Johan-Olsen argued that enough sour whey to put out a fire could only come from the production of sour milk cheeses. That may well be, but we still have no proof that it was anything like *gamalost*. Besides, the Möðruvallabók manuscript of the saga added that some of the women carried water and urine to the fire...[9]

Johan-Olsen was not a man who gave up easily. He had more arguments. The burning of Njal's farm at Bergthorsknoll (*Bergþórshváll*) in *c*. 1011 is considered a historical fact. Some of the buildings at the site have been excavated, and among other things, pieces of a peculiar white substance was found in 1885. The substance was sent to Copenhagen for analysis. The detailed report[10] concluded that the pieces were the remains of a soured milk product (*skyr*) and/or a sour milk cheese. In the opinion of Johan-Olsen, the report described a typical *gamalost*.[11] The saga of the sworn brothers related the story of the two enemies Butraldi and Thorgeir having to sit down together for a simple meal. There is a detailed report of what they ate:

> ...two platters were brought in; on one of them was some old short-rib mutton and on the other a large quantity of old cheese. Butraldi made a brief sign of the cross, then picked up the mutton ribs, carved off the meat and continued to eat until the bones were picked clean. Thorgeir took the cheese and cut off as much as he wanted, though it was hard and difficult to pare. Neither of them would share either the knife or the food with the other. Though the meal was not good, they did not bring out their own provisions for fear that it would be seen as a sign of weakness.[12]

This is actually the only time the phrase *forn ostr* appears in the saga literature.[13] Grøn made a comparison with *fornt vin* (old wine), arguing that 'forn' probably in both cases described something that improved with age.[14] To me it would seem that in this case, neither meat nor cheese had improved with age. Butraldi and Thorgeir were served a poor meal of old food. To interpret this as *gamalost* is at best wishful thinking.

Icelandic author Halldór Laxness wrote a novel, *The Happy Warriors*, based in part on the saga of the sworn brothers. His great tale of Bishop Grimkel bringing gold and silver hidden in a *gamalost* to the Pope in Rome is not part of the saga:

> Bishop Grimkel was in the guise of a pauper, with staff in hand. And the tale is told that he had in his bag no other treasure but a rotting fermented cheese, such as are made in the North and for smell surpass all created things in Christendom, so that thieves, robbers and cutthroats went far about a pilgrim carrying such an abomination....
>
> There was in [Rome] more filth, putrescence, leprosy and fetor of corpses and beggary than any otherwhere in the world in those times. Yet learned men hold that the stench rising from the cheese bishop Grimkel had with him from the North did little to better the Roman air, but rather the contrary.
>
> And here bishop Grimkel lifted his cheese from the bag, set it down at the Pope's feet and then stuck his knife in it, and out of the cheese poured a profusion of gold and silver like swarming maggots; and this hoard was both good and bright.[15]

The first descriptions

The first description of what seems actually to have been *gamalost* is found in Olaus Magnus' history of the Nordic people, published in Rome in 1555. He claimed that the people of Parma and Piacenza had Scandinavia to thank for their abundance of 'large and nourishing cheeses', described the cheeses of several Swedish districts and Finland and prized them for their quality. The people of Hälsingland and Norway, however, had what he called rotten cheese:

> Although they see, how it is filled with maggots, they get used to it, and appreciate it. It's bark-like rind, which is left after the inner parts have been eaten, is hard like tanned leather, and is used for shields in war.[16]

He does not have a name for this cheese, but both the area where it was made and the description suggest that it was *gamalost*. The bark-like rind fits the description of a common fault in *sæter* made cheese.[17]

Gunnerus is generally credited with being the first to distinguish between the two main methods of *gamalost* production. Even if the name itself is not much older than Gunnerus, the two methods are clearly distinguished, but described in less detail, by Christen Jensøn in his dictionary of Western Norwegian colloquialisms, published in Copenhagen in 1646:

> *Rør-Oest* or *Skiør-Oest* is made from sour milk by continuous stirring in a cauldron and heating over a slow fire till it reaches a temperature where you can just keep your hand in it. When the cheese has separated from the whey it is scooped up and into cloth-lined moulds where it is weighted to press out the remaining whey.[18]
>
> *Syde-Ost* is made from sour milk by bringing it to the boil in a cauldron, then removing it from the fire till the cheese sinks to the bottom. The whey is poured off and the cheese filled into moulds where it is weighted.[19]

He did not say whether, or how the cheeses were ripened. The dictionary was a collection of words from Askvoll in Sunnmøre only, and it is unlikely that *gamalost* was made locally by two different methods. The conformity in technique with the methods later described by Gunnerus is interesting.

Jonas Ramus (1649–1718) in his 'Description of Norway' listed several types of cheese made at the *sæter* including *Skiør-Ost* or *Suur-Ost* (both are synonyms of *gamalost*, according to Gunnerus), and *Knaa-Ost, Knøst-* or *Pult-Ost*. No further details are given.[20]

Bishop Erik Pontoppidan of Bergen wrote a 'Natural History of Norway', published in Copenhagen in 1752–53. He said that several kinds of cheese were made in Norway, naming some, but did not specifically mention *gamalost* or other sour-milk cheeses. He said that Norwegian peasants used to drink *blande*, made by mixing milk and water, or in winter, water and sour whey (*syre*). The peasants' wives boiled *syre* to preserve it through the summer for this purpose.[21] The presence of sour whey is an indication that sour-milk cheeses may have been made. However, when later authors[22] say that Pontoppidan mentioned sour-milk cheeses, they cannot have read his description.

Hans Strøm (1726–97) in his description of Sunnmøre[23], was the first to mention *gamalost* by that name. He said that *Skiør-Ost* or *Gammel-Ost* was made of curds or sour milk, boiled into cheese, pressed into moulds, and finally set aside to ferment, when it had to be stirred to get the fermentation going and to achieve the correct taste.

Christopher Hammer (1720–1804) published a 'Norwegian Household Calendar' in 1772. He described *skjørost* as being very dry and crumbly and recommended placing it close to the sauna for a day or so, to get the fermentation going. Depending on its further treatment, *skjørost* was made into *gamalost* or *pultost*. To make *gamalost*, he recommended steeping the cheese in wort.[24]

Johan Ernst Gunnerus (1718–73) was Bishop of Trondheim and one of the founders of the 'Royal Norwegian Science Society'. He wrote the first detailed work on the uses of milk in Norway, published in 1774. This is still the best available source on traditional ways with milk and milk products in this country. It included one of the most detailed descriptions of *gamalost* production.[25]

Gamalost is made of thick, soured milk. The milk should be skimmed first, unless 'a person of rank' wants a superb cheese, which is then made from whole milk. Gunnerus distinguished between two methods of *gamalost* making. By the first method, the thick, sour milk was heated in a cauldron over a slow fire, constantly stirring to break the curds into small, even grains. The milk was boiled for up to half an hour, whereafter it was removed from the fire. The curds should then sink to the bottom of the kettle, leaving the clear whey on top. After removing the whey, the still warms curds were filled into a mould. Some used a cloth to line the mould, which also facilitated the unmoulding.

By the second method, the milk was heated, but not boiled. The still floating curds were lifted out of the cauldron and into the cloth-lined mould. The cheese was then submersed in the whey again and boiled for half an hour.

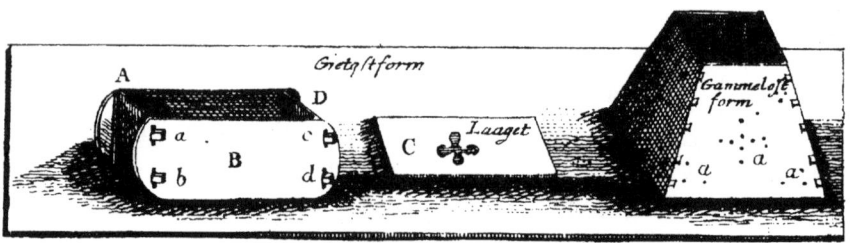

Cheese moulds as illustrated by Gunnerus, 1774. A whey cheese mould with lid to the left, an unusually square gamalost mould to the right.

Occasionally, cheeses were made by this method without the boiling, but this resulted only in poor cheeses, with poor keeping-qualities.

According to Gunnerus, the best cheeses were made by the first method, later to become known as the Sogn method after the district where it was used. These cheeses were held to be finer, more uniform and matured more easily. His second method would later be known as the Hardanger method.

The cheeses were left in the moulds for at least two, usually three or four, days. After unmoulding, they were placed on shelves in a warm room for a few days, until they started maturing. The signs of this were softening or some mould growth. They were then moved to a drier spot, where they were turned every other day, so as not to stick to the wood or birch bark on which they were standing. For keeping, they were transferred to a cooler place, stacked one on top of the other, but with pieces of wood or birch bark in between. In the autumn, cheeses were placed in wooden chests or barrels in a cellar, often wrapped in straw, especially oat straw. The longer the chests had been used for this purpose, the better the cheese was said to be.

Several tricks were used to improve a cheese that did not mature properly. Beer, wort, wine or brandy was poured into a cavity made in the cheese, or the cheese was wrapped in a cloth moistened in any one of these liquids.

A well-made *gamalost* was, according to Gunnerus, of a brown or brownish colour, soft enough to be sliced thinly, and having a pungent smell and a pleasantly sharp taste. Many people, even Danes and foreigners, he said, preferred it to any other cheese. He recommended it for being diuretic, and for 'warming and fortifying' the stomach. Perhaps drawing on his own experience, he said there was nothing better for those who had eaten too many oysters. Peasants used to boil a small piece of cheese in sour whey, drinking a few cups of this, hot, as a remedy against colds.

The best *gamalost* in Gunnerus' diocese came from Røros and Nordland, especially Vefsn. Generally, he said, persons of rank in the countryside, made a better cheese than the peasants did.

He described the common size as 8 kg[26], the shape cylindrical, or square. A square mould is also known by a description from Nordland, where it was known as a *gamalostlur*. It was 40 cm high, 20 cm square, somewhat wider on top.[27]

Gerhard Schøning (1722-1780) visited the farm of Storfosna in Trøndelag in June 1774, where he found that *gamalost* was made according to the Sogn method. It was wrapped in lukewarm, boiled straw after moulds had started appearing on it's surface.[28]

Nikolay Jonge wrote a 'Chorographic description of Norway', published in 1779. His description of cheeses was mostly word for word copied from Ramus, but he added *gamalost* as a separate type.[29] Possibly because Jonge (who was Danish) knew *gamalost*, but was not aware of the other Norwegian names for it.

The 'Norwegian journey' published by Johann Christian Fabricius in Hamburg in 1779 included a description of cheeses. His description of *gamalost* was a brief summary of Gunnerus. He added that it was known under the name of old Nordic cheese in Denmark, and much appreciated.[30]

The clergyman Wille described two kinds of *gamalost*, a sweet one, i.e. made from fresh milk, with rennet, as opposed to the one made from sour milk, *gammel Suur-Ost*.[31] More detailed descriptions and manuals on the production of *gamalost* were eventually published, but these are of less interest for the purpose of this paper.[32] *Gamalost* was considered of better quality and more valuable than *pultost* and *fatost*. The first was often made for sale, and the others for home consumption.[33] Hard pieces left over from a cheese were occasionally made into a sort of 'porridge', *gamalostgraut*, and eaten on bread. This is known both from Hardanger and Trøndelag.[34]

The travellers' tales

Several British and other travellers visited Norway during the eighteenth and nineteenth centuries and published their diaries. Naturally, their interest in the foods they were served varied. The fact that many encountered *gamalost* shows that it was a more common food in those days than it is today. Some of them liked it and some hated it, probably reflecting a very variable quality.

The English mineralogist and traveller Edward Daniel Clarke (1769–1822) visited Scandinavia in the summer of 1799, starting out in the company of three other young men, one of whom was Thomas Robert Malthus (1766–1834). They parted in Sweden, however, and did not go to Norway together. Clarke was served *gamalost* at Hov, south of Trondheim:

> We dined at *Hoff*; and for the first time tasted the old *Norwegian* cheese, called *Gammel Orse*, or *Norske*, of which the inhabitants are very fond. It resembles very excellent old *Cheshire* cheese, without any rankness. This kind of cheese is sometimes sent in presents to *England;* but the *Norwegians* themselves prize it so highly, that it is difficult to purchase any of it. The *Gammel Orse* is sometimes kept for ten years before it is brought to table. In making it, they use buttermilk, mixed with yeast.[35]

The Reverend Frederick Metcalfe, M.A., fellow of Lincoln College, Oxford, had two very different experiences with *gamalost* in 1856, but did not go into great detail. In Dalen (Telemark) he had a cheese not explicitly described as *gamalost* but it was, 'so sharp that it went like a dagger to my very vitals at the first taste...'.[36]

Later, he had more luck in Flåm (Sogn):

> Not sorry was I to darken the doors of Thorsten Fretum, whose house stood on an eminence, commanding a view up the valley and the Fjord. Bayersk Oel and Finkel[37] – old and good – raw ham, eggs, and gammel Ost – a banquet fit for the gods – were set before me.[38]

In the same year William Mattieu Williams (1820–92) ate *gamalost* in Romsdal. He was quite taken with it and praised it as 'a new sensation for epicures':

> My dinner at the Horgheim station consisted of 'smoerogbröd' and 'gammel ost', bread and butter and old cheese: charge 8 skillings, or 3½*d*. The gammel ost is a celebrated Norwegian dish, and this at Horgheim the finest example of it I have met with. It is a peculiar sort of cheese, made, I believe, with goats' milk mixed with herbs and sugar. When new, it is very detestable, but after many years' keeping it decomposes, and forms a sort of condiment rather than food. It is sprinkled in a moist powder, upon bread and butter. When in perfection, it is neither mouldy, moist, nor mitey; it is of a uniform pale brick colour, just capable of crumbling, and has a rich anchovy-pastish flavour with a faint suggestion of parmesan. If Fortnum and Mason, or Crosse and Blackwell, were to import some of this, put it into eccentric jars, and charge a sufficiently high price for it, our epicures would run into ecstacies about it, until they discovered that it is really a cheap and vulgar article.[39]

He had the opportunity to watch the cheese being made, but his description of this is not very accurate:

> In the corner opposite to the head of the bed, and almost within arm's reach, was the great stone hearth, covered with a stone and plaster dome. The other corners were occupied by benches on which the vessels for standing and mixing the milk with the other cheese materials were placed. There was also a second small apartment, or rather cupboard, for the stowage of pans, pails, &c. All was scrupulously clean in this particular saeter.
> Soon after sunset the sovereign mistress of the place came in, bearing heavy pails of rich milk from cows and goats. Some lumps of wood were taken from their store place under the bed, a crackling fire was soon blazing on the hearth, and the iron cauldron, filled with a mysterious mixture of goats' milk and other unknown ingredients, from

which the green cheese that ripens in time to "gammel ost" is made, was hooked to the black chain over the middle of the fire. For some hours after, every time I awoke the sticks were blazing, and the busy lass was there stirring, mixing, and watching till after midnight, or nearly to the dawn, when she disappeared.[40]

Williams later wrote several popular science books, among them *The Science of Cookery* and *The Chemistry of Cookery*, published in London in 1884 and 1885, respectively.

Francis Merrick Wyndham apparently would happily have done without the smell of *gamalost*, which followed him everywhere on his tour in 1859. First, in Kaupanger (Sogn):

> ... on a plate under a bell-glass were placed a few pieces of the strong-smelling *gammel ost*, or old cheese.[41]

Having supper at Bøvertun sæter, Sognefjell:

> ... our peace was much disturbed by some large sugar-loaf shaped cheeses, which, exhaling an extremely disagreeable odour, were far from desirable neighbours; however the *budeier*, at our particular request, removed them to the dairy.[42]

At a farm in Lom, he was shown around the store-rooms of the house:

> When we visited the dairy, I had the imprudence to venture into the cellar beneath, and to look into a coffin-shaped box where the *Gammel ost* was kept, and whence issued such an odour, that I quickly let go the lid and scrambled up the stairs again.[43]

In the mountains south of Lom, more *gamalost* was being made:

> In a very small room, which savoured strongly of cheese, a stout, rosy-faced girl was lustily[44] stirring with a stick a quantity of skim milk in a large iron cauldron, suspended over a blazing fire in one corner of the room. On some shelves against the wall stood a number of the large, white, conical gammel ost (old cheese), fully accounting for the unpleasant smell which had greeted us on entering. A door opened into the milk room, where was disposed a goodly array of flat wooden vessels filled with milk.[45]

He was not impressed by the quality of the white goat cheese and the 'very peculiar tasting' *mysost*, but:

The *gammel ost*, however, to my taste is the worst cheese of all, though not in the opinion of a native, who likes nothing better.[46]

The French explorer Paul Belloni Du Chaillu (1835–1903), known for his travels in West Africa, made several journeys in Scandinavia from 1871 to 1878. He encountered '*gammal ost*, the strongest old cheese one can taste' on a *smörgåsbord* in Gothenburg, Sweden[47], and later came across it several times in Norway. Of Sogndal, he said that:

> The district is celebrated for orchards of apples, and also for its *gammel ost* (old cheese), which, when old enough, is the strongest known, and, after one gets accustomed to eat it, an excellent appetiser.[48]

Du Chaillu is the only foreign traveller known to me who described our other main sour milk cheese, *pultost*:

> There are three peculiar kinds of cheese: 1. The *mysost* is made from the whey remaining from the common cheese, boiled till the water is evaporated; then it is shaped into square cakes, weighing from two to five pounds; the colour is dark brown. It must stand at least a day before it is fit to be eaten. It is made only at the sæters, where wood is plentiful, for it requires a great deal of fuel. It is eaten in thin slices, and with bread and butter; women and children are especially fond of it. The best is from goat's milk. It can hardly be called a cheese, as it consists chiefly of sugar and milk. 2. The *gammelost*, made from sour skimmed milk, is a fermented round cheese, which is kept for months in the cellar. 3. *Pultost* is also a fermented cheese, mixed with caraway-seeds, not formed into cakes, but preserved in wooden tubs.[49]

Valldalen used to be famous for its *gamalost*. Du Chaillu stayed here a couple of days, apparently having a good time with the young milk-maids. When he left, he was presented with a big cheese, provisions for his walk across Hardangervidda. One can only assume that it must have been a *gamalost*.[50]

The cookbook writers

Hanna Winsnes (1789–1872) in her famous cookbook described two methods for making *gamalost*. For the first one her description is very brief. The second method is more detailed, and unusual. The cheese, after being moulded, was wrapped in a cloth and boiled in the whey 'till fairly red'. She did not say how long this is going to take, but mentioned that the whey will begin to thicken. The two hours of boiling used in the dairy today hardly darkens the cheese at all, so it is more likely that something like 6 hours was necessary. After remoulding and a couple of days rest, it is wrapped in cloth again, and boiled once more. This time in beer, 'till [the beer] is almost absorbed.' She recommended making three cheeses at the same time, as they would be better, stacked on top of each other in a tall, narrow tub and wrapped in straw. To speed up the maturation, the straw should be boiled in juniper extract (*briskelåg*) and wrapped around the cheese while still warm. This moist straw would have to be replaced once a day, and for as long as you cared to. *Gamalost*, she said, must always be a year old when used.[51]

Elisabeth Undahl in 1893 included three recipes for making *gamalost*. One can be recognized as the 'Hardanger' method. The other two are word for word copied from Winsnes, except for two smaller changes. She added just enough to the description of Winsnes' first method to make it clearly distinguishable as the 'Sogn' method. For Winsnes' second method, she reduced the necessary maturation time to six months.[52]

Hulda Garborg (1862–1934), wife of the author Arne Garborg, highly recommended Norwegian products in her books. In her opinion, anyone who understood cheese, would most likely rate a perfect *gamalost* higher than a Roquefort:

> No cheese can be finer, and none worse than the *gamalost*. It is made in different ways, any way it *can* be good, any way it can turn bad. It has the temper of a fine wine, is capricious like an all too pretty maiden; it possesses secretive powers like no other Norwegian cheese.[53]

Another cheese she thought highly of, was the Western Norwegian *fatost*:

> I once got together a group of fully experienced connoisseurs (townies and globetrotters of the utmost description) and placed before them a crock of *fatost* from Jæren with Norwegian mountain butter and oatmeal flat-bread; when they had sampled this, they all agreed that this cheese outclassed both Roquefort and Gorgonzola.[54]

Dairy production and the nostalgic writers

Gamalost used to be made in the *sæter*, the mountain farm where cattle was kept during summer. Since the first Norwegian dairy was established in 1855, milk processing has gradually been taken over by the dairies, and the traditional cheesemaking slowly disappeared. Following the growing dairy production, several writers published test reports on the cheese production in the early twentieth century. The first was Olav Johan-Olsen in 1905, followed by Ludvig Funder (1915–17 and 1941). Schmidt-Nielsen and Benterud published two papers on the chemistry of *gamalost* in 1940. Sigurd Funder, son of Ludvig, wrote a thesis in English on 'The chief molds in Gammelost and the part played by them in the ripening process' in 1946.

There are no estimates of production on the *sæters*, but the dairy production in 1935–40 was about 260,000 kg per year. The *sæter* production was thought to add substantially to this figure.[55] By 1980, *sæter* production was nearly non-existent, and dairy production was down to 150,000 kg.[56] The start of the Norwegian Gamalost Society in 1983, along with a growing interest in food traditions, seems to have stopped the decline. The dairy production kindled an interest in the disappearing *sæter* made cheeses.

A soup in the sun, a novel by Leif Borthen[57], is the great account of a journey through Norway in search of the real *gamalost*. The author's Portuguese friend Sebastião da Silva remembered affectionately a cheese he had eaten at the house of the Norwegian consul in Lisbon in 1938. Years later he came to Norway with a *saudade*[58] in his heart, to find this cheese again. The friends soon realized that no dairy cheese would do. They found some local cheeses too, and agreed that the one made at a *sæter* near Voss was a great, unique cheese, but not *the* cheese. Sebastião said that 'You should know that the cheese served by your fellow countryman in Lisbon was an "angry" cheese, yes, not just angry, it was formidable. It was ruthlessly sovereign like a Genghis Khan. And *that* cheese, amigo, had a green centre.'

In 1987, the *gamalost* got it's own book, written by John Moberg, president of the Norwegian Gamalost Society. It is an irreverent book, jokingly describing the history of the cheese and it's present production and uses.

Since 1991 Vik dairy is the only *gamalost*-producing dairy.[59] The modern production still follows the general method described by Gunnerus in 1774. The dairy is located in Sogn, but the production follows the Hardanger method. Of course, stainless steel vats and moulds have replaced cauldrons and wooden moulds. The hygienic control is very strict, in fact more so than in any other cheese production in Norway. The source material is still skim milk. After souring, it is coagulated by heating to 63°C (145°F). A so-called decanter is used to separate the casein (curds) from the whey, whence the name 'decanter-method'. Centrifuge would probably be a more descriptive name for the contraption. A mill cuts the curds into a grainy substance, which is filled into the moulds. The moulds are lowered into vats of boiling whey for a couple of hours. After another few hours rest, the cheeses are unmoulded and placed on steel shelves, where they are sprayed with a pure mould culture, *Mucor mucedo* var. *racemosus*. After three days there is a thick mould growth, making the cheeses look like furry balls. The mould is flattened by hand, a process which is repeated after another two days. The ripening period is surprisingly short, 12–14 days for a standard size cheese of 1.7 kg. The small 380 g cheeses ripen in only 8–9 days. In order to maintain an even degree of ripening, all cheeses are frozen at the appropriate stage. Freezing apparently has no adverse effects on the quality. Today, 5 million litres of milk are made into 200,000 kg of *gamalost* each year.[60]

Does *gamalost* have a future? The pleasure of finding a particularly good cheese at the small dairy or *sæter* in a remote valley is gone. Wherever you buy your cheese today, rest assured that it is all made at the same location. This spring, just by chance, I found possibly the last person in Norway who makes traditional *gamalost* for sale. Gerd Lien from Røldal makes *gamalost* in a *sæter* at Haukelid every summer, and sells her produce through a shop in Bergen.[61] At a time of increased interest in farmhouse cheeses around the world, and desite growing imports of foreign cheeses to Norway, I think there is hope for *gamalost* too.

ACKNOWLEDGEMENTS

The following are some of the people who have helped making this paper possible: Øyvind Eltervåg, Sandnes Bibliotek, Sandnes; Ed Gunderson, Oregon, USA; Hallgerður Gísladóttir, Þjóðminjasafn Íslands / National Museum of Iceland; Gerd Lien, Fjellsjå sæter, Røldal; John Moberg, President of the Norwegian Gamalost Society; Henry Notaker, NRK, Oslo; Thea Nybø, Brattlandsdalen; John Inge Sørland, Tine Norwegian Dairies ba, Vik.

SELECT BIBLIOGRAPHY

Bang, J. S. 1802: Om Suurmelks Afbenyttelse til den saakaldte norske Gammelost og Syre. In: *Nordisk Landvæsens og Landhuusholdnings Magasin* Vol. 4, København pp. 169–175.

Benterud, S. J., J. Grude and E. E. Wold 1908: *Meieridriften i Norge indtil aaret 1905.* Grøndahl & Søns Forlag, Kristiania.

Bergfjord, Kjell 1997: «*Til jordbrugets ophjælp*» - *ei reise i soga til meieriet i Vik gjennom 100 år.* Sogn og Fjordane Meieri, Førde.

Björn K. Þórólfsson og Guðni Jónsson 1943: *Vestfirðinga sögur.* Íslenzk fornrit, VI. bindi. Hið Íslenzka Fornritafélag, Reykjavík.

Borthen, Leif 1961: *En suppe i solen.* Aschehoug, Oslo.

Brennu-Njáls saga. Íslenzk fornrit, XII. bindi. Hið Íslenzka Fornritafélag, Reykjavík 1954.

Carr, Sandy 1981: *The Simon and Schuster pocket guide to Cheese.* Simon and Schuster, New York.

Clarke, E. D. 1824: *Travels in various countries of Europe, Asia and Africa. Part the third : Scandinavia.* In three volumes: 9th, 10th, 11th. T. Cadell and W. Davies, London.

Du Chaillu, Paul B. 1899: *The land of the midnight sun.* New edition. [1st ed. London 1881]. George Newnes, London.

Fabricius, Johann Christian 1779: *Reise nach Norwegen mit Bemerkungen aus der Naturhistorie und Oekonomie.* Hamburg.

Fritzner, Johan 1954: *Ordbog over Det gamle norske Sprog.* Nytt uforandret opptrykk av 2. utgave (1883–1896). Tryggve Juul Møllers Forlag, Oslo.

Funder, L. 1915–17: *Undersøkelser vedrørende gammelost.* Statens Meieriforsøk, beretning nr. 3–5. Grøndahl & Søns Boktrykkeri, Kristiania.

Funder, Ludvig 1941: *Forsøk vedrørende gammelost i tiden 1938–1940*. Statens Meieriforsøk, beretning nr. 35. Grøndahl & Søns Boktrykkeri, Oslo.
Funder, Sigurd 1946: *The chief molds in Gammelost and the part played by them in the ripening process*. A. W. Brøggers Boktrykkeri a/s, Oslo.
Garborg, Hulda 1922: *Heimestell*. 3rd ed. [1st 1899]. Hans Martinussens Forlag, Bergen.
Gjertsen, I. P. 1854: *En paa 20 Aars Erfaring grundet Practisk Anviisning til at tillave fortrinlig god Gammelost samt at conservere samme*. Forfatterens Forlag, Bergen.
Grude, J. 1891: *Stølsdriften paa Vestlandet*. Dreyer Bogtrykkeri, Stavanger.
Grøn, Fredrik 1927: *Om kostholdet i Norge indtil aar 1500*. Oslo.
Guðbrandur Vigfússon & C. R. Unger (eds.) 1862: *Flateyjarbok*. Vol. 2. Christiania.
Gunnerus, Johan Ernst 1774: Oeconomisk Afhandling om alle de Maader, hvorpaa Melken nyttes i Norge. In: *Det Kongelige Norske Videnskabers Selskabs Skrifter, Kiøbenhavn* 5: 26–152.
Halldór Laxness 1958: *The Happy Warriors*. Methuen & Co., London.
Halldór Laxness 1983: *Gerpla*. [1st ed. 1952]. Helgafell, Reykjavík.
Hammer, Christopher 1772: *Norsk Huusholdings-Kalender*. 1st part. Christiania.
Jensøn, Christen 1646: *Den Norske Dictionarium eller Glosebog*. Faksimileutgave J. W. Eides Forlag, Bergen 1946. Kiøbenhaffn.
Johan-Olsen, Olav 1905: *Undersøgelser over ost og ostegjæring. I. Indledning - surmelksostene*. Udgivet med statsbidrag. Olaf Norlis Forlag, Kristiania.
Jón Jakobsson 1791: Um Miólkur-Not á Islandi. In: *Rit þess Konúngliga Islenzka Lærdóms-Listafélags*, vol. 11, pp. 193–241.
Jonge, Nikolay 1779: *Chorographisk Beskrivelse over Kongeriget Norge, samt Færøe, Iisland og Grønland*. Kiøbenhavn.
Konow 1878: *Tilvirkning af gammelost efter den paa godseier Knagenhjelms gaard Kaupanger i Sogn brugelige maade*. Meddelt af amtsdyrlæge Konow. Udgivet ved Nordlands landhusholdningsselskabs direktion. Trykt af Olaus Andresen. Bodø.
Metcalfe, Frederick 1858: *The Oxonian in Thelemarken*. 2 volumes. Hurst and Blackett, London.
Moberg, John 1987: *Gamalosten*. Det Norske Samlaget, Oslo.
Njal's saga. Penguin Books, London 1986.
Olaus Magnus 1976: *Historia om de nordiska folken, Vol. I–IV*. [Historia de gentibus Septentrionalibus, 1555]. Gidlunds, Stockholm.
Pontoppidan, Erik 1752–53: *Norges Naturlige Historie*. 2 volumes. Facsimile edition Rosenkilde og Bagger, København 1977.
Ramus, Jonas 1715: *Norriges Beskrivelse*. Kjøbenhavn.
Reinton, Lars 1969: *Til seters : Norsk seterbruk og seterstell*. Det Norske Samlaget, Oslo.
Ränk, Gustav 1956: Om äldre mjölkhushållning i Baltikum. In: *Svio-Estonica* XIII: pp 165–200.
Ränk, Gustav 1964: Gammal ost. In: *Gastronomisk kalender* 4: 47–56.
Ränk, Gustav 1964: Om äldre ostkultur i övre Sveriges fjälltrakter. In: *Norveg* 11: 35–44. Deutsche Zusammenfassung.
Ränk, Gustav 1966: *Från mjölk till ost : Drag ur den äldre mjölkhushållningen i Sverige*. Nordiska Museets handlingar 66. Nordiska Museet, Stockholm.
Ränk, Gustav 1971: Verbreitungsverhältnisse einiger Milchprodukte im eurasischen Raum. In: *Ethnologia Scandinavica* 1971: 120–126.
Schmidt-Nielsen, S. und Arne Benterud 1940: Über die flüchtigen Fettsäuren des norwegischen Sauermilchkäses "Gammelost". In: *Det Kongelige norske videnskabers selskabs forhandlinger bd XIII, nr 20* 83–86.
Schmidt-Nielsen, S. und Arne Benterud 1940: Zur Kenntnis des norwegischen Sauermilchkäses "Gammelost" : I Allgemeine Zusammensetzung. In: *Det Kongelige norske videnskabers selskabs forhandlinger bd XIII, nr 19* 71–82.

Schøning, Gerhard 1979: *Reise som giennem en Deel af Norge i de Aar 1773, 1774, 1775.* 2 volumes. Reissue of 1910 edition. [First ed. København 1778]. Tapir Forlag, Trondheim.
Setelarkivet til Norsk Ordbok. 1999a. The card files of the Norwegian Dictionary. Internet: http://www.dokpro.uio.no/cgi-bin/leksikografi/setelarkivet?base=NOB (accessed 25.05.99).
Setelarkivet til Trønderordboka. 1999b. The card files of 'Trønderordboka'. Internet: http://www.dokpro.uio.no/cgi-bin/leksikografi/setelarkivet?base=TOB (accessed 25.05.99).
Simon, André L. 1960: *Cheeses of the world.* 2nd ed. Faber and Faber, London.
Skrikrud, Eivind 1914: *Smør og ost : stutt rettleiding i laging av smør og ost paa vanlege gardsbruk.* Erik Gunleiksons Forlag, Risør.
Storch, V. 1887: *Kemiske og mikroskopiske undersøgelser af et ejendommeligt Stof, fundet ved Udgravninger, foretagne for det islandske Oldsagsselskab af Sigurd Vigfusson paa Bergthorshvol i Island.* [Also available in Icelandic]. Appendix to Árbók hins íslenzka fornleifafélags 1887. Kjøbenhavn.
Strøm, Hans 1762: *Physisk og Oeconomisk Beskrivelse over Fogderiet Søndmør, beliggende i Bergens Stift i Norge. Første Part.* Sorøe.
TINE Norwegian Dairies BA 1998: *Welcome to the World of Gamalost Cheese.* Bilingual, Norwegian and English. Internet: http://www.tine.no/gamalost (accessed 26.06.99).
Undahl, Elisabeth 1893: *Nyeste Husholdningsbog for husmoderen i byen og paa landet.* Chr. Stephansens forlag, Bergen.
Vi_ar Hreinsson (general editor) 1997: *The Complete sagas of Icelanders : including 49 tales.* 5 volumes. Leifur Eiríksson Publishing, Reykjavík.
Wille, Hans Jacob 1786: *Beskrivelse over Sillejords Præstegield i Øvre-Tellemarken i Norge.* Kiøbenhavn.
Williams, W. Mattieu 1859: *Through Norway with a knapsack.* With six tinted views and map. [1st edition]. Smith, Elder and Co., London.
Winsnes, Hanna 1845: *Lærebog i de forskjellige Grene af Huusholdningen.* C. A. Wulfsberg, Christiania. Facsimile edition, Damms Antikvariat, Oslo 1985.
Wyndham, Francis M. 1861: *Wild Life on the Fjelds of Norway.* Longman, Green, Longman, and Roberts, London.

REFERENCES

[1] *A Concise Encyclopedia of Gastronomy*, Overlook Press ed., New York 1981. To be fair, Simon in the later *Cheeses of the World* (Faber and Faber, London 1960) offered a detailed and correct description of *gamalost*, based on information from professor Knut Fægri, who many years later was made honorary member of INTERSPI.

[2] Carr 1981, pp. 114–115.

[3] Johan-Olsen 1905; S. Funder 1946.

[4] *Gamalost* is the 'nynorsk' spelling which seems more appropriate for this traditional cheese than the 'bokmål' spelling *gammelost*. In Swedish it is called *gammalost*.

[5] Setelarkivet 1999a, e.g. nos. 1501053311, 1501053313, 6505928027.

[6] Johan-Olsen 1905, p. 34; Grøn 1927, p. 99.

[7] *Njal's saga* 1986, p. 124; Icelandic: *Brennu-Njáls Saga* 1954, p. 125: '... ba∂ hann, at taka skyldi ostkistu Þorger∂ar, ok var svá gört; lag∂i han ∂ar í ni∂r snei∂irnar, ok stózk ∂at á endum ok ostkistan.'

[8] Grøn 1927, pp. 100–101.

[9] *Njal's saga* 1986, p. 265; *Brennu-Njáls Saga* 1954, p. 328 (sumar báru vatn e∂a hland); Johan-Olsen 1905, p. 35.

[10] Storch 1887.

[11] Johan-Olsen 1905, p. 36.

[12] Translation by Martin S. Regal in Vi∂ar Hreinsson 1997, vol. II, p. 341.

Icelandic: 'Fóstbræðra saga', in *Vestfirðinga sögur* 1943 p. 144–145: 'Frá verðgetum er sagt vandliga: Tveir diskar váru fram bornir; þar var eitt skammrifsstykki fornt á diskinum hvárum ok forn ostr til gnættar. Butraldi signdi skamma stund, tekr upp skammrifit ok skerr ok neytir ok leggr eigi niðr, fyrr en allt var rutt af rifjum. Þorgeirr tók upp ostinn ok skar af slíkt er honum syndisk; var hann harðr ok torsóttr. Hvárrgi þeira vildi deila við annan kníf né kjötstykki. En þó at þeim væri lítt verðr vandaðr, þá fóru þeir þó eigi til sjálfir at skepja sér mat, því at þeim þótti þat skömm sinnar karlmennsku.'

Old Norse: 'Fóstbræðra saga', in *Flateyjarbok* 1862, vol. 2, p. 103: 'Fra verdgetum er uandliga sagt .ij. voru diskar fram settir. þar var æitt fornnt skamrifstykki a diske huorum ok fornn ostr til gnættar. Butrallde signde skamma stund ok tok upp skamrifstyckit ok leggr æigi fyrr nidr en af var allt. Þorgeirr tok upp osthlutinn ok skar af sligt er honum syndizst. hann uar hardr ok torsottr. huorgi þeirra uillde æiga vid annan knif ne kiötstycki. en þo at þ eim uæri litt uanadadr nattuerdr þa foru þeir þo æigi j bur at skepia ser nattverd þuiat þeim þ otti þat suiuirding sinnar kallmenzsku.'

[13] It is easier to prove the existence of a phrase, than it's non-existence. I have digitally searched the full texts of some 80 sagas and tales (þættir) available on the Internet at http://www.snerpa.is/net/index.html and found several references to cheese, but nothing on particular types of cheese.
[14] Grøn 1927, p. 100.
[15] Halldór Laxness 1958, pp. 261–270.
[16] Olaus Magnus 1555, vol. 3, p. 85.
[17] Johan-Olsen 1905, p. 102.
[18] Jensøn 1646, p. 98.
[19] Ibid. p. 110.
[20] Ramus 1715, p. 19.
[21] Pontoppidan 1753, p. 432.
[22] Johan-Olsen 1905, p. 35; L. Funder 1941, p. 7; S. Funder 1946, p. 9
[23] Strøm 1762, volume 1, p. 377.
[24] Hammer 1772, vol. 1, p. 192.
[25] Gunnerus 1774, pp. 113–122.
[26] 1 Lispund = 7.97 kg.
[27] Setelarkivet 1999a, no. 1501053350.
[28] Schøning 1979, volume 1, p. 292.
[29] Jonge 1779, p. 33: 'Gammelost, som holdes rar af mange'.
[30] Fabricius 1779, pp. 250–252.
[31] Wille 1786, p. 191.
[32] E.g. Bang 1802, Gjertsen 1854, Konow 1878, Grude 1891 and Skrikrud 1914.
[33] Setelarkivet 1999a, no. 1501053313.
[34] Setelarkivet 1999a, no. 1501053348; Setelarkivet 1999b, no. 100091175.
[35] Clarke 1824, vol. 10, p. 292.
[36] Metcalfe 1858, vol. 1, p. 50.
[37] Bavarian type of beer and home-made liquor.
[38] Metcalfe 1858, vol. 1. p. 192.
[39] Williams 1859, pp. 191–192.
[40] Ibid. pp. 217–218.
[41] Wyndham 1861, p. 55.
[42] Ibid. p. 101. *Budeier* are milkmaids.
[43] Ibid. p. 116.
[44] The word *lustily* mysteriously disappeared in the Norwegian translation of 1970.
[45] Wyndham 1861, pp. 122–123.
[46] Ibid. p.228.

[47] Du Chaillu 1899, p. 7.
[48] Ibid. p. 201.
[49] Ibid. p. 342.
[50] Ibid. p. 253.
[51] Winsnes 1845, pp. 81–83.
[52] Undahl 1893, pp. 179–182.
[53] Garborg 1922, p.149 [Not in the 2nd ed. of 1903, 1st ed. 1899 not seen].
[54] Ibid. p. 147 [Identical in the 2nd ed. of 1903, p. 145, except that the cheese now was from Sunnmøre!]
[55] S. Funder 1946, p. 17.
[56] Moberg 1987, p. 13.
[57] Borthen 1961.
[58] The Portuguese word *saudade* comes from Latin *solitas* (loneliness) and denotes 'a feeling of nostalgic remembrance of people or things, absent or forever lost, accompanied by the desire to see or possess them once more.'
[59] Bergfjord 1997, p. 104.
[60] Jon Inge Sørland, TINE Norwegian Dairies BA, Vik, pers. comm. 1999.
[61] Gerd Lien, pers. comm. 1999.

The Origins of Taste in Milk, Cream, Butter and Cheese

Sarah Freeman and Silvija Davidson

Milk is composed principally of water and has a moderate pH (6.4–6.6), making it an ideal medium for the growth of natural flora. The other main components are fat, protein, lactose (a disaccharide composed of glucose and galactose) and citrate. This last occurs in small quantities but plays a particularly important role in the flavour of fermented products, butter and cheese among them.

Composition varies between species, and breeds within species (see chart in Sokolov's paper, below). Human milk is rather thin (low in protein) and sickly sweet (high in lactose) compared to cow's milk – a human baby takes twice the time to double its weight as does a calf. Sheep's milk, by contrast, is more concentrated than cow's in terms of both fat (almost double) and protein, though lactose levels are similar. Friesian milk (almost universally used outside the organic sector for both liquid milk and milk products) contains on average 37g fat, 34g protein, and 48g lactose per litre while Jersey milk has 51g fat, 38g protein, and 50g lactose – in other words, a considerably higher fat content, with a slight increase in protein and sugars.

However, composition can also vary considerably with the seasons. Mark Hardy, who produces the sheep's butter and cheeses that we have brought for tasting, says that just after lambing fat content may be as low as 40 g per litre, whereas six months later it has risen to about 85 g, so that he can make far more cheese now from the same volume of milk as he could in April. Citrate content also fluctuates dramatically: pasture grazing in the summer months will yield around 0.17 per cent, while winter feeding (silage, hay or other supplements) may give as little as 0.07 per cent This is likely to have a dramatic influence on the flavour of butter and cheese.

Feed and flavour

Feeding affects flavour in all sorts of ways. First, there is the grass: this may be organic or not, on permanent pasture or not, and new or old according to season. In my (Sarah's) view, the type of soil also influences the taste it produces, although the difference is more obvious in cheese than fresh milk;

it can also be argued that such difference is due not to pasture but atmospheric microflora. Anyone who eats organic vegetables will probably agree that organic grass is likely to have more flavour than conventional – and there is a reason for this: when chemical fertilizer is not used, plants have to work harder to gain nourishment. Also, in the absence of herbicide, there will be a wider variety of plants – perhaps dozens of different species, contributing to the spectrum of flavours; similarly, permanent pasture can be expected to yield more taste than a field which has been sown, most likely with monocultural fast-growing rye-grass, or maybe two sorts at most. Then there is the difference between spring and autumn grazing: in spring, the grass is higher in protein than in the autumn, and high-protein food makes for lower fat content (instrumental in taste perception). A further factor which affects the fat ratio is how long the cows have been in the field. As James Montgomery explains, they eat the leaf first, which is high-protein, and move on to the stem when no leaf is left. The stem is harder to digest, which raises fat content. In order to produce cheese of consistent quality, James moves the fences of the cows' field each day, thus ensuring that they eat a proportion of both leaf and stem.

Secondly, there is the consideration of winter and supplementary feed. The Montgomerys give their cows silage in the winter, which is higher in energy than hay; at Plaw Hatch Farm, which is biodynamic, the cows are fed chiefly on hay because the cheese-maker, Jo Schneider, finds that silage is sometimes too acid for them; he also says that he thinks he can detect the taste of it in the milk. Mary Quicke finds that ransomes give milk a garlic tinge and kale a cabbagy aroma. The Plaw Hatch cows are not given anything in the milking-parlour, but all conventional dairy-herds are given supplementary high-protein food at milking-time: at one time, fish-meal was popular but most farmers have now switched to mixtures which contain soya. At present, this is likely to contain a proportion which is genetically modified; one would hope that an alternative, GM-free sources will be found as soon as possible. (A somewhat more deleterious effect on flavour is likely to have resulted from the raw sewage recently employed by a number of French farmers to 'bulk out' the concentrate.)

Health and happiness
It seems obvious that if a cow is to give high quality milk, she must be in good health. Mastitis, i.e. infection of the udder, reckoned to affect around one to two per cent of cows at any one time, is known to give salty off-flavours. Equally important are issues of stress and over-milking. The modem dairy cow, producing around 2,000 gallons of milk a year, already yields more than eight times the levels naturally produced by a wild cow for her calf. A number of cows are being engineered to yield some 4,000 gallons a year and require

milking four times a day, if only to enable them to walk. An overmilked and stressed cow will produce milk that is low in protein.

It is also important to yield, if not to flavour, that she should be in a good state psychologically, i. e. happy and relaxed. Proverbially, milkmaids had to be calm and cheerful because if they were bad-tempered, the cow would withhold her milk: milkmaids may belong to history, but it has recently been shown that agreeable music significantly increases yield (it also encourages hens to lay regularly). I believe that research is now in progress to determine which kind of music is most effective. [I recall that Hugh Lillingston, erstwhile producer of Innes goat's milk and cheeses, kept his options open by piping through alternate doses of Mozart and heavy rock. (SD)]

Other factors
Factors known to affect milk composition and flavour range from milking practice variables (time of day; interval between milkings; stage of lactation; number of previous lactations) to environmental changes (whether the dairy building has received a fresh coat of paint).

Processing and flavour perception
Jo Schneider describes the processing of most of the milk sold today as 'taking it apart and putting it together again as people think it ought to be'.

All the milk available in liquid form, except a very little sold direct as 'green-top' from a handful of farms, is pasteurized. The method of pasteurization generally reckoned to destroy taste least is to hold it at 70°C for 15 seconds; however, because of concern about *Mycobacterium paratuberculosis,* the time has been increased to 25 seconds, which is why you may have noted an almost curdled texture and cooked taste or, to use Mary Quicke's description, 'stale, burn-on flavours'.

As the milk in the udder is sterile and today is usually piped straight from udder to dairy, with no exposure to the open atmosphere, the opportunities for bacteria to enter it are very limited: in spite of all precautions, however, a few succeed. Although relatively few in number, they include perhaps 40–50 different strains or variations of strains and make a startling difference to the flavour. Pasteurization does not of course discriminate between those bacteria which we might consider beneficial to flavour, and undesirable pathogens; it kills them all. The secondary effect of pasteurization, that of inactivating native enzymes, principally the fat-splitters responsible for rancidity in order to extend shelf-life, is almost certainly responsible for further flattening of flavour. [Despite these precautions, pasteurized milk reputedly sours more offensively, if less quickly, than raw milk and indeed is no guarantee of safety in products as any pathogen which subsequently enters the medium has a clear field and will reproduce very quickly – but this is a separate subject.]

Most milk sold is also homogenized: rather ominously, the name is derived from the Greek meaning 'of the same kind' and involves breaking down the fat globules by forcing them through a narrow opening onto a hard surface and reducing them all to the same size (about a quarter of the size of the larger globules). The resultant milk tastes more bland than the original, as there is a direct link between size of fat globule and taste perception: large globules give a creamier impression than small ones. Jersey milk, for instance, is actually less rich than sheep's but tastes creamier because the fat particles are much bigger. (We should stress, though, that this applies only to the taste of the liquid milk: sheep's and buffalo's yoghurt and cheese taste quite as rich as the composition of the milk would suggest.) Homogenized milk is also more sensitive to light spoilage, which gives two distinct types of off-flavours: one akin to rancidity, giving fishy and metallic flavours, is produced by photo-oxidation of unsaturated fatty acids, and another known as 'burnt feather' character, results from photo-oxidation of proteins.

To add insult to injury, the milk sold in supermarkets is standardized, which means that it has been separated and (unless sold skimmed), a proportion of the cream and milk solids returned. However, in order to maintain a consistent product, only the accepted average amount of cream and milk solids are put back, which means that the customer never receives the richer milk, with its fuller, creamier taste, which they might expect at this time of year, nor the full-bodied, protein-dense milk of the winter months.

BUTTER AND CHEESE
Aroma, bacteria and serendipity

Here we are in the realms of fermentation, or 'microbial modification of milk constituents', i.e. lactose, citrate, proteins and lipids. All fermented milk products involve the breakdown of lactose, the major source of carbon and energy for the microorganisms in milk, to yield a certain quantity of lactic acid. This provides the necessary environment for the crucial transformation of citrates into essential flavour compounds by so-called aroma bacteria. The principal bacteria are *Streptococcus lactis*, ssp *diacetylactis* and *Leuconostoc cremoris*. Of the two, only the *Streptococcus* can ferment lactose, a factor of some importance in determining starter culture cocktails. Their important role is the production of diacetyl, one of the most potent fragrances known to man. Diacetyl is what gives butters and cheeses (and indeed buttermilks and certain creams) their pleasant 'buttery' or 'nutty' character – as long as the proportions remain at around two parts per million; anything over four ppm begins to taste coarse and acrid. Mutant strains of bacteria have been known to produce a much higher proportion of diacetyl in a short space of time, resulting in considerable spoilage. Diacetyl can also be readily broken down into the flavourless, non-volatile acetoin. Indeed this happens as soon as citrate levels fall below a certain critical value, and is catalysed by an enzyme common to both flavour bacteria and other psychrotrophic bacteria usually found in dairy environs: in other words, flavour loss in, say, butter can occur in a cool dairy. Curiously, holding milk products at fridge temperature results in a gradual increase in diacetyl concentrations (our sources suggest that 'the reason for this phenomenon is not understood'). This is almost as serendipitous as the fact that the all-essential diacetyl production is a natural detoxification mechanism (involving conversion of pyruvic acid, accumulated from citrate metabolism, to neutral carbon compounds). In other words, 'flavour compounds are the products of necessity' (Marshall).

Butter: precision, uncertainty and spoilage
Selection of starter cultures is, however, a considerable step away from serendipity and involves such factors as balancing the acetaldehyde-producing *Streptococcus* strains (green, yoghurty flavours being undesirable in butter) with subspecies of *Leuconostoc* that are particularly adept at scavenging acetaldehyde; the addition of dextrose and yeast extract by way of *Leuconostoc* nutrients; even the creation of mutant bacteria lacking diacetyl reductase (to prevent breakdown into acetoin).

To produce ripened cream butter, pasteurized (usually) cream is cooled to 18.3°C, inoculated with starter culture, and held for five to six hours until the pH value drops to 4.9. The cream is then chilled to 12.8°C and kept until the pH value is 4.6. It is then cooled to 7.2°C and held until churning. Here the precision breaks down a little for, as Harold McGee points out *(On Food and Cooking)*, 'exactly how churning works is still unknown' (he nonetheless gives a fine disquisition on the theory of the process there is not room to reproduce here).

While ripened (cultured) butter has a much better flavour (from lactic acid and diacetyl) than simply churned cream butter, it also suffers from more rapid chemical deterioration, largely due to fat oxidation. There may also be spoilage from mutant subspecies of the two main aroma bacteria, namely ssp *maltaromicus,* which, as the name suggests, produce carbonyl compounds which impart a 'malty' flavour to the butter. Churning and washing themselves lead to variable loss of flavour as a result of the water-solubility of diacetyl.

Starter or paddle?
Two of the makers whose products we have brought for tasting, the Montgomerys and Keens, have farms within a few miles of each other, keep the same

breed of cow (Friesian, which is now almost universal outside the organic sector), and use almost the same processes, yet their butter and cheese tastes remarkably different. Part of the explanation for this lies with their starter bacteria. Both the Keen's and Montgomery's butter is made, not from the fresh but the whey-cream which is drained off during cheese-making and thus has already been treated with the cheese-making starter. James Montgomery (but not the Keens) adds a little extra starter to the butter. The Keens use a stainless steel churn; the Montgomerys are convinced that the taste of their butter is influenced by the fact that they make it with wooden paddles. Until not so long ago, they used wooden tubs too: these have now been replaced, but James's mother, Elizabeth, feels that not only wood but the right sort of wood is so important that she spent days searching for the kind she wanted.

Cheese: a very complex affair

Cheese is fermented and curdled to a greater degree than butter or other dairy products. Caseins coagulate, whey is drained or pressed out of the curd and in most instances fats and proteins, as well as lactose, are broken down. Starters are considerably more complex than in butter production: two to six types of bacteria are routinely used, according to the style of cheese, but others may be included to yield more complex flavours. Most cheese-makers do in fact use a basic laboratory product, but for others the starter they consider essential to the individuality of their product remains a closely guarded secret.

Mary Quicke notes that distinctive overtones are derived from different starters: 'DVI' starters give rise to rapid acid development and give a uni-dimensional flavour; Texel starters give a metallic back flavour; while 'bulk' starters (a wider spread of 'bugs') lead to more even acid development and more rounded flavour.

The taste of cheese depends not only on the quality of the milk and on the starter bacteria (plus the indigenous bacteria if the cheese is unpasteurized), but other microorganisms, i.e. yeasts and moulds, either deliberately introduced or atmospheric. The two best known moulds are *Penicillium roqueforti,* used in blue cheese production, and *Penicillium camemberti* for white mould rind cheeses such as brie. Another which is actually 'smeared' on at a later stage, but has a profound effect on taste, in *Brevibactenum linens* used, for example, on Gubbeen and Milleens. Many years ago, Eugene Burns, who makes Ardrahan, smeared his cheeses with *Br. Iinens* on a single occasion and has never needed to use it again as it established itself on his farm.

The different types of rennet used to set the milk also affect the final flavour. Microbial (fungal, i.e. vegetarian) rennet, notes Mary Quicke, gives a metallic, bitter back flavour, while the genetically-engineered Chymosin (a Vegetarian

Society-approved copy of the chymosin enzyme derived from a suckling calf's stomach) appears to give cleaner flavours.

Methods of salting produce another variable: in some instances salt is added directly to the curds, other cheeses are brined once formed, and a number of cheese-makers simply rub salt onto the surface of the cheese. Salting slows the activity of the starter bacteria and suppresses the growth of spoilage bacteria; to stop bacterial activity altogether salting has to be taken beyond the stage of pleasant palatability – *feta* cheese, for example, needs a certain amount of soaking before use. Reduction of water content is another inhibitor of bacterial activity. This takes place throughout the life of the cheese, from the draining of the curds to the moment of eating and again affects both texture and flavour.

When I (Sarah) told James Montgomery about this paper, he remarked that he hoped no one would ever be able to analyze exactly why a cheese (or anyway, his cheese) tastes as it does: I can reassure him that the subject is so complex and the changes which affect taste so small and subtle that they will probably never be fully detected by laboratory methods.

At the Symposium we demonstrated six types of hard cheese from the following makers, some of whose views are also quoted in the paper:

Mary Quicke (Friesian-Holstein cattle: cream, whey butter, Cheddar)
Stephen and George Keen (Friesian cattle: whey-butter, Cheddar)
James and Archie Montgomery (Friesian cattle .whey-butter, Cheddar)
Plaw Hatch Farm (biodynamic; Meuse-Rhine-lJssel (MRI) cattle: cream, Cheddar-style cheese)
Mark Hardy, Putlands Farm (organic; *fromage frais*, soft and hard sheep's cheese)
Chris Duckett (Friesian cattle: whey butter, Caerphilly)

ACKNOWLEDGEMENTS

Thanks particularly to: Jane Capon, The Village Farm, Diss (Domini Jersey Herd); Mark Hardy, Putlands Farm; Stephen Keen, Keen's Cheddar; Dr. Bernard C. Lamb, BSc, CBiol, FIBiol; James Montgomery, Montgomery's Cheddar; Matthew Organ, microbiologist, Lubborn Cheese; Mary Quicke, Home Farm (Quicke's Cheddar); Eurwen Richards, cheese consultant; Sir Julian Rose, Hardwick Estate; Rachel Rowlands; Jo Schneider, cheese-maker, and Jane Thomas, Plaw Hatch.

BIBLIOGRAPHY

Adams, M.R. and Moss, M.O., *Food Microbiology*, 1995, The Royal Society of Chemistry.
Davies, L.F. and Law, B.A., *Advances in the Microbiology and Biochemistry of Cheese and Fermented Milk*, 1984, Elsevier.
Freeman, S., *The Real Cheese Companion*, 1998, Little Brown & Co.
McGee, H., *On Food and Cooking*, 1984, Allen & Unwin.
Rose, A.H., ed., *Fermented Foods*, 1982, Academic Press.

Cato's Roman Cheesecakes: The Baking Techniques

Sally Grainger

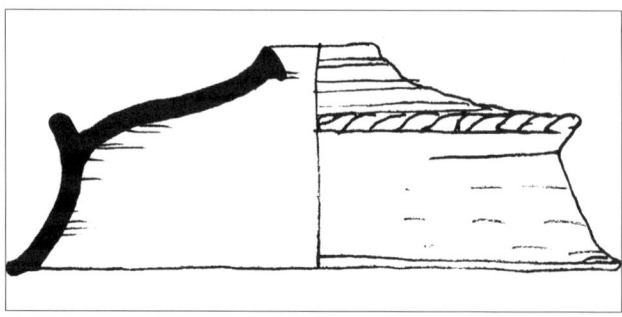

Mid-Imperial baking cover from Matrice. Drawing taken from Cubberly et al., 'Testa and clibani: The Baking Covers of Classical Italy', PSBR, volume 61, 1988.

I have been making and baking the cheesecakes from Cato's *De Agri Cultura*, namely *placenta* and *libum*, for many years. However, on most occasions, these cakes were baked in a modern oven and the cheese, so critical to the success of the cakes, was somewhat less then authentic, being an alternative to the freshly made sheep's cheese required in the recipe. The purpose of this article is to go back to the original recipes and follow the instructions to the letter, both for the cheese and in the baking method described, namely *sub testu*. Much of the ground work in identification of the *testum/clibanus* baking cover has been done by Tony Cubberly and others in a 1988 PBSR paper.[1] I took inspiration from this paper a number of years ago and sought a potter who could make me a *testum* that could withstand the heat of direct contact with the hot coals.

My experiments have proved very illuminating. Cato's cakes have been described elsewhere as rather stodgy and unappetizing and it has even been suggested that the recipes in Cato for *placenta* and *libum* were designed entirely for sacrificial purposes, i.e. that temple slaves and their like were the ultimate recipient of these particular cakes.[2] There is evidence for this but there is also evidence that these cakes were served as part of the *mensa secunda* (sweet course) and that they would necessarily be delicate, and appetizing as part of a fashionable Roman *cena*.[3]

It is clear from my experiments that a good deal of the success and appeal of these cakes is due to the baking process. They are very much more appetizing when served warm, straight from the oven, and when allowed to go cold, in my opinion, the altar is the best place for them. There is a quite considerable difference to the finished cakes when they are cooked *sub testu* as opposed to a large dry oven. It is not clear whether these cakes were mass produced in commercial bakers' ovens but I would suggest that as a dinner party sweet these cakes were always cooked *sub testu*.

A large wood-fired or charcoal-fired oven, used for roasting meat etc., could have been used to bake cakes but would have been uneconomical to keep hot all through the meal just to cook one or two cakes. The *testum/clibanus* baking cover is actually best used with the coals and ashes from the cooking platform after the rest of the meal is cooked. It is a very practical way of utilizing all the possible heat generated from a charcoal fire. The cakes are removed from the hearth and taken out to the diners almost immediately, soaked in honey. The whole process is both economical, practical and produces a superior product.

It is fairly well established that the Greek *plakous* and *placenta* are the same cake. *Placenta* is derived from the accusative form – *plakounta*. Athenaeus also tells us that the Greek was contracted from *plakoies*, an adjective form meaning 'full of or rich in' *plak* or in this case flakes.[4] Thus the cake should not be translated 'flat cake' but 'flaky cake', the flakes being the sheets of *tracta*. It is also apparent that the Greek term could also have a generic meaning 'cake' and other derived forms could mean 'cake maker' or 'baker'. This is so in Latin as well, hence *placentarius* means pastry cook. The cake appears frequently in Athenaeus:

> The streams of the tawny bee, mixed with the clotted river of bleating she-goats, placed upon the flat receptacle of the virgin daughter of Zeus, delighting in ten thousand delicate veils. (X. 449b-c)

This is a very elaborate, florid description but still resembles our *placenta* in all aspects. A further description taken from Athenaeus gives a more detailed picture of the finished cake. Phanias, a fourth-century BC writer on plants, writes that the cake looks like the seed pod of the domestic mallow with its ridges running out from the centre and its central pointed knob.[5] I am blest with a garden full of mallow bushes and can confirm the resemblance.

The recipe as written by Cato is rather confused at times and the ratios of ingredients somewhat inaccurate. The Roman pound is roughly equivalent to twelve of our imperial ounces. The ratio of ingredients in a recipe converted and reduced by a quarter to a manageable amount, works pretty well with some adaptation. (See appendix.)

But first we must turn to the cheese, for this is what we are here for. The fresh sheep's cheese is soaked, squeezed, beaten, sieved and beaten again. I found it quite strange that the cook had to go through all that rigmarole in order to get soft cheese to work with. There were two possibilities: soft cheeses like ricotta, mascarpone and the like were simply not made and the process of soaking, sieving and beating was the only way to make the cheese soft enough to take the honey; or they did make very soft cheese and it simply did not stand up – literally – to the recipe. I have used ricotta and mascarpone to make *placentas* and the result has invariably been flatter and less successful. The cheese mixture is very runny and invariably flows out as it bakes. It has not been easy to find fresh cheese in this country and I have had to resort to making it myself. I used the recipe in Varro, which states that two *congi* (twelve pints) of milk should take rennet the size of an olive – about a teaspoon.[6] I am grateful to Nepicar farms in Kent for the unpasteurized sheep's milk, rennet and the cheese moulds, which they were kind enough to lend me when I explained my purpose. The recipe, understandably, has no starter culture but I am reliably informed that it is not always necessary with fresh milk. It has also been suggested that a ready-made culture would have been present on the sides of the milk pail and on the container regularly used to make the cheese. I took advice on temperature, etc. from a very useful cheese book written for the farmer's wife,[7] but followed Columella's advice to expel the whey as quickly as possible and to soak the cheeses briefly in brine. I was then ready to proceed with the recipes for *placenta* and *libum* to the letter.

The cake *libum* is as difficult to pin down as *placenta* and for a similar reason: *libum* was often used as a generic term for cake. We would know very little about the specific cake were it not for Cato's recipe. However his recipe is at odds with the other evidence which always talks in terms of a dessert with honey.[8] With *libum* the cheese is simply broken down into a mash and formed into a dough with flour and one egg. There is no soaking to remove the salt and no sieving is required. Cato's cake seems to be a savoury item. It works very well with fresh salty cheeses such as feta but is totally inedible when cold and in that state may well have been used simply as a sacrificial offering. However if one uses fresh cheese soaked to remove the salt and pounded down to a paste then the result is quite delicious with honey poured over the finished cake. The recipe is a little corrupt but it is possible to rediscover the ratio of ingredients. It works out at 1 ½ pounds of cheese, and twelve or six ounces of flour to one egg. In this ratio, one egg is of little value but three gives just the right amount of aeration and moisture. It is difficult to conjecture a scribal error in this case as the 'one' is spelled out rather than as a Roman numeral, but I am assuming that the dinner sweet is going to be more delicate then the sacrificial cake. (I work to the proportions of eight cheese, two flour and one egg; cooking times are roughly 30 minutes.)

The finished cake is similar to a baked cheesecake with an open texture which sucks up the honey and renders the cake outstandingly good. The addition of sweet bay underneath leaves me speechless with praise of it. When cold this sweet version is palatable given that the honey has been fully absorbed while warm. However, it will not reheat. It is also of note that when this cake is cooked in a large, dry oven the resulting outer crust is drier and thicker than if baked *sub testu*. Baking *sub testu* creates a moist environment that prevents the outer crust from drying out before the cake is set. I will talk more on this phenomenon in relation to *placenta*.

Cato's recipe for *placenta* is confusing at times but the other cake recipes in the book are virtually unintelligible. I am inclined to blame Cato for this confusion. It is extremely unlikely that he had any conception of what he was writing about and certainly no experience of cooking or handling dough. I have plans to attempt to decipher the other recipes in the future.

It is necessary to define *alica* and *tracta* before *placenta* is attempted. The debate seems to be whether *tracta* is a precursor of filo or pasta. I am in neither camp, as *tracta* has characteristics of both, but it is insufficiently close to either of them for a firm definition.[9] It is unique. Pliny is quite clear as to the quality of *alica*.[10] It is naked wheat ground with a metal pestle to form a fine grain or groat. It is then sieved through various different sizes of mesh and the best *alica* is that which comes from the finest sieve. I cannot think of a commodity more like semolina than this.

This *alica* is softened with water and made into a dough with flour. However the numbers have become corrupt at this point.[11] The quantity of semolina mentioned will not absorb double its amount of flour. If the ratio is reversed, however, a very satisfactory dough is produced. Semolina has a textured appearance which remains even when it is made into a dough. It is this phenomenon which determines the appearance of the finished *tracta*. It has been suggested by Grandjouan that *tracta* were very fine, transparent sheets resembling filo.[12] My experiments have found that there is a limit to the thinness that can be achieved using a dough made with semolina. The grains of semolina soften and become pliable when made into dough but still retain their shape. Thus when the dough is pulled and stretched out into a disc it eventually pulls apart between the softened pieces. At this stage the sheets by no means resemble modern filo but are sufficiently thin to make the cake delicate and flaky. The sheets remind me of the home-made strudel dough I used to make as a chef and also of the rather spongy spring-roll pastry that can be bought at Chinese supermarkets.

The recipe distinctly says the *tracta* are to be dried and stacked prior to making the cake. We know that *tracta* were stored dry and were used to thicken sauces in the Apician recipes, but were they fully dried before being used in placenta? This is unlikely, partly because it is clear that one is expected to make fresh *tracta* each time and the sheets need only dry enough to be stacked.

Recently, I prepared a banquet for a society of hellenophiles in New York and individual *placentas* were on the menu. This meant making roughly 1,500 mini-*tractas* in a kitchen with a average temperature of 105°F. The drying process was very quick as you can imagine. It is not difficult to imagine similar temperatures in an ancient kitchen. It might be expected that, once fully dry, *tracta* would not cook inside the cake but, thankfully, this was not the case. It was not possible to detect which cakes had been made from dried *tracta* and which had been made from *tracta* left for a few hours only. The hard sheets are effectively steamed inside the cake.

It would be of interest to look at the other cakes that use *tracta* in Cato to see how they are used.[13] The other recipes are even more confused and little can be gleaned of any value. However the recipe for *spira* appears to say that the *tracta* are drawn out like a rope, implying that the dough is sufficiently soft for this to happen as the cake is being made.[14] A closer look suggests that the term *restis* – rope – could also mean the leaf of garlic or onion, i.e. that it is hollow and the sheets of *tracta* have not been pulled like a rope but rolled up and that these *tracta* are formed into what can only be described as cannelloni. From this I conclude that the *tracta* are rolled into tubes when fresh and allowed to dry slightly before being filled with the cheese mixture. They are then placed on the base in a spiral formation. This would not be possible if the tubes were fully dried as they would not bend into shape. It is my belief that the *tracta* are simply allowed to dry until they are firm enough for them to be handled. Any unused pieces of *tracta* would undoubtedly be fully dried and stored for thickening purposes.

If *tracta* do resemble filo and they are allowed to dry or rest for even a short time then they would be impossible to pick up or stack or spread with filling without crumbling into a thousand pieces. I have hideous memories, when I was a pastry chef, of inadequately covered filo that turned into powder in my hands. However *tracta* rolled or stretched to their limit and fully dried do not crumble without a certain force being exerted. They are about 1mm thick when fresh and as they dry they shrink, curl and thicken slightly. If we look at *tracta* use in Apicius, this dry pastry sheet is remarkably good at thickening sauces. The specific recipes where *tracta* are used to thicken are all stews of some kind with many ingredients.[15] The *tracta* are crumbled, probably in a *mortarium*, to the consistency of breadcrumbs and are lost amongst the other ingredients.

When sauces are made separately and poured over meat or fish they are invariably thickened with starch derived from wheat or rice which produces a smooth finish. I have conducted experiments with *tracta* sheets to thicken a number of different sauces and found them very efficient. A good deal more starch is expelled from the sheet then even I expected. The grains of pastry do not dissolve (that was too much to expect) but they do become transparent and are hardly noticeable.

Jon Solomon's paper on *tracta* questioned the purpose of spreading oil on them. This is a bit of a red herring as it is very common to do this to a ball of dough prior to working with it, especially in a pre-clingfilm society! The slightest amount of crust or dry patch on a ball of dough will interfere with the process of stretching and cause holes to appear at the edge of the dry patch. The recipe is confusing here as the instruction comes apparently after the *tracta* are formed. It is quite common, even today, to find that a cook, narrating a recipe, has forgotten a crucial aspect and adds it at the end or when it is remembered. It is a phenomenon found frequently in the Apician text too.

I turn now to the outer sheet, the cake's construction and the finished size. The outer sheet is a very simple flour and water dough pulled or stretched to double the size of the *tracta*. It should be very thin in order for it to be palatable and it resembles strudel dough. The cheese sieved and beaten loses little of its weight and absorbs the honey without becoming runny. I was able to make 12 *tracta* out of the reduced recipe and they are interlayered with the cheese mixture. The outer sheet is then simply pulled up and gathered in the centre and formed into a peak very like the central knob on the mallow seed. I try to take as much excess pastry away at this stage as it tends to sink down and form an unappetizing layer above the cheese filling. Here it is interesting to note that if the *tracta* were fully dried before assembly then earlier cooks may have, as have I, cut through the outer covering and caused a leak. Plenty of fresh oiled bay leaves are essential, as they prevent the cake from sticking to the board while it is carried to the hearth. A sharp shake of the board releases the cake and it slides off with ease. After baking the leaves also prevent sticking. It goes without saying that the presence of so many bay leaves gives the finished cake a wonderful flavour.

My quarter recipe made a nine-inch *placenta*. A full-size one is therefore roughly eighteen inches across. Cato talks of a twelve-inch board to make the cake. I would like to amend the figure to at least two to three feet. This would only mean that a Roman *i* or *ii* had been left off the text by a scribe. The most important issue in relation to the size of cake produced is of course the size of *testum* available. Cubberly gives a range of 30–50 cm (12–20 inches). It is apparent that the cake should not touch the edges of the *testum* but have at least four to five centimetres of free space. There should be a tight seal between the hearth and edge of the *testum* but there is still the possibility of ash seeping through and spoiling the finished cake.

It remains to describe briefly the baking method for these two cakes. The principle is that the *testum* is heated above a fire. The iron trivets that survive from Roman sites in Britain and Europe are ideal for this. If the cover is placed directly on the fire, it may maintain the embers over night, as the medieval covers were intended to do, but it also reduces the fire to a minimum and fails to heat the *testum* adequately. The small hole in the top of the dome (or holes in the sides in earlier examples), acts as a flue and helps to draw the fire. The *testum* takes about an hour to heat through on a charcoal fire, which has been used for other cooking tasks and has been alight for at least one hour before that. The cover, trivet and fire are removed. The surface is cleaned and the bread or cake is placed directly on the hearth with bay leaves beneath. The cover is placed over the cake and a makeshift lid or stopper is used to cover the hole in the centre and I suspect pieces of dough were used to close the side holes. The remains of the fire, which have only been pushed aside, are piled on the flange and around the sides. I have achieved a regular baking temp of 375–425°F while a cake is inside and the baking times are pretty much what they would be in any modern situation. These covers were ovens in the true sense, in that, it is the hot air inside that cooks the cake and not the hot sides touching the cake itself. I have been baking *sub testu* on a raised masonry platform, similar to the Pompeii hob, for over two years now and the technique has become second nature. I have become quite adept at judging the temperature inside from the look and feel of the fire outside. Occasionally more or less coals are added and the lid can be removed to adjust the temperature. I have had little trouble with ash on the finished cakes and if it does intrude it will simply blow off.

The *testum* used most frequently has now finally cracked. It was used at least 90 times to bake bread and cakes before the heat and stress caused the flange to fall off first.[16]

As an oven for bread, these covers are excellent as they create a steamy, moist atmosphere which ensures the dough is allowed to rise to its full extent before the crust is formed. I was unsure of the effect such an atmosphere would have on these cheesecakes. My conclusion is that it is just as useful. I have already stated that the *libum*, cooked *sub testu*, stays soft on the outside while when it is cooked in a larger oven it can dry out. The *placenta* has different requirements, yet the method is equally successful. A *placenta* cooked in a large oven can have a tough rubbery outer shell which has proved difficult to cut. A *placenta* straight out of the *testum* is wonderfully crisp and dry and when smeared in honey and taken to the dining room for immediate consumption one can only sympathize with the Roman and Greek gourmets who trained themselves to eat hot food so they could get at the cakes before their fellow guests.

REFERENCES

[1] Cubberly et al., 'Testa and Clibani: The Baking Covers of Classical Italy'. *Papers of the British School at Rome.* Vol. 61 1988.

[2] Horace Epi 1, 10, 10–11: temple slave running away from precinct for want of plain bread.

[3] Juv xi.59: dreaming of *placenta*. Horace Sat 2.8.24: guest swallows whole *placenta*. Martial Ep. ix. 90.18: quarter of a *placenta* as an offering to the gods. *Libum* appears frequently as an offering and is more often used generically for 'cake'. Martial Ep 3. 77, for *liba* and *placenta* as potential dessert cake.

[4] Athenaeus XIV.644b.

[5] above II.58d.

[6] Varro II.xi.4, Columella VII ff: 'A pail when it has been filled with milk, should always be kept at some degree of heat (not on fire), when the liquid has thickened it should immediately be transferred to wicker baskets or moulds, for it is of the utmost importance that the whey should percolate as quickly as possible and become separated from the solid matter....as soon as the cheese has become solid they place weights upon it; then when it is taken out of the moulds it should be placed in a shady cool place and it is sprinkled with pounded salt; when it is hardened it is still more violently compressed; again it is treated with salt and again compressed with weights.....Cheese which is to be eaten within a few days while still fresh is prepared with less trouble; for it is taken out of the wicker baskets and dipped in salt and brine and then dried a little in the sun'.

[7] Cheke, V. and A. Sheapard, *Butter and Cheese Making*, Granada 1985.

[8] Cato 75. Ovid Fasti 3. 725, 3 761, in which the cakes are infused in honey while hot. Here it is women who make and sell these *liba* as offerings. Athenaeus iii.126a, for honey in *libum* but this description is very odd as milk and *itria* are mentioned.

[9] See 'What was Tracta?' in *PPC* 12, 1982 for a different interpretation of *tracta*. Charles Perry has suggested that they are biscuit like in shape, basing this in part on a reference in Pliny (xviii.27.106) to a local bread made from soaked *alica*, which is kneaded '*ad speciem tractae*'. This would normally mean 'with the appearance of' but in this case the appearance is determined by the pots the bread is baked in and not the shape it is kneaded into. If the term is read as 'in the manner of' then the wine soaked *alica* is kneaded into a dough with flour just as happens with *tracta* dough.

[10] Pliny XVIII, 29, 112.

[11] Cato 76 line 2–3.

[12] Claireve Grandjouan, 'Hellenistic Relief Moulds of the Athenian Agora', *Hesperia*: Supplement 23 1989.

[13] Cato 76,77,78,79,81,82.

[14] Cato 77. Cato 82 has the *tracta* formed into balls with a cheese and honey mixture inside.

[15] Apicius IV,iii, 1–8: all dishes called *minutal*, basically stews.

[16] It would be of some interest to me to compare these pottery shards with archaeological finds of original *testa*.

APPENDIX

Placenta recipe, from Cato 'On Agriculture' (76). New translation by author's husband Dr Chris Grocock.

Placentam sic facito.
Farinae siligineae L. ii (unde solum facias) in tracta farina L. iiii et alicae primae L. ii. Alicam in aquam infundito. Vbi bene mollis erit in mortarium purum indito siccatoque bene. Deinde manibus depsito. Vbi bene subactum erit, farinae L. iiii paulatim addito. Id utrumque tracta facito. In qualo, ubi arescant, componito. Vbi arebunt, componito puriter. Cum facies singula tracta ubi depsueries panno oleo uncto tangito et circumtergeto unguitoque. Vbi tracta erunt focum ubi cocas calfacito bene et testum. Postea farinae L. ii conspargito condepsitoque. Inde facito solum tenue. Casei ouili P. xiiii ne acidum et bene recens in aquam indito. Ibi macerato, aquam ter mutato. Inde eximito siccatoque bene paulatim manibus siccum bene in mortarium inponito. Vbi omne caseum bene siccaueris, in mortarium purum manibus condepsito comminuitoque quam maxime. Deinde cribrum farinarum purum sumito caseumque per cribrum facito transeat in mortarium. Postea indito mellis boni P. iiii S. Id una bene commisceto cum caseo.

Postea in tabula pura, quae pateat P. I ibi balteum ponito, folia laurea uncta supponito, placentam fingito. Tracta singula in totum solum primum ponito, deinde de mortario tracta linito, tracta addito singulatim, item linito usque adeo donec omne caseum cum melle abusus eris. In summum tracta singula indito postea solum contrahito ornatoque focum bene primo temperatoque tune placentam imponito; testo caldo operito pruna insuper et circum operito.

Videto ut bene et otiose percoquas. Aperito dum inspicias bis aut ter. Vbi cocta erit eximito et melle unguito. Haec erit placenta semodialis.

Make placenta as follows:
2 lb wheat flour (this is for the base) and for the tracta 4 lb flour and 2 lb prime groats. Soak the groats in water. When they are soft put them in a clean mixing bowl and dry them thoroughly. Knead with the hands. When they are well mixed, gradually add 4 lb flour. Make tracta from the mixture. Put them in a basket to dry. When they are dry, stack carefully. When you make each tracta after kneading it wipe it with a cloth smeared in olive oil cover it up and oil it. When the tracta are finished thoroughly heat the hearth and testum where you are to cook. Next sift 2 lb flour and make a dough. Make a thin base from this. Put 14 lb of sweet and fresh sheep's cheese in water to soak. Macerate it and change the water three times. Take it out and dry it well gently in your hands and when dry place in a mixing bowl. When all the cheese is thoroughly dried, knead in a clean mixing bowl and break it down as much as you can. Then take a clean flour sieve and force the cheese through the sieve into the mixing bowl. Next add 4 ½ lb good honey and mix well with the cheese.

Next put the crust on a clean board measuring one foot across, placing it on top of oiled laurel leaves and make the placenta. First place an individual tracta over the whole base, then coat the tracta with the mixture from the mixing bowl, add the tracta one at a time and again coat them, continuing until you have used up all the cheese and honey. Place an individual tracta on top then pull the base together and finish it neatly. Make the hearth ready beforehand and place the placenta on it, cover with a heated testum and place hot coals on top and around it.

See that you cook it slowly and well. Uncover and check two or three times. When it is cooked remove it and coat it with honey. This is a half-peck placenta.

Dairy Food in the UAE

Philip Iddison

There is an abundant selection of dairy products both in the traditional *suq* and the modern supermarket in the United Arab Emirates. The products are both locally produced and imported. The substantial immigrant population, chiefly from other Arab countries and the Indian subcontinent, has an important role in ensuring this diversity but it is notable that the Emirati nationals have a long standing tradition of dairy product consumption. They have absorbed the new choices made available by the recent wealth in what is now a cosmopolitan and modern country.

Milk and milk products have been a major component of diet in the UAE since the domestication of stock in prehistoric times. Goat and sheep bones are recorded from archaeological sites dating before 4,000 BC. Whilst these may have been the remains of wild animals, contemporary accounts[1] of bedouin tribesmen taking wild ibex into captivity indicate the ease with which wild stock could be domesticated and the probable antiquity of dairy practices in the region. Dairy products were the second most important source of protein in the diet after fish.

Past and present dairy resources and products in the neighbouring countries are also pertinent as the concept of sovereign states is recent in Arabia and historically populations were relatively mobile. Sea trade was well established providing contacts with all the areas around the Gulf, the subcontinent to the east and Africa to the west. It has been suggested that a milk product, possibly a cheese, was being imported in pottery storage vessels from the Harrapan civilization in the Indus valley at the end of the third millennium BC.[2] Other evidence of milk processing is unlikely to be positively identified due to the simplicity of equipment and its ephemeral nature.[3]

Milk from goats and camels, and to a lesser extent sheep and cattle, was available to the population before oil revenues spurred recent development. The recent proliferation of dairy farms, dairies and the ready availability of milk products is therefore a natural development. The traditional and modern dairy practices of the country are reviewed in this paper and a glossary of dairy products from the past and present is included.

Prehistory of dairy in the UAE

In the pre-development era the population derived its food from a number of sources[4] of which herding and associated dairy farming was one of the principal activities. The country has diverse terrain ranging from flat coastal plains through sand desert and gravel plains to mountains. The population's economic activities reflected this diversity. As a simple model the population can be divided into three main groups with associated dairy food resources. Each dairy resource had particular attributes and restrictions on the production of dairy products.

Haleeb al jamal – *camel's milk – desert bedouin*

The domestication of the camel is thought to have been achieved in the second millennium BC. Camel's milk was a staple food for the bedouin and enabled small populations to occupy and make economic use of the extensive desert regions. The close association between camel and man gave the bedouin the freedom to travel extensively in the arid conditions of the Arabian interior. Their lifestyle, traditions and survival skills are of great antiquity but survive in very limited forms today.

The camel's physical adaptation and ability to adjust its metabolism to the sparse vegetation and limited water supplies in the desert is quite extraordinary. One of these adaptations is the ability to dilute its milk when in a dehydrated state.[5] This is thought to ensure an adequate water supply to the suckling calf when there are no alternate supplies.

Practically all human consumption of camel's milk is in the raw state as a fresh drink, *haleeb*. Commercial dairies in Saudi Arabia, Libya and Mauritania have started to pasteurize and package camel's milk but no initiative on these lines has yet been started in the UAE. Historically there was little need to process the milk as most consumption was by owners and herdsmen. With a lactation lasting up to eighteen months in a two-year reproductive cycle, permanent milk availability from a modestly-sized herd was ensured.[6] Camels were milked into a bowl supported on the milker's knee with the foot resting on the other knee, a finely balanced position. Traditional milking bowls from Oman were made of basket work with an external covering of goat hide. The baskets were made from the fibre of a palm tree, *Nannorrhops ritchieana*, which are rot-proof. Daily milk yields vary from as little as two kilos to nearly fourteen kilos for animals under station management in Saudi Arabia with supplemented feed. Average yields are about five litres per day.

There are technical difficulties in processing camel's milk into yoghurt, butter and cheese which have only recently been resolved.[7] This inability to process camel's milk was not a serious problem for desert bedouin who had to move regularly with their herd to find pasture. Some societies in other camel-rearing regions prepare a soured milk product but this does not seem to have been the case in the UAE.

Camel's milk varies in taste according to the pasture or feed and is generally more salty and acidic to the palate compared to the milk of other ruminants. It can also be the sweetest milk.

Travellers were entitled to satisfy their thirst with camel milk if they came across a milch camel, but presumably only if you were travelling in friendly tribe's territory. Milch camels generally have the udder covered with a cloth bag to prevent the calf suckling at will if the camel is to be milked for human consumption.

Interest in camel rearing is now chiefly concentrated on breeding racing camels as their traditional use as food supply and transport animals is now redundant in the UAE. However this does ensure that camel milk continues to be available for human consumption. In the 1990s it is still quite common for a UAE family to keep a camel for milk in the family compound.[8]

Haleeb ghanam, haleeb kharouf – *goat's and sheep's milk – plain, mountain and oasis*

In the north and east of the country the gravel plains with sparse vegetation and adjacent mountain ranges with a perennial water supply ensured a more settled existence and the potential for greater agricultural diversity.

Goats and sheep were ideally suited to the oases and the mountain terrain in the east of the country where daily access to water supplies was possible. Sheep and goats are recorded at fifth-millennium sites in Oman. The sheep were probably introduced from the Mesopotamian cultures to the north through well established trade links. Wild relatives of the goat are still present in the area.[9] As noted above the mechanism for their domestication could have been quite simple. Archaeological sites produce many arrow heads implying a hunting subsistence but this does not preclude the use of domestic animals. The wild prey provided meat whilst the domestic animals provided the reliable source of dairy products and other useful secondary products such as hair and dung.

In recent times there appears to have been little differentiation between the milk of sheep and goats.[10] Some milk was consumed raw but this was not the preferred fresh milk; camel's or cow's milk was the first choice. Most of this milk was processed into *laban, raab, labneh, chami, dihn/zibda, samn* and *kamil bathith*[11] for immediate consumption or storage and commercial purposes. With the mechanization of water supplies and ready availability of locally grown fodder, sheep and goat herds have increased and spread across the country in recent years. This has put additional pressure on the remaining natural vegetation which is seen as a free resource by the herdsmen, many of whom are now immigrant labour. Production now seems to be concentrated on the meat market which places a premium on Emirati reared meat and in particular sheep production seems to be exclusively for meat. A visit to a local market will however turn up some traditional milk products.

Haleeb baqar – *cow's milk – coastal entrepôt and oasis*

The coastal trading towns of the UAE have a long history which is recorded by archaeological research and sparse texts. The Shimal settlements of the third millennium BC were trading with Mesopotamia and the Indus valley cultures to the north and east respectively. Julfar was an important medieval city which traded with China, Thailand and East Africa.

Members of *Bos species* are recorded from the fifth millennium, probably introduced from Mesopotamia. However until the recent development era, cows seem to have been a relative rarity. Oxen were used principally as beast of burden; for instance to draw well water and would have had a secondary role as milk providers. An account of a merchant family in Dubai in the 1920s and 1930s records[12] that cows were kept to provide fresh milk for drinking and also for the preparation of *laban, dihn/zibda* and *samn* within the household. The cows were kept in the home compound, foraging on local desert plants supplemented by overripe dates and dried sardines. The quality of the milk must have been variable but Easa Al-Gurg attributes his survival to this key resource at a period when the Gulf economies had been devastated by the collapse of the natural pearl market.

The establishment of processing plants and dairy farms which started on a small scale in the 1970s has developed to satisfy practically every demand even though the population has increased substantially.[13] Initial investment was in reverse processing plants which converted imported dried milk and butter into a range of products, milk, cream, and yoghurt. This is still an important sector of the local dairy industry. The second stage of development brought the introduction of new dairy breeds which could cope with the climatic extremes and produce economic and consistent milk supplies to develop dairy product ranges. This second stage is now reaching maturity with market expansion being largely concentrated on widening product ranges to challenge a diverse import market.[14] Both these aspects are illustrated in the glossary of products which follows.

Traditional milk processing

Milk was processed in a number of ways to create products with desirable characteristics such as texture, taste and storage properties. The processing methods used in the UAE were ideally suited to the resources of small tribal units using basic equipment.

Fresh camel's milk was considered to be already 'cooked' if it was still at the temperature of the camel's udder.

The simplest processing was to heat the milk, a desirable objective in the mountains where temperatures can approach freezing point in the winter. An ingenious method was used in Oman. Selected round stones were heated in the fire and dropped directly into the milking bowl. This method has been in recorded use since Roman times[15] and is economical and requires no specialized implements. Hot milk was spiced with ginger, cardamom, fenugreek seed or saffron.

Milk was kept in animal skin bags, usually goatskins, for short term storage. In the typical summer temperatures of 30–48°C, initial fermentation or conversion to *laban*, a yoghurt-type product, was ensured by the build up of residues from previous batches with attendant bacteria. Storage bags were carefully selected and maintained to guarantee this process. There was no way to sterilize these storage bags.

The *laban* was consumed fresh and also made into three further products; *labneh*, a strained and thickened product; *chami*, a cooked and reduced soft cheese; and *dihn* or *zibda*, fresh butter. All these products had a limited storage life. *Chami* was a favourite breakfast dish and *zibda* was eaten with flat bread and the excellent local honey. *Zibda* was made by churning the *laban* in a goatskin, *sigga*, which was hung by a wooden tripod and rocked back and forth. Air was blown into the bag at regular intervals to ensure that it stayed inflated. The butter was collected and the liquid residue was either consumed as a drink, *sharab*, or may have been processed to a form of low-fat *kami*. Lizard skin bags were used to store *zibda*.

Again these products could be processed further to avoid waste as they had limited storage capabilities. There are a few references to cheese, *jibna*. This was probably *labneh* or *chami* which had been strained to reduce the moisture content to promote slightly longer storage. *Zibda* was converted to *samn*, clarified butter. This was a very important product as it could be stored for long periods and thus was an article of barter or commerce. It is still available in the traditional markets, usually sold in recycled Vimto bottles.[16] Goatskins were used in the past to store and transport *samn*.

Kami was prepared from *chami* or *labneh* flattened and pressed into small thin cakes and dried in the sun. The end product varies from broken granules to solid cakes, all rock hard and with a long storage life. This was a useful store food which was eaten directly, crumbled into dishes or pounded with water to make a form of fat-free milk. Like *samn* this artisanal product is still available in the traditional markets.

Imported dairy products

A large range of imported dairy products is available to the very cosmopolitan population of the UAE. Judging by the typical supermarket shelf, particularly popular products are butter and ghee, dried milk powder and a range of cheeses. Dried milk powder was an early introduction and was ideally suited to the very basic storage and transport conditions which prevailed until development. Ghee has effectively replaced *samn*.

The selection of Middle Eastern cheeses is particularly good. Local factories are now producing cheeses matching those of foreign origin. Most Middle Eastern cheeses are soft and relatively fresh. This reflects the climatic conditions which are generally unsuitable for maturing cheese. They are often kept in brine, oil or whey as a means of preservation. They are eaten directly as breakfast and mezze food as well as being used in cooked dishes. Unsalted cheeses are used in sweets such as *ataif,* a Ramadan favourite made by deep-frying pancakes stuffed with cheese or nuts.

There are few matured cheeses, *shankleesh* and *mish* being the main examples and *rumi* and *kashkawan* are semi-matured.

The *pasta filata* technique is used in the preparation of a number of these cheeses from northern Arabia such as *majouleh, halloumi,* and *mushalal.* The curd is kneaded in hot water, usually by hand producing a stringy or resilient texture and often a very characteristic shape.

Many of the cheeses have to be rinsed or even soaked before consumption to reduce the salt and enable the generally mild flavour to be appreciated.

Use of dairy products

The overwhelming use of dairy products is as primary food for direct consumption and this has always been the case. Studies in Gulf countries amongst national families show that there is a strong belief in the nutritional value of dairy food and also a strong sense of these being traditional foods. There has been an acceptance of imported products and modern packaging and some long established imports such as Nido brand milk powder have achieved icon status.

The traditional culinary repertoire consisted of robust dishes making good use of a limited scope of resources. Dairy products had a role in these dishes; nearly half the recipes in a book of substantially traditional UAE recipes contain at least one dairy product.[17] Additionally until the import of cooking oils became established in the 1970s, the main cooking fat medium was *samn* which would have been used for the substantial part of the cuisine which involves frying. A frequent use for butter or *samn* is as a liquid garnish on savoury and sweet dishes.

Colostrum was a valued product, a dessert, *alelbah*, was made from cow or goat colostrum in Bahrain. It was sweetened and spiced with cardamom and nigella seed.

Substitutes

Coconut milk does not have a significant role in Emirati cuisine although the coconut palm is regularly planted in coastal areas. This is not the case in neighbouring Oman where the influence of a once extensive empire including the East African littoral is noticeable. Coconut milk is a favoured cooking medium for savoury dishes which are often well spiced and also appears in many sweet dishes.

As noted above the use of *samn* as a cooking medium is declining due to the availability of cooking oils and also a perception of the health benefits associated with low cholesterol diet.

Conclusion

The pattern of dairy consumption in the UAE has probably changed little from the prehistoric era of the Umm al Nar and Shimal settlements of the third millennium BC until the discovery of oil as a major natural resource in the 1960s. People kept their own stock and had daily access to milk and straightforward techniques to prepare dairy produce. As a result dairy products formed a prime component of diet and an article of local commerce.

Over the last three decades the availability of cold storage for importing, merchandizing and home storage of products, and the consolidation of dairy farming into large commercial units is having a long term effect on the consumption patterns of dairy products. The quantity and range of chilled dairy products in the local supermarkets and neighbourhood stores is broadening choice and making some traditional products obsolete. *Samn* has been replaced by ghee a similar product originating from the subcontinent. Imported dried milk powder has replaced *kami* as a storage product that can be reconstituted to a form of milk.

At present the traditional products are still available in modest quantities in the local markets, but the young generation who has had no exposure to these old products is unlikely to want, to know how to use or be able to produce them in the future.

GLOSSARY OF DAIRY PRODUCTS IN THE UAE

This glossary details the dairy products currently available in the UAE, locally produced and imported, together with some products that are in the process of being consigned to history. Items which originate outside the Middle East region such as ice-cream and dried milk powder have not been included. A table giving some details of nutritional composition is attached.

AKAWI

Fresh white salted/unsalted cheese shaped by cloth wrapping into blocks of about 500 g with rounded corners, made from cow's milk and originating in the Lebanon. Production has spread to other countries and it is produced in several forms, salted hard versions for eating and minimum salt versions which melt readily for cooking. The low salt type is used for sweets, the salt being soaked out first. The cheese is firm with a slightly crumbly texture and a mild flavour.

ANARI

Cypriot cheese which is an import to the UAE markets. It is a medium fat soft cheese similar to *halloumi*.

ARRISH, ARISHI

A soft white cheese shaped into short cylinders with striated sides originating from Egypt. Low fat versions are made from skimmed milk. Versions are also prepared from the whey and precipitated proteins derived from making other cheeses such as *shankleesh* or *halloumi*. The flavour is mild, strengthening with time, and it has a quite dense crumbly texture. Also seen as a prepacked block cheese. Eaten fresh and also mixed with chopped vegetables or pickles to make a dip.

AYRAN

Another name for *sharab laban*, probably from north Arabia or Turkey.

BALADI

This cheese is shaped in soft round cakes 10 cm in diameter, off-white in colour and with a creamy paste. It is unsalted and has a very mild taste. For eating fresh as it does not melt on cooking. *Baladi* means local. (Dh. 13/kilo)

BATHITH

A Northern Emirates word for the sun-dried product of either *laban* or *chami*.

BULGARIA

White cheese with a dense smooth paste similar to feta, medium strength acidic flavour, now locally manufactured.

CHAMI

Laban cooked to a thickened consistency like curds, used as a breakfast food with bread and also to accompany dried dates.

DOMYATI, DOMBIATI, DAMMIETA

A white cheese from Egypt packaged in cuboidal half-kilo blocks, made from cow's or buffalo's milk. Salty, mild flavoured soft cheese with crevices in the dense paste. Unusually the salt is added before coagulation to control adverse bacteriological activity. Eaten fresh drizzled with olive oil and sliced cucumbers, tomatoes and flat bread. It can be matured under brine, darkening in colour and improving in flavour.

FETA

This well-known cheese is popular and widely available in the Middle East, loose as well as pre-packed.

HALEEB

Milk.

HALLOUMI

Another popular import from Cyprus, 1,000 tons are exported annually to Arab countries. A similar cheese, *hellim*, is made in the Lebanon and Syria. In Cyprus the cheeses were traditionally prepared from sheep's and goat's milk and after a minimum of 40 days in brine were considered to be ripe. Modern industrial production uses sheep's, goat's and cow's milk and the cheeses are not ripened. The cheeses are produced by the *pasta filata* technique and are moulded into a flattish block with a pronounced central fold. Dried mint is sometimes included in this fold and the individual cheeses are vacuum packed with a little whey. The high protein content ensures that this cheese does not melt when cooked making it ideal for grilling or frying as a mezze dish.

HALLOUMI LITE

This cheese is a recent arrival in the supermarket cold cabinet. Fat content is stated to be eight per cent, presumably quoted as a solids ratio. The reduction in fat results in a chewier cheese but there is little effect on the taste compared to *halloumi*.

ISTANBULI (JIBNA ISTANBULI)

A vacuum-packed half-kilo rectangular block of this traditional cheese from Syria. It proved to be a fresh white cheese packed in whey, quite dense in texture, with a slightly acid taste and it was lightly seasoned with nigella seeds, *habbat suda*. It was an import from Izmir in Turkey. I have not seen this name used in Turkey and assume it is an export name.

JADALA SURIA

Another name for *majouleh* indicating an origin in Syria.

JIBNA

The general word for cheese in Arabic often applied to local cheeses.

JIBNA BEYDAH

Generic term for the soft white cheeses produced throughout the Arab world from a variety of milk for immediate consumption. In the Gulf countries sheep and goat's milk were traditionally used and a crude form of rennet was prepared from the stomach of a suckling lamb.

KAMI

The UAE name for sun-dried yoghourt. It costs 10 to 20 Dirhams a kilo depending on quality, some is in small flattened cakes 1–2 inches in diameter, some is in granular pieces of varying size. It is hard and has a slightly musty aroma vaguely reminiscent of cheese.

KAREISH

An acid-curd cheese traditionally made from skimmed cow's or buffalo's milk in small-scale local production. It is the most common cheese made in Egypt and Sudan. The curd was allowed to coagulate naturally as the cream separated before being drained and pressed on reed mats and salted. Eaten fresh at breakfast or ripened in brine

KASHTA

Milk cooked with flour to form a thick cream for use in desserts. It is a north Arabian ingredient, used in restaurant desserts and not available commercially.

KASHKAWAN, KASKRAWALI

A semi-hard cheese in a flat disc shape about 20 cm diameter and 500 g in weight, originating from Syria or Lebanon, pale straw colour with no rind, semi-mature flavour improving with time in the fridge if not in a sealed container, rather stringy when cooked so best grated.

KUBROSI, KUBROST

A white cheese in the shape of a truncated cone with striations on the curved surface. Medium firm paste, a touch of salt and a mild flavour. Grills quite well and does not melt.

LABAN

Soured milk or yoghourt, an important element of the Arab diet. It used to be produced by storing milk overnight in a goatskin bag used regularly for the purpose by the Bedouin. It has a short storage life becoming more sharply flavoured with age. The word is used fairly loosely in the UAE to mean products with different consistencies. It can be like yoghourt as we are used to in the West with solid texture and it can also refer to a product of drinking consistency.

LABAN DAHAREEJ

Strained yoghurt formed into balls and dried enough to maintain their shape when stored in oil. They are served as a mezze dish with some of the storage oil drizzled over the top. Sold in jars in a number of variations made by rolling the balls in further flavourings, plain, with chilli pepper, mint, sumac etc.

LABAN KHAD

Buttermilk, correctly the liquid left after yoghurt has been churned to make *zibdeh/samn*, but also referring to a yoghurt drink.

LABNEH

Strained or thickened yoghurt, similar to a cream cheese. This is best eaten with a drizzle of olive oil and some fresh flat bread, *khubz*. It can be used with limitations as a substitute for cream and sour cream.

MAJOULEH, MAJDOULI

A string cheese from north Arabia where the curd strings are grouped together and formed into a hank about 15 cm long, kept in brine it has a rubbery texture and is salty and mild flavoured

MECHLALEH

The Omani name for *mushalal*, made from sheep's milk, hard and salty.

MISH

An Egyptian cheese, originally a grey salty peasant cheese with a very strong flavour. Fresh cheese curds and prepared cheeses such as *arrish/domyati* were poured into an earthenware crock (*ballas*) with some milk, salt, flavourings and a starter from a previous batch of *mish*. This was left to mature for one year, pasteurized and packed for consumption or distribution. There was a wide range of flavourings, sesame seed cake, milk solids from *samn* preparation, fenugreek, chilli/paprika, anise, cumin, cloves, nutmeg, thyme, nigella, etc. to taste.

It is usually encountered as a crumbled mass of curds in whey, generally pale red brown in colour but with some pure white curds. Red chilli pepper flakes betray the hot flavour which is also very salty. These seem to be prepared by the local dairies, imported *mish misri* from Egypt is also available and has a smoother more uniform consistency. The cheese has a strong spicy fragrance and is usually available in supermarkets in strong and milder forms depending on the chilli content.

MIZ

Small rock-hard marbles of cheese and salt, an Iranian product seen in Dubai spice suq.

MUSHALAL, JIBNA MUSHALAL, SHILAL

Another string cheese originating in Syria, the long curds left unbroken and ingeniously twisted into small bundle, usually with nigella seeds incorporated, salty and best rinsed before eating, a good breakfast cheese.

NABULSI, NABOLGHI

A white brined cheese, originally from Nablus in the West Bank, prepared from sheep's or goat's milk. The curds are pressed in small portions in cheese cloth and may be boiled in brine containing a mix of mastic gum and *mahaleb* to help preserve the cheese and improve its flavour and texture.

Samples purchased in the UAE were of Syrian origin and were white flat small tablets of dried folded roughly rectangular curd with salt crystals on the surface. Eaten thus they were very salty and with a chalky texture and flavour. After soaking in water for 24 hours, the paste had expanded to a creamy texture. Still quite salty and with a chalky flavour. It melts when cooked. This cheese is particularly used for sweet dishes. The heavy salting is a long term preservative which is removed by repeated soakings.

RAAB, ROB

The name in the Northern Emirates for a drink prepared from *laban* and water, similar to *sharab laban*. Elsewhere in the UAE it is referred to as *laban*.

RUMI, ROOMY

An Egyptian Cheddar-style cheese (*rumi* means Roman or Western), often sold in pre-sliced packets. The paste is firm and pale orange in colour, it has a full mature flavour and crumbly texture. Usually contains black peppercorns scattered rather infrequently through the paste.

SAMN

Clarified butter, the fresh butter is heated with flour and herbs or spices to remove the curds and impart a flavour, in this form it will keep without refrigeration for long periods. It was an important article of commerce and is still produced on a small scale locally.

SHANKLEESH, CHANGLISH

Well-strained sheep's yoghurt is de-fatted or fresh cottage cheese is salted and rolled into balls and dried thoroughly. These are stored until the paste darkens and a surface mould has developed. This is scraped off, the ball is coated with melted butter and then rolled in *za'atar,* thyme. It has quite a firm consistency but crumbles under the knife. The paste is pale brick red with spots of small white curds and it has a piquant salty flavour, one almost suspects the inclusion of a little chilli. This cheese is from Syria and Lebanon and is served as a mezze by slicing the cheese and drizzling olive oil over it. Sold in jars under oil and also loose or in vacuum packs. A country version *shankleesh baladi* is also available and is very similar with a softer, more mixed paste with more chilli flakes and inclusions of *za'atar* coating.

SHARAB

The word means drink and is applied to the buttermilk left after the churning to make *zibda*.

SHARAB LABAN

Yoghurt diluted with water and slightly salted to make a drink, ground fenugreek seed is used to flavour a version recently introduced into the local market, see also *raab*.

THALAJA

Fresh white cheese from Egypt with a soft creamy paste, moderately salted and with a well developed slightly acid taste.

ANIMAL, MILK PRODUCT AND HUMAN STATISTICS FOR THE UAE

	Camels	*Cattle*	*Goats*	*Sheep*	*Humans*
1976					
Population (no.)	39,416	15,803	198,142	73,159	250,000
Percentage milking (%)	–	36	46	42	N/A
Milk production (kg)	6,716,353	2,605,495	4,093,025	1,890,419	N/A
1992 (Abu Dhabi Emirate only)					
Population (no.)	155,071	7,686	605,873	2,300,000	
Milk production (kg)	–	18,465,000	–	N/A	

The 1976 statistics represent a pre-oil development picture of dairy use. Total milk production from all sources was 15,300 tonnes of which 3,500 tonnes was recorded as being processed in some way. The balance presumably was either drunk as fresh milk by humans or stock.

Processed dairy products and home consumption were recorded as:

zibda wa samn   butter and ghee   246 tonnes
(78% consumed by owner/household)
laban khadr   fresh curd drink   598 tonnes
(96% consumed by owner/household)
yaqat (jareed)   yoghurt   109 tonnes
(94% consumed by owner/household)
                           other products   7 tonnes
(84% consumed by owner/household)

By 1992, only milk production on commercial dairy farms was being registered with no records available for milk production other than cow's. There had been a fifteenfold increase in productivity per head.

COMPOSITION OF MIDDLE-EASTERN DAIRY PRODUCTS

Product	Fats %	Proteins %	Carbohydrate (chiefly lactose) %	Ash %	Salt %	Energy Kcal/100 g
MILK						
Camel	2.9-5.5	2.0-5.4	3.4-5.5	–	–	101
Goat	4.0-5.0	3.4-4.3	4.0-7.1	0.7	–	–
Sheep	5.0-8.0	4.0-7.0	4.0-5.0	0.9	–	–
Cow	4.0-5.2	2.8-3.6	4.5-4.6	0.7	–	–
MILK PRODUCTS						
Kami	20	50.0	–	–	3.0	–
Laban	3.2	3.6	2.0	0.7	–	50
Labneh	10.0	6.6-13.0	1.0	1.6	–	154
Samn	90.2	0.3	–	0.1	–	813
CHEESE						
Akawi	–	19.1	–	–	–	–
Arrish	–	17.6	–	–	–	–
Baladi	–	12.2	–	–	–	–
Domyati	23.4	21.1	–	2.6	–	511
Halloumi	24.0	19.0	–	5.0	3.0	
Kareish	6.0	17.0	–	6.0	4.5	122
Mish	2.7-11.0	9.7-12.6	3.1	11.9	11.5	150
Nabulsi	24.0	16.0	–	16.0	–	288
Shankleesh	5.6	35.0	3.0	12.2	–	215

BIBLIOGRAPHY

Al Gurg, Easa Saleh, *The Wells of Memory*, John Murray, London, 1998
Al Taie, Lamees Abdullah, *Al Azaf, The Omani Cookbook*, Oman Bookshop, Muscat, 1995
Al Zayani, Afnan Rashid, *A Taste of the Arabian Gulf*, Ministry of Information, Bahrain, 1988
Brock-Al Ansari, Celia, *The Complete United Arab Emirates Cookbook*, Emirates, Dubai, 1994
Dagher, Shawky M., *Traditional Foods in the Near East*, FAO, Rome, 1991
Devendra, C. and G.B. McLeroy, *Goat and Sheep Production in the Tropics*, Longman Scientific & Technical, Harlow, 1982
Dyck, Gertrude, *The Oasis - Al Ain Memoirs of "Doctora Latifa"*, Motivate Publishing, Dubai, 1995
Ghareeb, Edmund and Ibrahim Al Abed (editors), *Perspectives on the United Arab Emirates*, Trident Press, London, 1997
Gouin, P, 'Râppes, jarres et faisselles: la production et l'exportation des produits laitiers dans l'Indus du 3e millénaire', *Paleorient*, 16/2, 1993
Heard-Bey, Frauke, *From Trucial States to United Arab Emirates*, Longman, London, 1996
Hobbs, Joseph J., *Bedouin Life in the Egyptian Wilderness*, The American University in Cairo Press, Cairo, 1990
Iddison, Philip, 'Arabian Traveller's Observations on Bedouin Food', *Proceedings of the Oxford Symposium on Food and Cookery {1996}*, Prospect Books, 1997
Keohane, Alan, *Bedouin – Nomads of the Desert*, Stacey International, London, 1994
Lancaster, William and Fidelity, *Draft Commentary and Archive compiled for the National Museum Ras Al Khaimah UAE*, 1997–1998

Musaiger, Abdulrahman O., *The State of Food and Nutrition in Bahrain*, UNICEF, Riyadh, 1993

———, *Traditional Foods in the Arabian Gulf Countries*, FAO/RNEA/Arabian Gulf University, Bahrain, 1993

Nesbitt, Mark, 'Archaeobotanical Evidence for the Early Dilmun Diet at Saar', *Arabian Archaeology and Epigraphy*, Denmark, 1993

Ramet, J.P, *La Technologie des Fromages au Lait de Dromadaire (Camelus dromedarius)*, FAO, Rome , 1993

Smith, Sylvia & Richard Duebel, 'Mauritania's Dromedary Dairy', *Aramco World*, Houston, Nov–Dec 1997

Wilson, R. Trevor. *Camels*, Macmillan Education, London, 1998

Food Composition Tables for the Near East, FAO, Rome, 1982
The Technology of Traditional Milk Products in Developing countries, FAO, Rome, 1990
1977 Statistics Yearbook, UAE Government, 1977
1994 Statistics Yearbook, UAE Government, 1994

Unpublished interviews with Saudi Arabian nationals, Gulf News, product labels and ephemera.

REFERENCES

[1] Hobbs recounts from his travels with the Ma'aza tribe of Egypt in the early 1980s the capture of a young ibex which would be 'introduced to the hunter's herd to be reared as a goat'.

[2] Gouin.

[3] See later for processing equipment and contrast the firm evidence for processing of dates to produce date syrup, *dibis*, as early as the second millennium BC in the Gulf.

[4] Heard-Bey.

[5] An ability shared only with humans and cattle amongst the mammals.

[6] It has been estimated that the nutritional needs of a human would be met by between 3 and 8 milch camels, Wilson.

[7] FAO Study by Ramet.

[8] Heard-Bey.

[9] The Arabian *tahr*, a small goat-like ungulate may still survive in the Hajar mountains of Oman and was hunted in the UAE until recent times.

[10] In a personal communication, William Lancaster states that the inhabitants of Ras Al Khaimah still make little differentiation between goats and sheep, they are more often identified by their colour e.g. *bir aswad* are black-coated sheep or goats. This attitude also applies to their milk which is often mixed for consumption or processing.

[11] See Glossary.

[12] Easa Saleh Al-Gurg's family lived in Lingah on the Persian side of the Gulf until they emigrated to Dubai in the 1920s and may have brought the tradition with them.

[13] The milk production statistics appended show the change in nature of the local dairy industry over the last two decades.

[14] Recently introduced products are *laban* flavoured with ground fenugreek seed and date flavoured milk.

[15] The method is mentioned in Pliny, Dioscorides and others.

[16] Vimto is a fruit cordial drink, originally from the north of England, now locally packaged in Saudi Arabia. I have never met anyone who has admitted to drinking this product and although it is on the local supermarket shelves, I still wonder where all the empty Vimto bottles come from.

[17] In Al-Ansari's book, 48 per cent of the recipes have a dairy product and in a further 27 per cent oil is used as a cooking medium. Comparative figures for Saudi cuisine are similar.

What's in a Name?
Some Thoughts on the Origins, Evolution and Sad Demise of Béchamel Sauce

Cathy K. Kaufman

One culinary legend[1] attributes the invention of the eponymous béchamel sauce to the marquis de Nointel, Louis de Béchameil (1630–1703), *maître d'hôtel* to Louis XIV. A handsome, wealthy gourmet in charge of the finest table in seventeenth-century France, Béchameil's high profile allegedly provoked the duc d'Escars to carp: 'that fellow Béchameil has all the luck! I was making breast of chicken *à la crème* more than twenty years before he was born, but I never had the chance of giving my name even to the most modest sauce.'[2] The good duke had a point: dairy sauces did not suddenly burst forth in the seventeenth century as the inspiration of the marquis. Milk sauce appears in Apicius and again in thirteenth-century France, although infrequently. Cream sauces and cream-enriched dishes gained popularity in the eighteenth century, reigned over the finest cooking in the nineteenth century, only to be deposed in the late twentieth century by that tyrant, nouvelle cuisine.[3]

Ask any contemporary cook to define béchamel and the response will be a 'white' sauce made from milk whisked into a roux (a gently cooked mixture of fat and flour), and boiled until thickened, with a frequent, but not mandatory, enrichment of cream.[4] Although we often think of béchamel as a sauce, it has a second, albeit related, meaning: that of a category of dishes in which savoury titbits are bound in béchamel sauce, that is, dishes *à la béchamel*.[5] As we shall see, this second meaning has clear antecedents in medieval and Renaissance recipes and is how the term béchamel is first used in cookery. As currently used, béchamel is the mother of little sauces,[6] a binder for made dishes such as lasagne, moussaka, gratins and vol-au-vents, and a base for stable soufflés. Its bland versatility – nutmeg is now its only dependable seasoning other than salt and pepper – is its greatest drawback, having earned it the sobriquet 'glue'.[7]

Milk and cream sauces were not always made by adding those liquids to a roux. Nor were 'white' sauces historically roux and dairy-based. From at least imperial Rome through the eighteenth century, when the name béchamel first appeared, a white sauce was generally understood to be a wine, vinegar or verjuice-based sauce bound with eggs and possibly an additional thickener.[8] Some of the earlier white sauces contained pounded nuts, while modern versions are mounted with butter. These modern egg-acid sauces are still very much with us in hollandaise, béarnaise and sabayon, and the shift in white sauce nomenclature from these egg-acid sauces to roux-bound sauces begins in the nineteenth century.[9]

Although béchamel is ubiquitous in twentieth-century classical cookbooks and cooking school curricula, animal milk and cream sauces[10] were uncommon until the eighteenth century. What accounts for their relative scarcity, given the wide availability of milk and its broad use in varying guises throughout Europe? Certainly milk's high perishability explains some of the preference for cheese and other cultured products, although béchamel flourished prior to refrigeration and pasteurization.[11] Another possibility is that milk was considered unhealthy for consumption and subject to adulteration. The evidence is mixed, but even when milk was viewed suspiciously, cooked milk was usually given the medical green light.[12] A third, and perhaps most significant, element is the changing tastes among the arbiters of culinary fashion during the seventeenth and eighteenth centuries. As the spicy and acid flavours of medieval and Renaissance cookery become more widely available and thus less of a social marker, the trendsetters found new flavour combinations right in their own backyards with butter and cream flavoured foods. Whether their delight in the unctuous mouthfeel and mildly sweet taste of milk and cream was a cause or an effect of larger dairy herds, milk and cream became frequent ingredients in sauce-making.[13]

Milk sauce in early cookery: the Apician 'béchamel'

Milk was drunk and processed into curds, butter and cheese in Neolithic and early civilizations, although I have found no evidence that milk was used as an ingredient except possibly in porridges or other grain-based dishes.[14] Athenaeus gives us greater culinary uses of milk, listing cakes, breads and puddings made with milk. In a fleeting passage, he remarks, 'how the fecund miscarried matrix [of a sow] rounds out cheese-like in the stew, covered with white sauce!'[15] Nothing hints that this white sauce is made with milk, and all of Athenaeus' explicit references to milk in its liquid form are as a beverage.

Milk finds its way into several Roman sauce recipes. Although most of Apicius' several white sauce recipes, specifically called either *ius album* or *ius candidum*, do not contain milk,[16] *De Re Coquinaria* contains a recipe (8.6.6) for lamb or kid that employs milk bound with wheat starch for the sauce. The cook is instructed to 'pour [on the cooked meat] the following hot sauce: Milk, ground pepper, *liquamen, caroenum*, a little *defrutum* [both forms of reduced grape juice], and also a little oil; when this comes to the boil add *amulum* [wheat starch] to thicken.' While Apicius' technique differs from a modern béchamel, we have all of the characteristic elements spelled out: milk, flavourings, fat and starch. It will be more than 1,000 years before we see these elements reunited in a single sauce recipe.

Milk sauces in medieval and Renaissance cookery
Milk sauces form only a small minority of the many different sauces documented for the medieval and Renaissance cook. They exist in France and England, and I have found indirect evidence of milk sauces in other regions.[17] Why are they so infrequent, given that medieval and Renaissance cookery books emphasized the importance of having a pantry of sauces available both to improve the savour of a dish and to correct the diner's temperament under the doctrine of humours?[18] Animal milk and almond milk both had virtually identical humoral qualities that matched the ideal human temperament.[19] During the millennium stretching from Anthimus[20] to Andrew Boorde,[21] commentators generally favoured the use of milk when it was either warm from the critter or cooked. Thus no medical theory impeded the enjoyment of milk sauces. Milk's rapid perishability and the Catholic Church's fasting constraints also seem insufficient to explain the preference for almond milk over animal milk. Many foodstuffs found in the cookery books were highly perishable. All but the meanest city dweller could have access to fresh milk for at least a portion of the year. And therein may lie the problem. Almonds were imported to non-Mediterranean Europe, while milk was local. What status could attach to the liberal use of milk in sauces? Perhaps this explains milk's infrequent appearance in books written for upper class households.

The earliest post-Apician milk sauce comes from the anonymous author of the *Traité du XIIIe Siècle*, written *ca*. 1290. Among the several treatments for roasted capon and hen, depending upon the season, is the following: 'The flesh of capons and hens is good roasted with a sauce of wine in the summer and in winter, a sauce flavoured with garlic and cinnamon and ginger, mixed with either almond or ewe's milk.'[22] This winter sauce seems an early version of the ginger-based Jance sauces that were popular in France during the fourteenth and fifteenth centuries. The Jances were a family of cooked sauces characterized by their pronounced ginger flavour. The liquids used to make the sauces varied and included wine, verjuice, almond milk, cow's milk and bouillon. Taillevent's *Le Viandier* (*ca*. 1375), the *Goodman of Paris* (*ca*. 1390), Chiquart's *On Cookery* (*ca*. 1420) and *The Vivendier* (*ca*. 1450) all contain multiple recipes for Jance sauces, and, with the exception of Chiquart, all have cow's milk variations. The Jance sauces were frequently recommended for poultry, a highly relevant usage in light of the duc d'Escars grumblings over Béchameil's chicken in cream.

The Vivendier's recipe is most interesting for our study, as it takes the unprecedented step of combining both milk and verjuice in a single recipe: 'Jance of Cow's Milk. White bread tempered in milk, strained egg yolks, and white ginger distempered with verjuice. Boil everything together.'[23] Assuming, as Terence Scully cautions, that there is no scribal error in this combination, we have a rather daring technical challenge and an interesting flavour combination. As the *Goodman of Paris* notes, adding egg yolks to bind a hot liquid is tricky, for one always risks the yolks' curdling.[24] Moreover, adding acid in the form of verjuice could also wreak havoc with the eggs. The use of bread in the hands of a skilled cook provided some protection against the yolks' curdling when exposed to the combined forces of the fire's heat and the verjuice. The bread also helped to thicken the sauce, almost as elegantly as the modern roux would. We will revisit this technical issue of binding sauces with eggs in the early modern recipes, where we will also see the addition of a few drops of lemon or bitter orange juice to brighten the sauce's savour, anticipated here by the little bit of verjuice distempering the ginger.

Another family of medieval milk sauce recipes comes from a now-lost manuscript composed in the third quarter of the fourteenth century called *Le Grand Cuisinier de Toute Cuisine*, but surviving in a printed edition of 1540. *Le Grand Cuisinier* gives us two recipes for *dodine blanche*:

> White Dodine. Take some good cow's milk, put it to cook in a frying pan beneath your roast with some white powder [sweet spices], two or three egg yolks; strain your milk and let it cook altogether with a little sugar, some salt and a little bit of parsley leaves. If you want, put in some minced marjoram. Place your roasted waterfowl on top.

> Another White Dodine. To make white dodine, take cow's milk, hard-cooked egg yolks, white powder, fried rings of onion. And strain and cook it in a frying pan, and don't forget to support it with sugar, a little salt and the fat from your duck.[25]

The first thing to note about these recipes is that a *dodine* is a sauce that is ordinarily 'made from breast of capon, almonds, garlic and eggs, and served with goslings'.[26] Thus our anonymous *Grand Cuisinier* deviates from normal culinary practice of the era. Second, the use of fat and roast drippings in both recipes presages the seventeenth- and eighteenth-century masters' turn to fattier sauces. Third, were it not for the use of *poudre blanche* in the first recipe, that sauce would be quite at home in some of the early eighteenth-century recipe collections, especially in their reliance on eggs, rather than starch, for thickening.[27] Lastly, both of these *dodines* are designed for use with poultry.

English contributions to the evolution of béchamel are in the *Forme of Cury* (*ca.* 1390), although they are much farther removed from the essential elements of béchamel than the French recipes of the era. One is for capon cooked in milk and results in a dish suggestive of béchamel in its second meaning:

> Douce iame. Take gode cowe mylke and do it in a pot. Take persel, sawge, ysope, saueray, and ooþer gode herbes; hewe hem and do hem in the mylke and seeþ hem. Take capouns half yrosted and smyte hem on pecys, and do þerto pynes and hony clarified; salt it and colour it with safroun, & serue it forth.[28]

Typical of the English manuscripts, the recipe is vague on the final texture of the dish, and one might conclude that the addition of cut-up capons and pine nuts leads to a sort of blancmange. However, given the failure to grind the capon, the use of small, but intact pine nuts rather than ground almonds, and the lack of typical blancmange spices, this is probably a looser, saucier dish than the typical blancmange.[29]

The other noteworthy English recipe is for poached eggs with a custard sauce:

> Pochee. Take ayren and breke hem in scaldyng hoot water, and whan þei bene sode ynowh take hem vp and take zolkes of ayren rawe & mylke, and swyng hem togydre; and do þerto powdour gynger, safroun and salt, set it ouere the fire, and lat it not boile. And take ayren isode & cast þe sew onoward, & serue it forth.[30]

This custard sauce, thickened by the addition of egg yolks to hot milk, admonishes against boiling the sauce, necessary because the recipe lacks any starch.

Guiliano Bugialli asserts that *balsamella* was known in fifteenth-century Italy and brought to France in 1533 by one of Catherine de' Medici's cooks, called Pantanelli.[31] Bugialli traces the term *balsamella* to Florentine ladies' use of a milk and flour mask, or *balsamo*, to beautify their complexions,[32] and further points to a fifteenth-century manuscript in the Biblioteca Universitaria di Bologna that ostensibly explains the technique of roux and *balsamella*. The recipe, while a mixture of flour, milk and egg whites, is for thick dough that is rolled out and fried.[33] It is not a sauce and seems closer to a variation on the fritter recipes found throughout Europe at that time, including another Italian recipe for 'ravioli'.[34] Not to slight the Italian contribution to the art of cooking with milk and flour: Platina has several recipes for milk, meal and egg white combinations that are made into fritters, and these recipes are more delicate than their English and French counterparts in which fritters are made from a combination of flour and eggs.[35] These particular recipes, however, had reached France a quarter of a century before Catherine de' Medici[36] and are not the grandparents of béchamel, given the older starch thickened milk sauces found in France and the milk, bread and egg-thickened made dishes found in France (see note 29, supra) and England.

The beginning of modern uses of milk and cream in sauce-making

The seventeenth century begins the shift in France from the spice-acid palate, which had dominated cookery from the late Middle Ages through the Renaissance, to a herb, cream and butter palate, which required new culinary techniques. This change, of course, did not take place all at once, and vestiges of the old flavour combinations and sauce-making techniques are found even today. Vexing to the chef searching the works of François de la Varenne, Pierre de Lune and L.S.R. for the ur-béchamel sauce is the way in which these seventeenth-century masters flirt with roux but never expressly use them to thicken milk or cream sauces. These masters know how to make a pastry cream from milk, butter, flour and eggs, yet they never unequivocally alter the proportions of these same ingredients to create a sauce. Nonetheless, I believe that béchamel is found in the day-to-day workings of these kitchens.

La Varenne's *The French Cook*[37] contains a handful of recipes that may be béchamel sauces. First, it bears repeating the well-known point that *The French Cook* contains perhaps the earliest recipe for roux, called a 'Thickning of flowre' which, in La Varenne's version, is really a very thick velouté, béchamel's sister sauce. La Varenne instructs the cook to melt lard and add flour, stirring the mixture until it boils, add some onion and some 'good broath, mushrooms, and a drop of vinegar; then after it hath boyled with its seasoning, passe all through the strainer, and put it in a pot; when you will use it, you shall set it upon warme embers for to thicken your sauces'.[38] La Varenne also provides thickenings of almonds, mushrooms and truffles, leaving to the cook's discretion the choice of binder in each dish. He advises the cook to prepare batches of these liaisons, ready to incorporate whenever the need arises. Few, if any, of his recipes direct the cook to employ a specific thickening in a dish; certainly none is specified in any La Varenne recipe containing milk or cream.

La Varenne's recipe for 'Mushrooms with Cream' illustrates how these thickeners might have been used. The cook is to sauté some mushrooms in butter with minced parsley and green onions, 'and then you may put some creame to them, which when it hath boyled a little while, and the sauce being thickned, you may serve'.[39] We do not know whether La Varenne intends the sauce to be thickened merely by reducing the cream or whether the cook was to employ his 'Thickning of flowre' or another of the liaisons; in actual practice, given the longevity and popularity of *The French Cook* and the exigencies of the kitchen, I suspect cooks did both. Because La Varenne relies so much on the cook's judgment, the best we can say is that his recipe for 'Mushrooms with Cream' is consistent with our definition of béchamel.

La Varenne is very aware of the binding power of milk cooked with flour. His recipe for 'Tourt of Massepin' calls for a 'creame of milk' filling, made by allaying 'a very little flowre with a quart of milk, seeth it well, and let it be very thinne; then put a little butter in, four yolks of eggs, and two whites well beaten…'.[40] After sweetening the mixture is baked in an almond paste shell. The instructions to add only a 'very little' flour to a quart of milk and to keep the mix 'very thinne', relying on the eggs to bind the tourte in the oven, indicates that the master had made an unappealingly pasty milk 'cream' at some point in his career and hoped to spare other cooks from this mistake.

Writing three years after La Varenne, Nicolas de Bonnefons presents a recipe for chicken fricassee that gives the cook the choice of finishing the sauce with a yolk-verjuice or a yolk-cream liaison, reminiscent of the *Vivendier's* Jance sauce. After browning a tender chicken in fat 'to give it a fried taste', the cook is to stew the bird with bouillon or water, add some wine, butter and:

> whatever you believe will be tasty, such as mushrooms…[W]hen it is cooked, you can throw in some egg yolks mixed with verjuice, or, if you like, some cream, giving it just a little bouillon and returning it carefully [to the fire] so that the broth doesn't coagulate.[41]

No wonder the duc d'Escars was infuriated by all the credit going to Louis de Béchameil for his chicken in cream! The utter casualness with which Bonnefons suggests the substitution of cream for the medieval favourite verjuice implies that many seventeenth-century kitchens were accustomed to adding cream to sauces, and indeed, L.S.R.'s 1674 recipe for *Fricassée de pigeonneaux*[42] offers the cook the same choice. Yet Bonnefons' admonition about carefully reheating the sauce shows that, at least in Bonnefons' case, we are still dealing with that old but tricky technique of tempering eggs into hot sauces.

Pierre de Lune's 1656 recipe for *Poulets fricassés à la crème*[43] is simplicity itself: the cook is to sauté chicken parts, braise the browned chicken in bouillon and wine, degrease the sauce and enrich it with cream at service. Although the specific recipe gives us no instructions for thickening, de Lune's introductory remarks explain how to make a roux, called *farine frite*. His formula follows La Varenne's 'Thickning of flowre, and he tells the cook that this mix will be useful 'for sauces and other things'.[44] Like La Varenne's 'Mushrooms with Cream', this recipe is consistent with our definition of béchamel.

François Massialot expressly adds the flour that is implied in some of the earlier recipes, although it is not designed to be the sole binder for the sauce:

> Filets of a fat Pullet, with Cream. Take the Filets of large fat Pullets roasted, and cut them into pieces: Then put into a stew-pan a little lard and Parsley, and having toss'd it up with a little Flower, add Artichoke-bottoms…, and a little clear Broth, all well-season'd. When they are sufficiently stew'd, put your Filets to them, and a little before they are serv'd up, pour in a little Milk-cream; taking care to keep them hot. To thicken them, let one or two Eggs be beaten with Cream, and having brought it to a due consistence, let it all be set on the Table…[45]

John Middleton has a similar recipe,[46] and Henry Howard comes even closer to creating a béchamel in a recipe for 'Frigasee White' by binding his cream sauce with the uncooked sister of roux, beurre manié:

> [After braising chicken in broth] put in half a Pint of good Cream, and a piece of Butter rowled up in Flour and the Yolk of an Egg; stir it all about till it is as thick as Cream; wring in the Juice of a Limon, take care it don't [sic] curdle it.[47]

The first dish named béchamel

Kudos for the earliest named béchamel goes to Vincent La Chapelle, who publishes a recipe for *Turbôt à la bechamelle* in the 1733 first edition of *The Modern Cook*. The cook is to have ready a turbot poached in a court bouillon; then place minced parsley, green onions and shallots in a casserole with a good bit of butter, season with salt, pepper, and nutmeg, and add a small pinch of flour. The turbot is sliced and placed in the casserole 'with some cream, or some milk, or equally well some water'. The sauce is cooked until bound, seasoning adjusted, and served with lemon.[48] Remarkably, this béchamel does not require either milk or cream as the liquid for the sauce, as La Chapelle claims that the sauce can be made *equally well* with water. Imagine the horror overwhelming the duke d'Escars! Although La Chapelle calls his turbot dish a béchamel, the roux-cream combination clearly is not the defining characteristic for him. *The Modern Cook* contains a recipe for 'A Quarter of Veal with Cream,' in which the cook is instructed:

Make a Sauce as follows; Take a good piece of fresh butter, put it into a Stew-pan with a little chopped Parsley, grated Nutmeg, a little Dust of Flower, and at least a Pint of Cream; set that over the Stove, and take care that it does not turn to Oil: make it savoury, take the Quarter of Veal off the spit, put it into the Dish, take off the Bacon, pare the handle, pour the Cream Sauce over it, serve it hot.[49]

If this isn't a béchamel, what is?

The name 'béchamel' spread rapidly to denote specific dishes cooked with cream in the eighteenth century. François Marin publishes in 1739 a recipe for *Filet de Poularde à la Béchamel*,[50] using a mix of roux and egg yolk to bind the cream. By 1759 William Verral pens *Poitrine des poulardes a la Benjamele*,[51] following the beurre manié technique employed by Henry Howard. Verral notes that '[t]his sauce may serve for any sort of white meat, and is now very much in fashion', and offers recipes for sweetbreads in pastry, eggs and sherdoons, all *à la Benjamele*.

Notwithstanding Verral's enthusiasm, béchamel is still a type of made dish, and not yet an independent sauce. The exhaustive *Dictionnaire portatif de cuisine* has no entry for béchamel, either in its long section on sauces or as a type of made dish. Certainly there are sauces that are related to béchamel, such as *Sauce à la morue* (onions, mushrooms and garlic infused into cream, thickened with flour, strained and seasoned with reduced stock), and others which rely on the older techniques of thickening with bread or yolks, such as *Sauce à la reine* (veal, ham, mushrooms, onions, sautéed in oil, simmered with stock and Champagne, degreased, bound with bread soaked in cream, strained, and finished with butter and more cream), or *Sauce à l'ivoire* (*sauce à la reine* bound with egg yolks).[52] Béchamel simply has not yet arrived as a chic and fundamental preparation.

This continues in Menon's *La Cuisiniere bourgeoise*.[53] The sauce chapter is silent on béchamel, although the poultry section contains a recipe along the model of Howard's for *Poularde à la Béchamel*. That the dish has not reached exalted status is suggested by the fact it could be served as an entrée or an hors d'oeuvres, rather than as a major dish.[54]

The apotheosis and tragic fall of béchamel sauce

Carême single-handedly changes the culinary world's view of béchamel in his 1833 opus *L'art de la cuisine française au XIXe siècle*. In his introductory remarks to his sauce chapter, he proclaims that there are four great sauces, espagnole, velouté, béchamel and allemande, from which every variation can be quickly made. Carême's béchamel is more sophisticated than any of the recipes heretofore cited, and requires both a generous budget and meticulous kitchen staff to execute. In brief, the cook is to create a blond 'quintessence' of meat juices by gently simmering ham, several cuts of veal and fowls with a preexisting meat bouillon in a covered vessel. When the first reduction is made, the meats are pricked to release the juices, the sides of the vessel cleaned, the evaporated bouillon replenished, and the whole thing goes back to the slow fire to extract more juices, with ever vigilant skimming of grease. Eventually the meats are cooked, the liquid is strained, and one proceeds to make velouté, a prerequisite of Carême's béchamel, adding in several stages the rich broth to a pale roux. The velouté simmers for hours, with aromatics added and attentive degreasing. Finally the liquid is 'perfectly clarified', at which point it is divided, and enriched either as an allemande (with egg yolks) or as béchamel, by adding unboiled heavy cream. Carême drops a footnote here that helps perpetuate the marquis de Bechameil's presumed invention of the sauce:

> It is essential that the reduction of this sauce be done by a man who loves his station: for an insouciant man will let it take colour, and that is the worst thing that could happen to this sauce which, moreover, honours the marquis de Béchamel, who thought of the addition of cream to velouté and gave his name to this excellent sauce.[55]

This extraordinarily indulgent feat of sauce-making could not last, if, indeed, it ever was the common practice. The 1836 *Dictionnaire de cuisine* has two béchamel entries, one following the model posited by Carême, and the other for *béchamelle maigre*. The *maigre* version calls for boiling cream with flour, taking care that it does not colour, and adding a milk and water mixture that has been infused with aromatics.[56] We have reached the beginning of the end of béchamel as fine cuisine.

By 1871, Urbain-Dubois[57] has removed even the possibility of a meat base for béchamel. Although Urbain-Dubois has a lovely recipe for a roux-bound 'plain white sauce', made by gently stewing meats and reinforcing their savour with bouillon, this concoction does not form part of his béchamel recipe. Urbain-Dubois' béchamel is a quickly executed affair made by cooking a roux with a bit of ham for ten minutes, diluting same with milk, infusing some aromatics for an additional twenty minutes, and finishing the sauce by straining, reducing and mounting with raw cream.[58] Escoffier continues this pale imitation, substituting veal for the ham, and estimates that the sauce can be made in one hour.[59]

These modern béchamels are dreary compared with the complex flavour of Carême's sauce. Perhaps La Chapelle and the eighteenth-century cooks following his example were correct in limiting béchamel to one of many minor preparations. Our late-twentieth-century reaction to combining flour thickeners with cream in sauce-making is akin to showing garlic to a vampire: we cringe. Yet we cannot escape the desire for cream sauces, even if only in name. Much like La Chapelle's advice that water works as well as milk or cream in the original *Turbôt à la bechamelle*, we now are confronted with recipes for 'Old-Fashioned Cream Gravy' to grace American fried chicken that instructs the cook to make a roux from chicken fat and add milk or water to create a 'just fabulous' gravy.[60] I am sure that the duc d'Escars is rolling in his grave.

REFERENCES

[1] Other charmingly undocumented tales include Waverley Root's recitation of Cardinal Albornoz dining on Italian *balsamella* in fourteenth century Emilia-Romagna and, of course, Catherine de' Medici's cooks.

[2] *Larousse Gastronomique* (Lang, Jennifer Harvey, ed.) New York: Crown (1988), p. 81.

[3] Gault, Henri and Millau, Christian, 'The Ten Commandments of Nouvelle Cuisine,' *Guide Nouveau* (1978). Commandment 7 'recognizes the pretension, the inanity, the mediocrity of those rich and heavy sauces. THESE TERRIBLE BROWN AND WHITE SAUCES... that have assassinated so many livers and have covered up so many insipid pieces of meat.'

[4] *Grand Larousse Universal,* Paris (1989); *Trésor de la langue française,* Paris: Centre National de la Recherche Scientifique (1975); *Dizionario Encyclopedio Italiano*, Rome (1955); Oliver, Raymond, *Classic Sauces and Their Preparation*, London: Wine and Food Society (1967).

[5] *Trésor de la langue française*, t. IV, p. 341.

[6] Antonin Carême describes over 70 'petites sauces' derived from béchamel in *L'art de la cuisine française au XIXe siècle*, (1833) Paris: Éditions Payot & Rivages (1994).

[7] The derisive 'glue' is not limited to Anglophone kitchens; the Italians refer to *salsa colla* in lieu of *balsamella* or *besciamella*. *The Magazine of La Cucina Italiana*, (June 1999), p. 91.

[8] One exception is François Massialot in *The Court and Country Cook*, London (1702), in which he defines white sauce as a sort of blancmange: 'White Sauce, a Sauce made of blanched Almonds and the Breast of a Capon, pounded together, adding Cinnamon, Cloves, Ginger, Rose-Water and Sugar.'

[9] Burnet, *Dictionnaire de cuisine et d'économie ménagère*, Paris, à la librairie usuelle (1836), p. 639.

[10] A few of the large number of medieval and renaissance recipes for almond milk expressly permit the substitution of animal milk; whether actual culinary practice engaged in more frequent substitution within the fasting constraints is open. I also will treat milk and cream interchangeably, as the boundary between the two in cookery books is quite fluid.

[11] Preserving milk's freshness continued to be an issue into the twentieth century, even with the availability of ice. A small pamphlet published by the Hood Dairy in 1906 promoting the wholesomeness of its milk bragged that in a competition among dairies, the Hood milk, although 'nearly <u>four days old</u>, ... was perfectly sound and in excellent shape after having taken the <u>longest</u> journey of any of the twenty-one lots of milk exhibited.' (Emphasis in original.) Hill, Janet McKenzie, *Recipes for the Use of Milk,* Boston: H.P. Hood & Sons (1906).

[12] The medical theories about the healthfulness of milk varied with time, location and age of the drinker. As but one example, the *Tacuinum Sanitatis* calls sweet milk 'good nourish-

ment' although dangerous for those with fevers. Combining milk with seedless raisins was thought to neutralize the dangers.

[13] See, e.g., J.-L. Flandrin, 'La goût et la nécessité: sur l'usage des graisses dans les cuisines d'Europe occidentale (XIVe–XVIIIe siècle),' *Annales* 38 (1983), pp. 369–401.

[14] Brothwell, Don and Brothwell, Patricia, *Food in Antiquity*, Baltimore: Johns Hopkins University Press (1998); R. J. Forbes, *Studies in Ancient Technology*, III, (3rd ed.), New York: E. J. Brill (1993).

[15] Athenaeus, *Deipnosophistae*, (C. B. Gulick, trans.), I:435, Cambridge, Mass.: Harvard University Press (1993)

[16] Apicius, *De Re Coquinaria*, (Barbara Flower & Elizabeth Rosenbaum, trans.), London: Harrap (1958), see, e.g., Recipes 5.3.2, 6.5.3, 6.5.5, 7.6.4, 7.6.5, 7.6.6, 7.6.7, 7.6.9. The exception is Recipe 6.9.11 for Chicken à la Varius, which has a milk sauce bound with egg whites and is 'known as white sauce.'

[17] The *Diuersa Servicia* (contained in *Curye on Inglysch*, see note 28, infra,) has a recipe for a White German Broth that permits the use of either almond or animal milk for cooking morsels of chicken. The dish is finished by adding spices to the pot. The name of the dish suggests that some form of milk sauce was known in Germany. Rudolph Grewe's translation and synthesis of four related manuscripts written in Danish, Icelandic and Low German has several sauce recipes, although none contain milk; the milk recipes are for custards and porridges. *Current Research in Culinary History: Sources, Topics and Methods,* Boston: Culinary Historians of Boston (1985) pp. 27–45.

[18] See, e.g., Scully, Terence, The Opusculum de saporibus of Magninus Mediolanesis', *Medium Ævum* 54 (1985); Chiquart, *On Cookery*, (T. Scully, trans.), New York: Peter Lang (1986); Platina, *On Right Pleasure and Good Health*, (M.E. Milham, trans.), Tempe, Arizona: Renaissance Society of America (1998).

[19] Scully, Terence, *The Art of Cookery in the Middle Ages*, Woodbridge: The Boydell Press (1995), p. 49.

[20] Anthimus, *De obseruatione ciborum*, Devon: Prospect Books (1996), §§ 9, 75–76, 79–81.

[21] Boorde, Andrew, *The Dyetary of Helth*, (1562), 'Cream the which doth not stand long on the milk, and sodden with a little sugar, is nourishing. Clouted cream and raw cream put together is eaten more for a sensual appetite . . .[and] hath put men in jeopardy of their lives.', quoted in Simon, André L., *The Star Chamber Dinner Accounts*, London: Wine & Food Society (1959).

[22] 'Traité ou l'on enseigne a faire et appareiller tous boires comme vin, clairet, mouré et autres, ainsi qu'a appareiller et assaisonner toutes viandes selon divers usages de divers pays' (ca. 1290), manuscript from the Bibliothèque Nationale (F. lat. 7131), excerpted in Guégan, Bertrand, *La Fleur de la Cuisine française*, Paris: Editions de la Sirène (1920–21), at pp. 1–2. All of the translations from the French are mine unless otherwise indicated.

[23] *The Vivendier*, (Terence Scully, trans.), Devon: Prospect Books (1997), p. 66.

[24] *The Goodman of Paris*, (Eileen Power, trans.), London: Routledge (1928), p. 290.

[25] 'Le grand cuisinier de toute cuisine,' (ca. 1350–1380),, excerpted in Guégan, supra., at p. 9.

[26] Godefroy, Frédéric, *Dictionnaire de l'ancienne langue française*, Paris: 1880 (Kraus Reprint Corporation 1961).

[27] In fact, these recipes seem so 'modern' for the fourteenth century that I am wondering whether the 1540 printed edition may be a hoax in its dating of these recipes; even so, the recipes are relatively modern for the sixteenth century.

[28] Hieatt, Constance B. and Butler, Sharon, *Curye on Inglysch*, London: EETS (1985), *Forme of Cury 64*.

[29] *The Vivendier* and *Le Viandier* also have dishes suggestive of béchamel's second meaning, in which peas are puréed, bound with milk, bread and egg yolks, and served with cut-up chicken. *The Vivendier*, § 46; *Le Viandier*, § 11.

[30] *Curye on Inglysche, Forme of Cury* 92.
[31] Bugialli, Guiliano, *The Fine Art of Italian Cooking*, New York: Quadrangle (1977), pp. 51–54; see especially Ms. #158 at p. 54.
[32] Kamman, Madeleine, *The New Making of a Cook*, New York: William Morrow & Co. (1997), p. 277.
[33] My thanks to Dr. Vincent J. Bartolini for translating this recipe for me.
[34] *Curye on Inglysch*, p. 189.
[35] Ibid., at *Diuersa Servicia 19*.
[36] The first French edition of Platina was published in Lyons in 1505.
[37] De la Varenne, François, *The French Cook*, (2d ed.) (1654). The work was originally published in 1651 as *Le Cuisinier françois*. All of my references come from the 1654 English translation.
[38] Ibid., p. 120.
[39] Ibid., p. 214. I do not consider that the addition of roux to hot cream, instead of Adding cream to a roux, to be a significant difference in technique for our purposes.
[40] Ibid., p.242
[41] Bonnefons, Nicolas de, *Les Délices de la campagne*, quoted in Guégan, *La Fleur de la Cuisine*, p. 96.
[42] L.S.R., *L'Art de bien traiter* (1674), reprinted in *L'art de la cuisine française au XVIIe siècle*, Paris: Éditions Payot & Rivages (1995), p. 99.
[43] De Lune, Pierre, *Le Cuisinier* (1656), reprinted in *L'art de la cuisine française au XVIIe siècle*, p. 327.
[44] Ibid., p. 242.
[45] Massialot, *The Court and Country Cook*, p. 124.
[46] Middleton, John, *Five Hundred New Receipts on Cookery*, London: 1734, pp. 125–126.
[47] Howard, Henry, *The British Cook's Companion*, (5th ed.) London: 1729, p. 44.
[48] The complete recipe is found in Volume 3, page 138 of *The Modern Cook*, which was privately published in London by the author. My thanks to Gilly Lehmann for generously confirming this citation.
[49] *The Modern Cook* (1733). Volume 2, pp. 184–85.
[50] Marin, François, *Les Dons de Comus* (1739), excerpted in Guégan, *La Fleur de la Cuisine*, p. 251.
[51] Verral, William, *A Complete System of Cookery*, London (1759), pp. 135–6.
[52] *Dictionnaire portatif de cuisine* (1767) Paris: Éditions Payot & Rivages (1995), see esp. pp. 460–64.
[53] Menon, *La Cuisiniere bourgeoise*, (6th ed.), Brussels: 1783.
[54] Ibid., at 188–89.
[55] Carême, *L'art de la cuisine*, at pp. 292–93.
[56] Burnet, *Dictionnaire de cuisine*, pp. 54–55.
[57] Urbain-Dubois, *The Household Cookery Book*, London: Longmans, Green & Co. (1871).
[58] Ibid, at 62.
[59] Escoffier, Auguste, *The Escoffier Cookbook*, New York: Crown Publishers (1969), P. 21.
[60] McDermott, Nancie, 'Southern Fried Chicken' in *Fine Cooking*, September 1999, pp. 27–29.

Low-temperature Cheese-making: Ancient Wisdom not Outdated

Lidia Kitrilakis and Sotiris Kitrilakis

Soft, brine-ripened and preserved cheese is probably the first type of cheese ever made. Like every other activity of importance Homer makes note of it. We are referring to cheese made of whole milk from sheep and goats using natural rennet from the stomachs of suckling animals and preserved in sea-salt brine. This method of making cheese is still in use in Greece at farmsteads and in small dairies. Characteristically, warm milk obtained immediately after milking the herd is filtered into the vat, rennet is added and the curd is ready for cutting within an hour. There is very little time for harmful bacterial growth to develop. The natural rennet containing slurry from young animals' stomachs supplies not only the curd-forming enzymes but also the bacterial cultures.

The cut curd is placed in basket or metal moulds to drain overnight and is salted the next day. Depending on weather (temperature and humidity) the cheese is placed in brine after a few days and allowed to ripen. Never is the milk or curd temperature raised above body-temperature. Prior to refrigeration, this method preserved for many months valuable milk nutrients in the form of cheese. The salt content was considerably higher but the cheese kept well at cellar temperature for the entire year.

This type of cheese has always been an important food for the Greeks. The average person consumes 50–100 gr during several meals every week. Unlike hard cheeses, which are consumed sparingly and with special dishes, *feta* is an everyday food. Soft brine cheese (fat content about 30 per cent) is a staple in the Greek diet and has been for many centuries. Milk on the other hand is not consumed in significant quantities after childhood. We are all aware of the Cretan diet with 40 per cent calories from fat (olive oil and soft cheese primarily) which is associated with very low incidence of heart and circulatory diseases.

Commercial soft cheese production in Greece has adopted certain technical innovations but continues to follow traditional techniques in most important aspects. Milk comes from free-ranging herds of sheep and goats. There are about 4,000 herds with an average count of 300 animals. They graze on wild grasses and plants on private, communal and public lands. When weather prevents field grazing they are fed hay and dried grasses. Some are given formula feed on such occasions but this is the exception. These conditions keep the animals healthy without the need to add medication to the feed, which would not be practical in any case.

About 1,500 small independent dairies and a few large corporate units operate in Greece. The small producers are located near the herds and receive the milk within one to three hours after milking. Some are able to make cheese with the still-warm milk without reheating. Most dairies do pasteurize the milk at low temperature. Lower temperature and longer time produce better flavour. Of course, pasteurization of the milk is the major departure from tradition. There is no question that some of the richness of flavour in the mature cheese is lost in pasteurizing.

Indicative of the high quality of the milk and relatively gentle nature of low temperature pasteurization is that no addition of calcium is necessary. Even large dairies are able to make cheese without it.

Some of the smaller dairies still make their own rennet, but most use laboratory products adding cultures that they grow and maintain. The origin and processing of the culture is the guarded secret of every cheese-maker. The large dairies tend to use the standardized products from laboratories. Small dairies frequently use modified yoghurt cultures. It is surprising how close the flavour of carefully made cheese can be to the traditional products made with unpasteurized milk. Cutting the curd and draining the whey is a key step in the process. Most cheese-makers will kneed the curd in the mould to obtain the right 'feel'. The elasticity and texture are repeatedly tested by squeezing the curd between thumb and index-finger. Using such techniques each batch is custom-made. The milk of free-ranging animals varies constantly depending on weather conditions, the relative abundance of various plants and many other factors. It is up to the cheese-makers to compensate for these changes as much as possible by adjusting the variables he controls. In the end, each batch of cheese varies in flavour and texture. But traditional *feta*, like good wine or farmstead Cheddar, is enjoyed for the special character each cheese-maker produces and the endless variety of complex flavours. Ageing in wood barrels (birch) adds many nuances of flavour as the cheese matures for several months. No substitute has been found for that lengthy process.

Traditional *feta* is facing a major challenge from industrial dairies in Northern Europe and the United States where a continuous process has been developed based on ultrafiltration of cow's milk. 'Standardized' cow's milk (3.5 per cent fat, 3.5 per cent protein, 12.5 per cent total solids) is kept in refrigerated tanks. The milk is warmed to 50°C and processed in hyper-filtration units where it is concentrated 1 to 5.1 and to a total solids content of 39.5 per cent.

This concentration of solids is about the same as that of formed curd. When acidification takes place this slurry will expel two to three per cent of whey to reach the final solids concentration of 43 per cent. The slurry is pasteurized at 77°C for one minute and homogenized at pressures of 50–75 bar. The objective of severe pasteurization is to kill bacteria formed during hyper-filtration. Homogenizing at high pressure restructures the proteins to resemble changes that occur during normal curd formation. Acid-cultures are added and frequently lipolytic enzymes. The slurry forms curds quickly in this acid environment with much less rennin than traditional cheeses. Within 24 hours the curds develop a pH of 4.8 and are ready for packing in 8 per cent NaCl brine.

The '*feta*' is ready. With this technique one kilo of '*feta*' can be made with 5.3 litres of cow's milk while 7.3 litres are required when traditional methods are used. That is a 35 per cent improvement in yield. The whole process is continuous and equipment manufacturers offer turnkey installations.

As one might expect, the flavour and texture of this type of *feta* are very different from the traditional product made with sheep's and goat's milk at low temperature. We have not been able to find any studies on the digestion of these cheeses, which one would expect to be closer to high temperate hard cheeses. What is of critical importance is that ultrafiltration of cow's milk and continuous processing brings the cost of these cheeses to half the cost of sheep's milk traditional *feta*. In the hands of major Danish, German and American cheese-producing firms these cheeses are marketed aggressively through mass-market outlets. It has been our experience that given a chance and a taste many consumers will recognize the difference and pay the higher price for traditionally-made cheese.

Most consumers are never given the choice and the pressure to reduce costs is driving processors to more 'efficient' methods. There is also pressure to stable the sheep and to feed them formula in order to increase milk production and to 'standardize' the characteristics of the milk year-round. This will have a profound effect on the flavour of the raw material. Even when using sheep's milk there is pressure to adopt continuous hyper-filtration and to abandon barrel ageing altogether.

BIBLIOGRAPHY

Anifantakis, Em., *Greek cheeses a tradition of centuries*, National Dairy Committee of Greece, Athens 1991.

Bogiatzoglu, Em., 'High temperature effects in the production of brine cheese from hyper-filtrated cow's milk', Study XII, Athens University 1983.

Kyriakopoulos, Pan., *Cheese making in Action*, Triaina, Athens 1995.

The Artisanal and Regional Cheeses of Greece

Diane Kochilas

Ios is celebrated for its flocks and herds, and of all the islands Ios is the most celebrated for its *mysethra*, 'food of the gods', as they call it here. It is simply a curd made of boiled sheep's milk strained and pressed into a wicker basket called tyrobolon, just as they are spoken of in the *Odyssey*; from this basket it gets a pretty pattern before being turned out onto a plate. When eaten with honey it is truly delicious.

> Theodore J. Bent, *Early Voyages and Travels in the Levant*, London, 1893

The subject of Greek artisanal cheeses is vast. Most Westerners associate Greece with just a few cheeses: feta, of course; kasseri, and perhaps graviera, most likely from Crete. These are the most popular cheeses for export, but they do not even begin to represent the gamut of Greek regional cheeses. Every island, every mountain town, indeed, sometimes even a particular village, produces its own, unique cheese.

Probably nowhere else in the world has one particular food enjoyed the kind of unbroken continuity that cheese-making has in Greece. According to the myths, it was Aristaio, the son of Apollo, who taught man how to make cheese. Homer recounts the story of Odysseus and his encounter with the Cyclop Polyfimos in the cave where the one-eyed monster kept and coddled his little bundles of white sheep's milk cheese, the description of which sounds remarkably similar to a whole range of simple sheep's and goat's milk cheeses that Greek shepherds still make today. In Ancient Sparta, one of the rights of passage for a young man was to steal a head of cheese from someone's house without, of course, getting caught. And in the ancient Athenian Agora, there was a whole section devoted to the sale of cheeses; the fresh ones sold by weight, the hard, aged cheeses sold by the piece. Some of the cheeses of the ancient Greeks were made by curdling milk with fig sap, something still practised here and there by individual, home cheese-makers in villages throughout the country. The ancients also savoured the combination of cheese and honey, and they prepared myriad cheese pies, both sweet and savoury. To this day, cheese and honey are one of the great, albeit simple, dishes of Crete, where both *graviera* and fresh and soured whey cheeses are served together with the island's famed thyme honey. As for the existence of cheese pies, from the farthest reaches of Epirus to the remote mountain villages of Crete, cheese pies of every shape and form are prepared ubiquitously.

Greeks have always been shepherds, so cheese-making has always been a part of the culinary culture here, a way, certainly to preserve milk in a hot climate. Most regional cheeses are still made on a small scale, and many are not available outside their particular regions of origin. Greece is a mountainous country and grazing land has always been scarce. Sheep and goats, though, survive easily in precipitous places, hence the majority of Greek cheeses is made with sheep's or goat's milk, or a combination of the two. Cow's milk cheeses are made in a few places, namely in some of the Cycladic islands, such as Tinos, Syros and Naxos. These islands were Venetian strongholds in the fifteenth, sixteenth and seventeenth centuries. The cows came with the Venetians and so did the knowledge of how to make certain cow's milk cheeses.

The art of making cheese might be an ancient one in Greece, but the industry is young. Some of the most popular Greek cheeses were not made even half a century ago.

Greeks eat an inordinate amount of cheese – about 40 pounds a year, each. Roughly 70 per cent of the estimated 1,000,000 tons of sheep's and goat's milk produced annually is used to make cheese, as is about 30 per cent of the 700,000 tons of cow's milk. Brine-cured cheeses – essentially *feta* – account for more than two-thirds of 190,000 tons of sheep's and goat's milk cheese produced in Greece each year.

There are at least 70 distinct regional Greek cheeses, obviously too many to list here. Instead, I have given a brief explanation of the main categories of Greek cheeses and then go on to describe in more detail those cheeses that for me represent the most unusual and obscure, the real artisanal, regional gems that are the soul of the Greek dairy business.

Greek regional cheeses fall into several main categories. There are first the soft, naturally fermented cheeses. These are the simplest cheeses because they are naturally fermented. They go by various names and come from various places: from Hania, in Crete, comes the *piktogalo,* literally 'thick milk'; from Domokou on the mainland, we get *Katiki;* from the mountains of Roumeli comes the *tsalafouti,* and from Epirus in the north-west, one savours the delicious *galotyri,* literally 'milk cheese'. Because the flora differs from place to place, the cheeses all have distinct, unique flavours. Generally, though, they are very thick and either smooth like yoghurt or somewhat grainy, tart and very sour. The method for preparing them varies slightly but generally, the fresh milk is heated to about 80°C, cooled overnight, salted, and then left to ferment and thicken on its own, without the addition of rennet, either in goatskins or barrels. These cheeses usually take at least three months to ferment fully.

Secondly, there are soft mould-cheeses. They generally go by the name *kopanisti,* or 'beaten' cheese, and there are many varieties throughout Greece, but especially from the islands of the Aegean. Syros, Tinos, Andros, Mykonos, Lesvos, Ikaria, Samos and more all produce *kopanisti. Kopanisti* is essentially sheep's or goat's milk cheese which, once set (with rennet), strained, and salted is left to ferment and sour for a period of time. The cheese is made in batches, sometimes over several weeks or months, and each new batch is mixed in with the previous lot, kneaded, and pressed into either a clay or plastic jug or a goatskin in such a way so that as little air as possible gets in. This kneading and ageing process enables the *kopanisti* to develop bacteria – mould – which is what makes it so strong. *Kopanisti* is extremely peppery and pungent, usually eaten with a little bread as a *meze* for strong, local eau-de-vie. Its colour ranges – depending on the strain of bacteria that forms the mould – from light salmon to pink, to light greenish blue.

Thirdly, there are whey cheeses. There is a long and glorious litany of such cheeses in Greece, made from the residuals after the production of *feta, graviera* or other hard semi-hard cheeses. Some whey cheeses are fresh; others are dried and eaten as grating cheeses or with fruit as table cheeses. The main cheeses in this category are: *Manouri,* a delicious, creamy, log-shaped cheese made with sheep's milk whey and the addition of fresh milk. *Manouri* is generally made on the Greek mainland and in western Macedonia. Some of the best comes from Kozani. It is eaten as a table cheese and is delicious with honey and fruit. Then there are *myzithra* and *anthotyro.* Generally, these cheeses are made by heating the whey until the curds rise to the top, at which point they are collected, and drained, and salted lightly. *Anthotyro* is richer than *myzithra* because whole milk and sometimes cream are added to the whey as it simmers. *Anthotyro* is usually drained in baskets, which gives the cheese its characteristic, fez-like shape, and as it ages it develops a natural rind. *Myzithra* is usually drained in cheesecloth. As it ages it becomes a rock-hard, alabaster-white ball, which is usually savoured for grating. In some parts of Greece, especially Crete and some of the other Aegean islands, fresh *myzithra* is salted and pressed and then left to ferment slightly. The result is a creamy, spreadable cheese called *xinomyzithra,* or 'sour *myzithra*'.

The fourth category is brine cheeses. These are, of course, the white cheeses that are aged and preserved in salt brine. King among them is *feta,* which is made with either sheep's or goat's milk, or a combination of the two. *Feta* is kept either in barrels or large tins. There are other brine cheeses, too. Among them: *telemes*, a speciality of Thrace, in the north-eastern corner of Greece, which is like *feta* except that it is made exclusively with cow's milk and always aged in tins; *sfela*, a speciality of the Peloponnese, especially the area around Kalamata, which is made with sheep's or goat's milk, shaped into thin bricks and almost always aged in tins. It is much harder and saltier than *feta*. Finally there is the *batsios* of western Macedonia, another hard white brine cheese made from sheep's or goat's milk whey. It is low in fat and has a sour, sharp taste. Unlike *feta* or *telemes* or *sfela*, *batsios* is filled with air pockets and is often floured and pan-fried.

Soft table cheeses come next. The list is too long to detail here. Suffice it to say that soft table cheeses are made in every corner of Greece and that they vary tremendously from place to place. They are young cheeses, usually not more than a few months old. When left to age, they become different cheeses, either hard table or grating cheeses.

Then there are the semi-hard cheeses. There are at least half a dozen semi-hard table cheeses, too many to detail here. Probably the best-known among them is *kasseri* and the smoked cheese *Metsovone*, from Metsovo in Epirus. Other regional semi-hard cheeses include the cylindrically shaped *formaella*, which has a mild flavour and a chewy texture not unlike that of *mozzarella*, and the Cypriot *haloumi*.

Hard cheeses are the final group. *Graviera*, from Crete, Tinos, Naxos, and Agrafa among other places, made with either sheep's milk or cow's milk; *kefalotyri*, the oldest Greek hard cheese, made in many parts of the country; *vlachotyri*, from Epirus; *ladotyri* from Lesvos and Zakynthos; *kefalograviera*, again made in many parts of Greece. These and other hard Greek cheeses vary tremendously. Since most are produced on a large scale, I have opted to omit a detailed description of them.

My favourite cheeses – cheeses steeped in wine

It was in a small *cafeneio* in Appolonia, Sifnos that I first discovered that Greeks do something with cheese that no one else in the world, to the best of my knowledge, is known to do: as a way to preserve it, they steep their little heads of sharp goat's and sheep's milk cheese in lees, in must or in wine. There we were, having the usual non-locals' platter of mixed *mezedes* when an avuncular old man, cane, moustache, baggy clothes and all, ordered a little *ghilomeno* with his *raki*. It was the smell that first seduced me. Strong and sweet and singular. Then it was the looks of the thing: round and about two inches high, ridged from the basket in which it had been drained, and covered with an inky pigment. 'What's that?' I asked Margarita, the proprietress. 'Oh, that. You wouldn't like that. It's something that only Sifniotes like.' I went back to Athens stinking like an unscrubbed wine-barrel, having talked Margarita into parting with six whole heads and having packed them in my suitcase. My life was never the same after that.

It takes a leap of fate to dunk small heads of precious home-made cheese into the thick black mud at the bottom of the wine barrel, but it was a jump taken long ago.

Lees-steeped cheeses go by different names. In Sifnos, where the best-known are to be found, the cheese is called either by its genteel name, *manoura* (nothing to do with *manouri*) or by its local name *ghilomeno,* the word for wine dregs in the island's dialect. The cheese isn't exclusive to Sifnos. A very similar, although somewhat milder, *ghilomeno* is also made in Folegandros and I would bet in a few other of the Cyclades, too. In Folegandros it is sold both with the wine-dregs still clinging to it and scraped clean and gentrified for urban palates.

In Santorini, the practice of preserving aged *chloro* (the local cheese) in the wine dregs is not a tradition, but at least one island gastronome, a restaurateur named Yiorgos Hatziyiannakis (Seline Restaurant), took the bold step of 'marinating' local goat's milk cheese in the sweet golden mud at the bottom of the Vinsanto barrels. The result was a cheese a little at odds with itself, salty and sweet at once, but nice.

In island cultures, habits travel fast. Wine-soaked cheeses are not unique to the Cyclades. We encounter them in the Dodecanese as well. The cheese of Leros – known locally as simply *Leriko tyro* – is a hard *myzithra* also preserved in wine mud, but here the technique for preserving it differs from that of other islands. For one, the milk itself used to make the *myzithra* is mixed with a little sea water for flavour before the cheese is set. Then, once salted and drained in the requisite baskets, then air-dried to the proper consistency, Leros' cheese is rubbed with a salve of wine dregs, savory, pepper and oil. The cheese, thus swaddled, is left to mature and then used almost exclusively as a grating cheese, over fresh pasta and accentuated by a sprinkling of dried mint.

The two other Dodecanese islands known for their wine-aged cheeses are Kos and Nyssiros, but in both places the cheeses are entirely different from the dried, aged dregs-steeped cheeses of the Cyclades.

The wine-steeped cheeses of Kos and Nyssiros are called either *kokkinotyri, possia* or *krassotyri.* I had gone looking for these, too, and was surprised that even in touristy Kos local producers still make this regional speciality.

Whereas in the Cyclades the cheese is usually made into small round heads, in the Dodecanese it is usually made into oblong, log-shaped loaves. *Krassotyri* is almost always a goat's milk cheese, but in Kos, where it is also made and sold commercially by several large local producers, cow's milk is also used. Typically, unpasteurized goat's milk is set with rennet, strained in a long, cylindrical straw mould for two to three days, sun-dried for another day, then placed in brine for about a week. After that, the cheese is sun-dried again for several days and, finally, steeped in another, unusual, brine: boiled vinegar and salt, or a combination of boiled wine (or the lees), vinegar, and salt. It acquires a deep brick-coloured exterior as it steeps, but the colour doesn't exactly come from the boiled wine or wine and vinegar mix. It comes from a handful of sea-washed pieces of pine bark that are boiled along with the wine mixture and that impart a deep red colour.

The cheese is delicious, and of all the wine-steeped cheeses in Greece the *krassotyri* has the most pronounced winy flavour. The commercial variety tends to be uniform and semi-hard, but if you stumble upon the artisanally made cheese it will be quite different, smaller and harder and more strongly flavoured. In Kos, the village of Kefalos on the far end of the island, away from the capital, is the place to go looking for the real thing.

A cheese lapped by the sea

Limnos, June 1998. The ride over the pebbles to the end of a beach near Kaminia in Petros Honas' Skoda was bumpy, to say the least. But this was a reconnaissance mission for the one thing that had eluded me more than once on Vulcan's island: the *melipasto*, a kind of young *feta* which instead of being preserved in brine is washed in the sea and dried in the sun.

I had been intrigued ever since I first heard of the cheese several years ago. Once on the island, I had chased the process from cheese-maker to cheese-maker, watching the beginning, when the curds are set, in the village of Thanos, at the home of one Kyria Foteini, who spends an intense month each summer making about five or six kilos of cheese a day in her kitchen so that her large family will have enough to last all winter long. I had witnessed stage two at a home-grown facility in Karyolakas, near Kaspako. There, the little rounds of *melipasto* – or *melichloro*, as it also called – were sitting in *kafasia* (makeshift wooden cupboards hung on a post outside), where they would stay covered but in the sun to mature for about two weeks. The washing is the last stage before *melipasto* is ready to eat. We stopped short on the rocks. 'Take your camera,' said Petros.

There she was, Kyria Astera, kerchief'd against the sun and wet from the knees down, bending into the sea. Brush in one hand, cheese in the other, one by one she scrubbed the fat little disks of her *melipasto* that were afloat in an orange crate wedged between some rocks. Others, already brushed with sea water, were drying on a towel. Watching her I realized I was witnessing something which probably hadn't changed in a few thousand years.

Touloumi, or skin-aged cheese

To the uninitiated, the sight of a 160-odd pound old goatskin spread bellyside up and inside out, and burgeoning with chunks of soft white cheese, is startling at best. But *touloumi*, for all its eyebrow-raising possibilities, is, in fact, one of the oldest and rarest cheeses in Greece. According to some sources it is the forerunner of feta. The cheese, rich and pleasantly sour, is named for the goatskin in which it is aged. Ikaria, Samos, and Mytillene all produce *touloumi*, and in each island the flavor of the cheese is quite distinct.

The skin has to be cured first. It is salted with about six pounds of coarse salt, then folded and weighted down for about two weeks. After that, it is washed and shaven, the feet tied together, and the bag turned inside out, so that the hairy part is on the inside. This is the strangest thing about *touloumotiri*. One local cheese-maker in Samos explained that otherwise the skin would sweat, which would keep the cheese from drying properly.

The cheese that goes into the bags is made with unpasteurized sheep's or goat's milk, which is coagulated, cut, drained and salted just like *feta*. The chunks of cheese that are cured in the bag vary in size depending on how wide the opening is. The secret to making *touloumotiri* is to fill the bag without leaving any empty spaces between the pieces. Sometimes soft *mizithra* is added to the bag, to help fill it in.

Touloumi is usually exceptionally pungent and peppery and quite often as it ages over the winter, it develops a pinkish (edible) mold. Most *touloumi* ages for at least three months. It is classified as a soft cheese in Greece and has to have a maximum moisture content of 56 per cent.

(The last two sections are taken from my forthcoming book *The Glorious Foods of Greece, Region by Region*, published by William Morrow.)

Fresh From the Cow's Nest: Condensed Milk and Culinary Innovation

Rachel Laudan

A small boy comes across a pile of condensed milk cans. He turns to his mother and says, 'Look Mother, there's a cow's nest'. That, at least, is the joke among anthropologists.

The cow's nest and its contents – evaporated milk, sweetened condensed milk, and dried milk powder – are not looked on favourably by temperate-land foodies.[1] Condensed milk is scorned as an unacceptable alternative to fresh cream, one of a flood of substitute foods invented in the late nineteenth and early twentieth century, as bad as margarine (spurious butter), Spam (second-rate ham), custard powder (a feeble imitation of egg custard), and canned vienna sausages (rubbery stand-ins for real German frankfurters).

First, a little history to understand why in England and the United States condensed milk is held in such contempt. In part, of course, it is simply that it comes in a can. We live in a period where fresh and natural are the watch-words and foods in cans are the mark of our unenlightened mothers or grand-mothers and of the benighted poor. In part, though, it is because condensed milk came in on a wave of enthusiasm for fresh milk and failed to live up to expectations.

That there was such enthusiasm for fresh milk in the nineteenth century is interesting in itself. Milk is a wonderful substance and all that, but it does have its disadvantages. So rich that all kinds of disagreeable organisms like to breed in it, it goes bad in a matter of days or hours depending on the temperature, and indeed without high standards of cleanliness can be down-right dangerous. Besides it is so obviously the excretion of the cow. Growing up on a dairy farm, I would never have dreamt of a nice glass of milk fresh from the cow. A disguise of tea or cocoa, a milk pudding or a béchamel sauce was absolutely necessary. In most parts of the world for most of history, people have wisely avoided fresh milk, instead promptly transforming it into cheese or yoghurt or butter, for example.

One striking exception to this general rule is to be found in the late nine-teenth century in north-west Europe and the countries settled by peoples of north-west European extraction: Australia, New Zealand, Canada, the United States. There fresh milk was elevated to the very peak of the alimentary pyramid. 'Milk', proclaimed the editor of my aunt's 1888 edition of Mrs. Beeton, 'is the most complete of all articles of food.... It contains all the elements that enter into the composition of the human body.'[2] Scientists, such as John Lawes and Joseph Gilbert, at the pioneering Rothamstead Research Station, opined that the three great principles that gave stamina – saccharina, oleosa, and albuminosa – could all be found in milk. Louis Pasteur suggested a way of making it safer by pasteurization, a technique that was gradually adopted in the first part of the twentieth century. Railway lines meant that milk could be sped from the dairy to a distant town at rates unheard of before. By 1900, 5,000,000 Londoners were consuming 60,000,000 gallons of milk annually, a quart a week per person, not much by today's standards but a huge increase over earlier consumption.[3] In 1926, Americans were consuming a third more dairy products than they had just seven years earlier.[4]

What an incentive to invent a substitute for fresh milk for reasons of profit, philanthropy or a judicious combination of the two. After all, all kinds of people – overlanders making their way to the West, soldiers on campaign, sailors aboard ship – were out of reach of this precious fluid. In 1856, Gail Borden in the United States patented a process for condensing and canning milk so that it would have a long, sanitary shelf life and quickly went in to business. His timing was just right; canned condensed milk was taken up by the military in the American Civil War.[5]

A decade later, the Anglo-Swiss Condensed Milk Company opened the first condensery in Europe. By the 1880s, it was turning out 25 million cans a year. Before long, condenseries were springing up everywhere, especially where, as in Switzerland or Wisconsin, cows produced more milk than could be sold fresh. On the opposite side of the world, the settlement of remote south Queensland in Australia led J. H. McConnel to set up a condensery at Toogoolawah in the 1890s. In 1907, Anglo-Swiss, now merged with Nestlé, took over the Toogoolawah condensery and set up warehouses in Singapore, Bombay and Hong Kong.[6] World War I further spurred the growth of condenseries. Nestlé bought several in the United States to keep up with demand for government contracts in Europe, where fresh milk went directly to towns. Condensed milk was now big business.

One of its greatest growth areas was in milk for babies and young children; Henri Nestlé had invented his infant formula in 1867.[7] With the germ theory of disease the latest thing in medicine, a host of experts concerned with public health, from doctors and nurses to home economists urged cleanliness and sterility, including clean and sterile canned milk, as the way to good health. In the 1910s and 1920s, the big companies began marketing aggressively to mothers and children, first in Europe and the United States and then in tropical Asia, Africa, and Latin America. In 1924, the Borden Company produced a text for school children, *Nutrition and Health: With Twenty Suggested Lessons for Nutrition Classes*, including one class devoted to Eagle Brand condensed milk. Fierce merger wars began and from them emerged the huge corporations that now dominate the world market. Nestlé is the world's largest food company.

By the 1960s, condensed milk was on the way out. Soldiers in two wars had grown sick of what they dubbed the Armoured Cow. Housewives accustomed to fresh milk rejected the thick unctuousness of sweetened condensed milk and were tired of trying to reconstitute powdered milk (the new technology, developed in the 1950s, that made this easier came too late). Mothers heard of books called *Milk and Murder* and became reluctant to use canned or powdered milk under any circumstances, let alone as their babies' major food.[8]

In the 1970s, the big companies turned their attention to ultrapasteurized milk and yoghurt. Ultrapasteurized tasted much closer to fresh than canned or dried had ever done. Yoghurt, a strange new taste, was sold by adding sugar and arguing that it really was good for you. It was preserved not by chemistry and machines but by natural processes. Consumers gained a vague notion that people who consumed lots of yoghurt, people who lived in villages somewhere in the Balkans or the Caucasus or other remote areas, lived to tremendous ages. No matter that the commercial yoghurt was thickened with gums and gelatines, sweetened with fruits to make their taste more acceptable, and, on careful inspection of the label, often made out of powdered milk anyway, it was yoghurt that seemed the way of the future. Yoghurts spread across the shelves of the world, first in the advanced countries, then elsewhere. In Mexico now, sweetened yoghurt is well on its way to replacing the traditional *atole* (thick corn drink) as a quick breakfast.

A quick search on the web confirms just how low condensed milk has sunk. Borden's *Nutrition and Health* can be downloaded from a website entitled 'Wall O' Shame'. Dried milk is used commonly as an ingredient in home-made dog biscuits. Its other destination is a home-made version of the currently fashionable milk paints (pastel, slightly chalky colours like those of the nineteenth century). Condensed milk is apparently made at home by that curious group of people who swop recipes for imitations of industrial foods: Kentucky fried chicken batter, McDonald's special sauce, and Bisquick. (The recipe for condensed milk is a cup of skimmed milk powder, a third of a cup of boiling water, two thirds of a cup of sugar and a tablespoon of margarine by the way.)[9] Armed with the grocery-store variety or this cheaper product, you can proceed to green slime. Add a little cornstarch, a smidgen of green food colouring, and condensed milk becomes an edible finger paint to use at Halloween should you have the urge to disguise yourself as a creature from the film *The Exorcist*.[10]

That, at least, is the situation in the United States and Great Britain. In many other parts of the world, though, condensed milk is honoured for its own virtues. Many of these lands do not have a tradition of eating dairy products, so that they come to condensed milk without the presuppositions born of long familiarity with fresh milk. Other areas have dairy traditions but have used their milk in ways largely unknown in north-west Europe. Most are semi-tropical or tropical where a can on the shelf is still more reliable than the 'fresh' milk hawked door to door on the back of a donkey or in a pick-up truck. Cooks there have discovered that condensed milk has some very interesting culinary properties. Most of these, naturally enough, are in sweets, desserts and pastries simply because condensing increases the proportion of milk sugars, even when additional sugar is not added.

To start with the simplest uses, many people find condensed milk delicious by itself, sweet unctuousness not being so easily come by in many parts of the world.[11] In Hawaii, people of all classes and races pour it over unsweetened ship's biscuit or crackers and leave them to soften. Father Damien allowed himself these as a treat when he worked with the lepers on the bleak, inaccessible peninsula of Kalaupapa. Sue Yim, editor of the food page of the *Honolulu Star-Bulletin*, was amazed as a child that something so simple could be so delicious when her grandfather made them for her. The Filipinos have their own version. Toast white bread, dip in condensed milk, and sprinkle with grated coconut and cinnamon.[12] This love of straight condensed milk is not limited to the tropics, I might add. An Oxford Symposiast, expert on the cooking of the Italian Renaissance, told me that every Christmas he treats himself to a whole can of condensed milk, dipping his finger in the can and licking it until every last drop has gone.

Almost equally simple is condensed milk frozen to make a simple dessert. Again in Hawaii, plantation workers mixed a can of condensed milk, a can of syrup, and five cans of water and froze it in aluminium ice cube trays. The cubes (called ice cake) slipped easily down a throat sore from hours in a dusty, scratchy field of sugar or pineapple.

Only slightly more complicated is the use of condensed milk to sweeten and soften drinks. In India, it goes into hot, spiced tea. In Thailand it goes into iced coffee and into fruit shakes. With avocado it makes a drink as rich and suave as heavy cream. I have no idea about the commercial version, but one of the favourite underground recipes of the 1970s in the United States was for homemade Irish Cream Liqueur. It saw the light of day in the early 1980s in one of

Maida Heatter's marvellous dessert cookbooks. A combination of a can of evaporated milk with a nearly equal proportion of Irish whiskey, three or four eggs, and extracts of vanilla, coconut, coffee and chocolate (a formula purportedly passed on to a wine-tasting group in San Diego by a lady from Ireland), defeated the real thing in a blind tasting to the surprise of the originally sceptical Heatter.[13]

For the Chinese, the fact that condensed milks don't taste like fresh milk, far from being a curse, is a blessing. A common dish in Canton and Hong Kong is 'fried milk'. Evaporated milk, or a mixture of evaporated and fresh milk, is thickened with cornstarch and cooled so that it forms a creamy solid. Both the basic idea and the name must have come originally from Macao where the Portuguese make the popular Iberian dessert *leche frita*, a paste of milk, flour, sugar and egg yolks, cooled, breaded and deep-fried. The Chinese, though, give this classic a new twist. They add pieces of the creamy solid to stir fries or they flavour it with salty minced ham, cut it in diamonds, dust with flour, and deep-fry it to accompany spare ribs or to eat with sugar.[14]

Cooks in the tropics discovered that condensed milk sets to a rich, satisfying cream when it is mixed with acids, opening up all kinds of possibilities for creamy fruit desserts. The type example of these, at least in the United States, is key lime pie. It takes its name from the Florida Keys, the string of islands that stretch south from the tip of Florida into the Caribbean. Cream pies were an American staple in the 1930s, a pie crust filled with an egg custard flavoured with vanilla or banana or coconut, and topped with whipped cream. In the Keys, though, without either refrigerators or dairy cows, a pie that could be made without fresh milk and with a meringue topping was welcome.[15] Lime juice from the typical small lime of the Keys (*C. aurantifolia,* though any lime or lemon will do) was mixed with condensed milk and poured into the crust. A few hours later creamy tangy pie was ready to eat. The principle, though, is used much more widely. In India and Latin America, for example, all kinds of tropical fruit are pureed with condensed milk to make what we would call fools or mousses.

Indeed, many of the most interesting uses of condensed milk come from India and Latin America. Unbeknownst to Gail Borden, Henri Nestlé and other pioneers, condensed milk was not, in fact, their invention. It was a way of preserving milk that had been practised for centuries in certain parts of the world, though rarely in Europe.[16] In fourteenth- and fifteenth-century India, the cooks at the royal courts of the Hindu kingdoms boiled down milk to produce a series of ever thicker substances until a solid milk fudge (now called *khoya*) was obtained. All the products were flavoured with fragrant flowers or fruits and eaten as sweetmeats.[17] In the late sixteenth century, the Mughal Emperor Akbar enjoyed *kulfi*, an ice-cream made by freezing khoya with pistachios. Traditionally these milk products were the basis of many sweet dishes and even some vegetarian ones. Grains of milk fudge are fried in the valley of Kashmir, then cooked with green peas in a spiced tomato gravy to make an equivalent to *keema matar* made with ground meat.[18]

Across Latin America, milk, usually with sugar added, has been boiled down to the point where it turns a caramel colour and just pours since the beginning of the colonial period.[19] Presumably this is a technique that came from Iberia and perhaps ultimately from the Islamic world, though this is not a path I have had time to trace. In most places, the condensed milk is called *dulce de leche*. In Mexico, it is now almost always called *cajeta* (after the little boxes in which it was, and sometimes still is, sold), though in the nineteenth century, it was part of the families of *antes* and *postres*. Boiled down further, again with sugar, it becomes one of the family of *jamoncillos* (sweets resembling little hams, a resemblance that escapes me).[20]

When the condensed milk companies began producing all those millions of cans, they had to find somewhere to sell them. The obvious market was the tropics, particularly the parts of the tropics that already had a tradition of condensing milk and liked the product and the flavour. As early as the late 1930s, India was importing 24,000 hundredweight each of whole and skimmed condensed milk.[21] By 1947/48, it was 105,000 hundredweight of skimmed condensed milk and nearly 100,000 hundredweight of milk powder. Not all this went into infant formula though that is the use we hear about. Much of it was seized on as a handy substitute for traditional milk products, just as it was in Latin America.

Recipes developed in these countries are actively promoted by the condensed milk companies. Grocery check-out lines in Mexico right now display the handsome, hard back, full colour *Sweet Moments with the Milkmaid (Dulces Momentos con la Lechera*, Lechera being the name of Nestlé condensed milk in that part of the world).[22] Housewives since the 1930s have been happy to play their part by sending in entries to competitions. The competition for new recipes run by Nestlé in Mexico in September 1998 netted 15,000 entries. The recipes get around too. I had thought that the marketing departments of the multinationals met at intervals to swap and adapt the mountain of recipes sent in by their competitors, though Josué de la Maza, of Nestlé, Mexico, tells me that it is rare for this to happen.[23]

In India today full cream dried milk powder, made by the roller drying not spray drying method, is much used instead of khoya.[24] It is used for all the many flavours of milk fudge (*burfee*), for a variety of cakes, balls and dumplings, and for many of the sweets in syrup, such as *gulab jamun*, that staple of overseas Indian restaurants.

In Latin America, condensed milk is turned into an easy home-made version of *dulce de leche* or *cajeta* by immersing the can in boiling water for four or five hours. (The technique is now being promoted in Britain by BBC cookshow host, Gary Rhodes. He describes the result as 'toffee cream' and suggests using it for crumbles, creams, ice creams and sauces.)[25] Condensed milk is also the basis of *flan magico*. Some time in the 1950s or 1960s, an inventive soul tried making crème caramel with condensed milk and popping it in the pressure cooker instead of in a *bain marie*. (Pressure cookers are very popular in Latin America; large populations live at very high altitudes making cooking very slow, and expensive as well because so much fuel is required.) The resulting *flan* was ready in just fifteen minutes. It has become the dessert of choice across Latin America. Rich and delicious, it is robust enough to be flavoured with pear, sweet potato, mango, sesame and a host of other flavourings including the best of all, coconut. Flan, though, is now being challenged for first place in the dessert line by *pastel de tres leches* (three milk cake), a sponge cake soaked in a syrup of fresh, evaporated and condensed milk.

To my mind, these constitute a worthy panoply of recipes using condensed milk. Why are they not more widely known? Because, I believe, a cookbook author writing for audiences in the United States or England is going to think long and hard before admitting that the cuisine they are promoting makes abundant use of condensed milk. Indeed, I would bet that many recipes using milk or cream in books on Asian or Latin American cooking for the English or US market are back adaptations of recipes originally using condensed milk. Rick Bayless, in his prizewinning *Mexican Cooking*, gives a recipe for what he calls 'celebration cake' (*pastel de tres leches*). He explains that 'the classic "three milks" from the title are sweetened condensed milk, evaporated milk and whole milk; I've replaced the first with *cajeta* (the goat milk caramel so famous in Mexico), the last with heavy cream (to give it a little more custardiness).'[26] Presumably he suspected that the kind of person who shells out $40 for a book on authentic Mexican cuisine was not going to be happy finding recipes for condensed milk. Only the brave and the confident leave these ingredients in, though you will find a steamed yoghurt, coconut pistachio sweetmeat, and a mango pudding in Madhur Jaffrey's *Flavors of India* (1995).

Meanwhile in the Asian Pacific region, the demand for condensed milk is rising again. Production within the region is now 1,500 tons a day or 3.8 million cans.[27] India exports condensed milks with names that carry on the grand tradition: Rosa, Happy Cow, Super One, Queen Cow, and adds Golden Dragon for good measure.[28] New kinds of condensed milks are available, most notably condensed soy milk and condensed coconut milk.[29] Is it too much to hope that we might one day have condensed almond milk? Or donkey milk? Or even yak milk?

It is impossible to resist drawing a few morals. First, the oft-repeated remark that in the past preserving methods created interesting new textures and flavours in a way that new methods (freezing and canning) do not, is simply not true. Condensed milk is just one of many canned products that has a texture and flavour different from the fresh product. Each new product has to be judged on its merits.

Second, dismissing new foodstuffs as mere imitations or poor substitutes is a mistake. Of course new inventions are likely to start out life as imitations. Our imaginations, after all, are limited. Aluminium was originally imitation silver, sugar was imitation honey, and I'd be prepared to bet that *tofu* was originally cheap and nasty *paneer* for Buddhist missionaries from India. But just as aluminium came into its own when it turned out to be more useful for airplane bodies and food wrapping than for tea services for the Czarina of Russia, and sugar turned out to have interesting properties that honey did not when heated or mixed with other ingredients, so too condensed milk comes into its own when it is not thought of as second-rate fresh milk.

Third, discovering uses for new foodstuffs takes a while. We are still discovering new uses for sugar and for soybeans after hundreds of years. Condensed milk has time. Time, that is, if we don't relegate it to dog biscuits and green slime, and miss out on a new culinary ingredient fresh from the cow's nest.

REFERENCES

[1] .The three major kinds of preserved milk – evaporated, sweetened and condensed, and dried – all depend on extracting water from milk (condensed milk adds sugar too). Hence I shall refer to them collectively as 'condensed milk'.

[2] *Mrs. Beeton's Book of Household Management* (London: Ward, Lock & Co., 1988), 1156.

[3] William Fream. Dairy and Dairy-Farming. Quoted from the British Dairy Farmer's Association, who in turn quoted it from a magazine called *Household Words*. Encyclopedia Britannica. 11th edition. Volume 7 1910-1911.

[4] Elmer Verner McCollum, *A History of Nutrition* (Boston: Houghton Mifflin, 1957), 120. He credits this to government urgings about diet.

[5] Joe B. Franz, *Gail Borden, Dairy Maker to a Nation* (Norman, Oklahoma: University of Oklahoma Press, 1951); Martin L. Bell, *A Portrait of Progress: The Business History of the Pet Milk Company from 1885 to 1960* (St. Louis, 1962); Jean Heer, *Nestlé: Ciento Veinticinco Años de 1866 a 1991* (Vévey, Switzerland: Nestlé, 1991). Some information may be gleaned from Harvey Levenstein, *Revolution at the Table: The Transformation of the American Diet* (Oxford: Oxford University Press, 1988), chs. 10 and 12.

[6] www.ozemail.com.au/~gmcbryde.toogool.html

7 www.nestle.com

8 Dr Cicely Williams drew attention to the problems associated with the use of infant formula in *Milk and Murder* (1939). Naomi Baumslag, *Milk, Money and Madness: The Culture and Politics of Breastfeeding* (Westport, Conn.: Bergin & Garvey, 1995); Valerie A. Fildes, *Breasts, Bottles and Babies: A History of Infant Feeding* (Edinburgh: Edinburgh University Press, 1986).

9 ftp.neosoft.com/recipes/eggs-dairy/condensed milk

10 php.iupui.edu/~kcarmony/slime.html

11 Rachel Laudan, *The Food of Paradise* (Honolulu, Hawaii: University of Hawaii Press), 61-65.

12 Barbara MacPherson, personal communication.

13 *Maida Heatter's New Book of Great Desserts* (New York: Knopf, 1980), ix.

14 Ken Hom, *The Taste of China* (London: Macmillan, 1992 for 1990), 66; *Fragrant Harbor Taste* (New York: Fireside, 1992 for 1989); Maite Manjon, *The Gastronomy of Spain and Portugal* (New York: Prentice Hall, 1990), 174.

15 That, at least, is the story told by Maida Heatter who has lived in Florida for many years. I have been unable to confirm it elsewhere. *Heatter,* 169.

16 Though I do have a French recipe for a *confiture de lait,* made by boiling 3 litres of milk with 2.5 kilograms of sugar and a pod of vanilla until it is the consistency of honey and the colour of creme caramel. Suzanne Fonteneau, *Les Confitures a l'Ancienne* (Paris: Dargaud, 1988 for 1979), 91.

17 K.T. Achaya, *Indian Food: A Historical Companion* (New Delhi: Oxford University Press, 1994), 94, 116, 132.

18 Julie Sahni, *Classic Indian Cooking* (New York: Morrow), 1980, 55-56.

19 *Libro de cocina. Convento de San Jerónimo. Selección y transcripción atribuidas a Sor Juana Inés de la Cruz* (Mexico: Instituto Mexiquense de Cultura, 1996). *Recetario de Dominga de Guzmán,* ed. Guadalupe Pérez San Vicente (Mexico: Sanborns, 1996). Beatriz Rossels, *La gastronomía en Potosí y Charcas, Siglos XVIII y XIX, 800 recetas de la cocina criolla* (La Paz: Editora 'Khana Cruz,' 1995).

20 *Nuevo Cocinero Mexicano en Forma de Diccionario* (Mexico: Porrúa, 1992 for 1888).

21 K.T. Achaya, *The Food Industries of British India* (Delhi: Oxford University Press, 1994), 78.

22 There are just too many of these to list. For Brazil, for example, there is the lavish, beautifully illustrated volume of over 300 pages with recipes using Nestlé products *O Doce Brasileiro* (Nestlé Industrial e Commercial Ltda, 1990).

23 Personal communication, 10 June, 1999.

24 Jack Santa Maria, *Indian Sweet Cookery* (Boulder: Shambhala, 1980), 26; Yamuna Devi, *Lord Krishna's Cuisine: The Art of Indian Vegetarian Cookery* (New York: Bala, 1987), 297.

25 *Great Food: Over 175 Recipes from Six of the World's Greatest Chefs* (Seattle, Washington: 175 West, 1998),178.

26 Rick Bayless, *Mexican Kitchen: Capturing the Vibrant Flavors of a World-Class Cuisine* (New York:Scribner, 1996), 394.

27 www.spv.com

28 www.indomilk.com

29 www.prosoya.ca

The Rise of the Cream Sauce, 1660–1760

Gilly Lehmann

When we think of milk and cream in cookery, we tend to associate it more with sweet dishes than with savoury ones, and when we look back to the cookery of the seventeenth and eighteenth centuries, what springs to mind is the creams, jellies and syllabubs which have already been explored by Peter Brears and Ivan Day.[1] But part of the seventeenth-century shift to modern cookery, as well as the separation of sweet and savoury, was the increased use of butter, and later cream, in savoury dishes. As Jean-Louis Flandrin has pointed out, medieval sauces tended to be fatless as well as sharp[2] or, in the fifteenth century, sweet and sour, a tendency which was more marked in England than in France.[3] By the end of the fifteenth century in France, and a little later in England, butter was beginning to be used in some of these sauces, often replacing bread as the thickening agent. Late sixteenth-century cookery-books show a mix of these two types of sauce: for instance, A.W.'s *Book of Cookrye* (1591), gives six receipts for sauce and 31 receipts for boiled and stewed meats served with sauce; of these, 78 per cent contain vinegar or verjuice, 76 per cent contain sugar and/or dried fruit; in 59 per cent the sharp and sweet elements are combined. Bread is still much used to thicken sauces; 41 per cent of the receipts contain bread as a thickener, but 41 per cent now contain butter.[4] Butter had the effect of tempering the acidity given by the vinegar or verjuice, and its use thus modified the flavour as well as the texture of the sauce. The use of butter marked the first phase of the elimination of medieval flavour combinations.

The next phase was the separation of the late-medieval and sixteenth-century sweet-and-savoury flavours, and this went hand in hand with an enormous increase in the use of butter. The separation of flavours, associated with the increased use of butter to smooth out acidity, began earlier in France than in England. The dearth of new cookery-books in France between 1560 and 1651, the date of La Varenne's *Cuisinier françois*, makes it difficult to say when this started, but one English book does provide some evidence of what was happening. John Murrell's *New Booke of Cookerie* (1615) contains a section in which most of the receipts are described as French (plus a few others scattered elsewhere), and another entitled 'London Cookerie'. In the next edition (1617), the latter is described as 'English Cookerie', and the title-page of the book indicates that the directions are set out in the 'now, new, English and French fashion'. A comparison of the two groups of receipts shows that the French were beginning to discard the sugar and dried fruit, still much more present in the English receipts (sugar in 47 per cent, dried fruit in 21 per cent of the French receipts, as against 57 per cent and 53 per cent of the English receipts); verjuice or vinegar, with pepper to season, plus butter added to the sauce are typical of the French receipts (in 89 per cent, 84 per cent, and 74 per cent of the receipts, compared to 43 per cent, 30 per cent, and 43 per cent of the English receipts).[5] These are the flavours which Sarah Peterson ascribes to the new French cuisine which was to usher in modern cooking in the seventeenth century.[6] What is notably absent is the use of salt (in only 16 per cent of the French receipts, and 20 per cent of the English ones), although Peterson sees salt as another characteristic of the culinary revolution. Butter is an important presence in the French receipts, but cream does not appear, except in one receipt where the cook is directed to thicken the sauce with rice boiled in cream.

By the time La Varenne published his receipts in 1651, French cookery had virtually eliminated the sweet element in savoury dishes, and the presence of acid and butter, while less marked than in the 'French' receipts given by John Murrell, is still considerable: 39 per cent of La Varenne's sauces contain butter.[7] In fact, La Varenne's method of thickening a sauce usually uses a roux made of flour and lard or melted bacon fat, a method given in the receipt to make a 'Liaison de farine' and used frequently in the receipts for entrées.[8] In these receipts, cream is never present, but it does make its appearance in the receipts for 'entremets' for flesh days, used for small raggoos of vegetables such as asparagus or peas.[9] The court cooks who followed La Varenne continued along the same lines, and the number of cream sauces gradually increased: in 1656 Pierre de Lune gave several receipts for cream sauces with vegetables, plus one for 'Poulets fricassés à la crème'.[10] French cookery seems to have used cream principally for small dishes of vegetables. But the use of butter continued to rise: in 1674 L.S.R.'s *L'Art de bien traiter* uses butter in 55 per cent of all its receipts and in 80 per cent of its sauces (but this is exceptionally high), and the 1777 edition of *La Cuisinière bourgeoise* 55 per cent of the sauces contain butter.[11]

In England, La Varenne's work was translated in 1653 and was widely copied, but the most popular book of the second half of the seventeenth century, *The Compleat Cook* (the cookery section of the three-part *Queens Closet Opened*, 1655), looks a little old-fashioned compared to La Varenne's manual. While many of the made dishes have the new flavours, there are also 11 per cent which still contain sugar, looking back to an earlier period. Butter is present in 37 per cent of the made dishes, but cream sauces were not taken up: a receipt where one might expect to find cream, 'To Fricate Champigneons' uses a lear of mutton gravy, egg yolks and citrus juice as a sauce for the mushrooms.[12] The only exception is the cream and egg yolk sauce for artichokes, but this is still a sweet dish, with sugar used twice during the cooking. Later books, such as Robert May's *Accomplisht Cook* (1660), are even more archaic: in three groups of receipts ('boiled meats', 'hashes', and 'pottages'), 24 per cent of the made dishes use sugar, 30 per cent dried fruit. But butter appears in 64 per cent of these receipts, and there are three per cent containing cream; the cream sauces are always associated with sugar.[13] May's book (and the same is true of other Restoration books, such as those by Rabisha and Hannah Woolley) looks back to cookery before the French culinary revolution: his 'French' receipts are not notable for the elimination of sweet and savoury combinations.

However, the manuscript receipt books of the Restoration period show that the separation of flavours was being adopted even while printed cookery books continued to cling to the old style. As the new flavours came in, the use of cream increased, and these MSS show cream being used with meat more often than in France, perhaps because vegetables were less fashionable in England. Rebecca Price's manuscript, begun in 1681, is remarkably modern in its flavours: sweet elements, in the form of sugar or dried fruit, are present in only six per cent of the fish and meat receipts, while butter appears in 73 per cent. The typical flavour combination in this manuscript is the association of anchovy and wine, tempered by the frequent use of butter. Cream sauces appear too, used for veal, rabbit and poultry.[14] Other manuscript collections of the period also show the importance of the anchovy-wine combination, and of the use of butter in sauces: examples are found in the MSS of Mary Watts and of Katherine Windham, but neither of these MSS gives any cream sauce.[15] By the turn of the century, however, cream sauces were becoming more frequent: Diana Astry's collection, which dates from the first decade of the eighteenth century, has a cream sauce for a 'white frigcacee [sic] of chicken', and there is one cream sauce, for 'White Scotch Collops', in Ann Blencowe's receipt book, dated 1694.[16] A somewhat later manuscript, belonging to Margaretta Acworth (1727/8–1794), has four cream sauces, all for chicken or veal.[17]

Cream sauces reached the printed cookery-books in the early years of the eighteenth century. As far as I know, the first receipts appeared in Henry Howard's book, *England's Newest Way* (1703); five such receipts appear in the second edition (1708), and they are in the early part of the text rather than in the 'additions' which were grafted on to the original text in this and subsequent editions.[18] Howard's book follows the trends of the ladies' MSS: in his receipts for meat, fish or vegetable in or with sauce, 56 per cent contain anchovy, 71 per cent wine. The flavour combinations are those of the Restoration, but sugar and dried fruit have entirely disappeared. Another court cook, Patrick Lamb, uses cream sauces for braised lettuce, and for chickens and asparagus.[19] The women authors were not far behind: in 1714 Mary Kettilby included three cream sauces in her book, and in 1727 E. Smith gave eleven, for items as diverse as soles, tripe, ox-palates, pig, and peas, as well as the usual chicken and veal.[20]

The result of the increased use of cream in sauces was to produce blander flavours. Often books gave receipts for brown and white fricassees, using the same meat: in E. Smith's book, the brown has the strong anchovy-wine flavouring, while the white is flavoured only by broth, and garnished with mushrooms or oysters. Many of these early cream sauces use an egg-yolk liaison to thicken the sauce, sometimes helped along by, or more often replaced by, 'a little bit of butter rolled in flour', the eighteenth-century term for 'beurre manié'. Of E. Smith's eleven sauces, five use flour and butter, five eggs; the sauce for the peas is simply reduced cream. Eighteenth-century 'white sauces' were not based on milk, but were made either with cream, or with butter, as in today's 'beurre blanc'. Although the inspiration for E. Smith's book is a modified version of court cookery, adapted to the potential of the domestic kitchen, the sharp increase in the use of cream anticipates the developments of 'nouvelle cuisine' in the 1730s and 1740s. The delicate sauces were more appropriate to the refinement of the small dishes of this style, as opposed to the strong meaty flavours of the grand dishes (olios, bisques and terrines) of court cookery.

Soon the term 'à la crème' began to be used in England, although at first in a very restricted context. In 1730 Charles Carter's glossary of technical terms includes 'A la Creame, i.e. to force Fowl, and dress with a Cream Sauce'. The receipt is very elaborate, with the roasted fowl half carved and boned, reassembled with the white meat used as a farce interleaved with layers of oysters, baked, and served on a cream and egg-yolk sauce enriched with a garnish of sweetbreads, cockscombs, morels, truffles, mushrooms and forcemeat balls, a typically court-style dish.[21] This receipt is not the only one in the book to use a cream sauce (there is a white fricassee, and cream sauces for turkey with celery, tench, oysters, salt fish, mushrooms, asparagus, and artichokes[22]), but it is the only one to mention the fact in the title. Since Carter tends to use French terms, the implication is that English cookery had already gone further than French cookery in its use of cream sauces.

French 'nouvelle cuisine', however, made use of them for its smaller side dishes. William Verral's cookery-book gives an idea of the dishes produced by one of the masters of the new style, Pierre Clouet. Clouet made elegant cream sauces with prawns for fish, and dishes of fowls and eggs were sauced 'a la

Benjamele', since Béchamel sauce started life made with cream rather than milk.[23] Later English books had as much difficulty as Verral with the term: Mary Smith gives a receipt for 'Polard Besh-a-mell'.[24] Although the name was garbled, the dish was not debased, and even when used as a sauce for leftovers, Béchamel still contained cream and egg, as in Charlotte Mason's receipt for 'Chickens hashed, called bichamele'.[25] In cookery less influenced by French terminology, the white and brown fricassees remained standard items in English cookery-books for the rest of the eighteenth century: in 1769, Elizabeth Raffald gave two such pairs, along with twenty-four other cream sauces, not to mention the cream added to bread sauces and onion sauces.[26] Everything was now a candidate for cream sauces: fish, meat, vegetables, and, in Raffald's book as in most others after 1760, leftovers.

The prevalence of cream sauces in the English cookery-books from the early eighteenth century onwards, especially in the books by women authors, suggests that the use of cream to sauce meat and fish as well as vegetables was as much a native development as an imitation of the French style. Two of the important practitioners of the French 'nouvelle cuisine' of the 1730s, Pierre Clouet and Vincent La Chapelle, both worked in England, the first for the Duke of Newcastle, the second for the Earl of Chesterfield, and their use of cream may well have been influenced by their observation of English cookery. As has already been noticed, E. Smith's cookery-book makes more extensive use of cream sauces than Charles Carter's, and Carter's cookery was essentially French. Elizabeth Raffald's book shows just how ubiquitous cream had become in gentry and even middle-class cookery.

With such a long history in English cookery, one might have thought that the cream sauce was here to stay, but the rot began to set in during the nineteenth century. Mrs Beeton gives two receipts for béchamel, one using stock, cream and arrowroot, the other water and milk, thickened with beurre manié and egg yolks; out of three receipts for white sauce, only one contains cream.[27] Gradually milk and flour thickening ousted cream, and wartime rationing finished the job. When Dorothy Hartley published her account of traditional cooking, *Food in England*, in 1954, she did not envisage using it for anything other than sweet dishes, saying that cream was now 'forgotten'; and in 1956 Constance Spry commented rather defensively that using cream in a chicken sauce might be thought extravagant, but 'this seems at least as good a way of enjoying cream as by using it to enrich cakes and gâteaux.'[28] Savoury cream sauces had disappeared, and have returned only recently albeit in a lighter form threatened, however, by the current trend towards healthier eating, one wonders how long the cream sauce can survive in contemporary English cookery. The eighteenth century really was the 'golden age'.

REFERENCES

1. See, for instance, Peter Brears' two-part essay on 'Transparent Pleasures' in *PPC* 53 and 54, and Ivan Day's essay on syllabubs in *PPC* 53.
2. See Flandrin's two essays, 'Pour une histoire du goût' (particularly p. 14) and 'Et le beurre conquit la France', in *L'histoire* 85 (1986), pp. 12–19, 108–111. See also the shorter version of the second essay, 'La cuisine au beurre', in Flandrin, *Chronique de Platine*, Paris: Odile Jacob, 1992, pp. 237–253.
3. See Bruno Laurioux, 'Spices in the Medieval Diet: A New Approach', *Food and Foodways* 1 (1985), pp. 45–47.
4. A.W., *A Book of Cookrye* (1591), facs. repr., Amsterdam: Theatrum Orbis Terrarum, 1976, ff. 3–14v.
5. John Murrell, *A New Booke of Cookerie* (1615), facs. repr., Amsterdam: Theatrum Orbis Terrarum, 1972, pp. 1–13, 48–9, 70–72 (French receipts), pp. 53–78 (English or 'London' receipts). For these statistics, I have excluded from both groups the receipts for sousing, which automatically contain acid and would thus give a false impression.
6. See Peterson, *Acquired Taste*, Ithaca/London: Cornell UP, 1994, pp. 183–86. Peterson claims that the salt-acid combination was the basic framework for the new cuisine, but is obliged to notice that 'it was not until well into the eighteenth century that salt per se (and not simply a pickle) began to be added to dishes as a matter of course' (p. 186).
7. See the table in the introduction by Jean-Louis Flandrin, Philip Hyman and Mary Hyman to La Varenne, *Le Cuisinier françois* (1651), repr. of the Bibliothèque Bleue ed., Paris: Montalba, 1983, p. 25.
8. See the liaison receipt, p. 198; the entrées are found pp. 135–158. The flour-based thickening is found in 33 per cent of the receipts, and an egg yolk liaison in 11 per cent.
9. Receipts 88 and 96, pp. 193, 194–5.
10. See P. de Lune, *Le Cuisinier* (1656), repr. in G. & L. Laurendon (eds), *L'Art de la cuisine française au XVIIe siècle*, Paris: Payot, 1995, pp. 275, 276, 278, 428, 429, 327. Pierre de Lune's 'sauce blanche', contrary to what one might expect, uses broth and an egg yolk liaison, but no cream.
11. See the tables in Flandrin, Hyman and Hyman, pp. 24–25.
12. See W.M., *The Compleat Cook* (1655), facs. repr. of 1671 ed., London: Prospect, 1984, p. 9. The made dishes are scattered throughout the text.
13. These receipts are found in May, *The Accomplisht Cook* (1660), facs. repr. of 1685 ed., Totnes: Prospect, 1994, pp. 1–18, 38–80
14. See Price, *The Compleat Cook*, ed. M. Masson, London: Routledge, 1974, pp. 92, 94, 94–95, 96. The fish receipts are pp. 62–70, the meat receipts pp. 79–97.
15. Mary Watts' MS, dating from the second half of the seventeenth century, is now in Kent RO (U49 F15); Katherine Windham's MS, unusually written out towards the end of her life rather than shortly before marriage, and begun in 1707, is now in Norfolk RO (WKC 6/457).
16. See B. Stitt (ed.), 'Diana Astry's Recipe Book', *Publications of the Bedfordshire Historical Record Society* 37 (1957), p. 93; *The Receipt Book of Mrs. Ann Blencowe, A.D. 1694*, London: Guy Chapman, 1925, p. 34. Interestingly, in both MSS these receipts are attributed to titled ladies.
17. Margaretta Acworth's MS is now PRO C107/108. The cream sauces are scattered throughout the MS, with one in the earliest group of receipts, which the modern editors of a selection from the MS suggest was copied out for her by her mother. See A. & F. Prochaska (eds), *Margaretta Acworth's Georgian Cookery Book*, London: Pavilion, 1987, p. 15. Although the receipts were collected over a long period, some of them date from the mid seventeenth century or even earlier, and none seems to be later than about 1750. Loose leaves contain dated receipts from the 1720s and 30s. I am grateful to Dr Amanda Bevan of the PRO for her comments on the handwriting in this MS.

[18] See, for instance, 'To Ragow a Neck of Veal', Howard, *England's Newest Way*, London: Chr. Coningsby, 1708, pp. 49–50. The receipts in or with sauce are pp. 26–62.
[19] Lamb, *Royal Cookery*, London: Maurice Atkins, 1710, pp. 86, 90.
[20] See E. Smith, *The Compleat Housewife* (1727), facs. repr. of ed. 16 (1758), London: Studio, 1994, pp. 42, 46–47, 48, 49, 51, 52, 65, 68, 78–79, 93–94, 95.
[21] Carter, *The Complete Practical Cook* (1730), facs. repr., London: Prospect, 1984, pp. 209 (glossary), 57 (receipt).
[22] Ibid., pp. 74, 43, 67, 69, 73, 75, 94.
[23] For the receipts, see A. Haly (ed.), *William Verrall's [sic] Cookery Book*, Lewes: Southover, 1988, pp. 41, 91, 111–12; there are other cream sauces, for instance for peas, cardoons, and lettuce pp. 115, 119, 120–21, and the Béchamel sauce is suggested as a variant several times. For the history of the term, see Claudine Brécourt-Villars, *Mots de table, Mots de bouche*, Paris: Stock, 1996, p. 58.
[24] M. Smith, *The Complete House-keeper, and Professed Cook*, Newcastle: T. Slack for the Author, 1772, p. 127.
[25] Mason, *The Lady's Assistant*, London: J. Walter, 1773, p. 242.
[26] See Raffald, *The Experienced English House-keeper* (1769), repr. ed. A. Bagnall, Lewes: Southover, 1997, pp. 49 and 70–71; also 10, 15, 16, 20, 27–28, 29, 30, 31–32, 37, 38, 44–45, 46, 50–51, 52, 53, 57, 62–63, 69–70, 143, 145, 146.
[27] I. Beeton, *The Book of Household Management* (1861), facs. repr. as *Beeton's Book of Household Management*, London: Chancellor, 1997, pp. 181–82, 256.
[28] See Hartley, *Food in England* (1954), London: Futura, 1985, pp. 470–76; Spry & Hume, *The Constance Spry Cookery Book* (1956), London: Dent, 1967, p. 1016.

Finnish-American Milk Products in the Northwoods

Yvonne R. Lockwood & William G. Lockwood

This paper concerns two milk products and their role in the lives of Finnish Americans in the Upper Peninsula of Michigan: *viili*, a type of clabbered milk, and *juusto*, a fresh cheese.

The state of Michigan consists of two great peninsulas defined by the Great Lakes. They were not connected until 1957 when they were finally joined by the Mackinac Bridge. Even today, they are distant and remote entities; it is a good twelve-hour drive from Detroit, Michigan's major city, to the furthest corner of the Upper Peninsula. Important concentrations of population, the major industrial sites and financial institutions, and the centre of state government are all located in the southern half of the Lower Peninsula. The Upper Peninsula was principally the location of extractive industries, first lumbering and, later, copper and iron mining. As these declined in the early post-World War II period, they were replaced by tourism which is currently the major industry of the region. It is largely an undeveloped area of dense woods and extensive wetlands where hunting and fishing still provide a significant part of the local diet.

Between the last decade of the nineteenth century through the first decades of the twentieth, Finns, as well as Italians, Poles, and South Slavs, immigrated to the Upper Peninsula in large numbers to work the mines. Many Finns, once they had established a foothold in America, purchased farms on marginal land left after loggers had cut away the trees, leaving stumps and debris. Some tried to survive by farming on these so-called 'stump farms' (cf. Alanen 1981; Gough 1997). It was said that with ten cows, one could make a living. However, many farmers had to work in mines and lumber mills several months a year to pay debts, leaving their wives to manage the agricultural work (Hoglund 1978: 9). In other cases farms provided subsistence to supplement ones regular job in the mines or lumber industry. Even those who settled in the small mining towns of the area usually kept a garden and cow, supplemented by hunting and fishing (in and out of season). Many small-town Finnish-American families kept a milk cow until the 1950s.

The majority of Michigan's Finnish population are now third- and fourth-generation Americans. It is the largest concentration of Finnish Americans in the United States. Nearly 112,000 or seventeen per cent of all Finnish Americans live in Michigan, the vast majority in the Upper Peninsula (*Finnish American Reporter*, November 1993: 7; March 2000: 10). With the exception of the oldest generation, they do not speak Finnish, other than perhaps a few bits of vocabulary having special symbolic importance. Most of the food they cook and eat is midwestern American. Perhaps the most distinctive foodstuff they call their own is pasty, learned from Cornish immigrants who preceded them in the mines. This was diffused by Finnish Americans until it is today a speciality of the region (Lockwood and Lockwood 1991, 1983). Pasty became so much a part of the local diet that some Finnish Americans came to consider it of Finnish origin. But the community has also maintained a few dishes of genuine Finnish origin. Among the most important of these are *viili* and *juusto*.

In Finland milk products are ubiquitous. After bread, milk and sour-milk products are the most important staples in the Finnish national diet (Käkönen 1974: viii). According to statistics, dairy accounts for two-thirds of Finnish agricultural income and the consumption of milk is almost 300 litres per person a year (Rajanen 1981: 104). A recent cursory survey of shops in Tamperi showed eight different kinds of milk, ten different types of buttermilk, nine different types of *viili*, and too many kinds of yoghurt to count (all with fruit or other additions). Both the emphasis on dairying and the high consumption of milk products are interesting, given that seventeen per cent of the Finnish national population is lactose intolerant, which is higher than for Europeans generally. For this reason, a good number of the milk products sold in Finnish markets are treated for lactase deficiency.[1] Most lactose intolerant adults can consume up to a pint of milk a day without serious repercussions. Moreover, the various types of cultured milk products are acceptable because the fermenting bacteria uses the lactose as fuel and the final product (cheese, yoghurt, *viili*) is practically lactose free

Viili is an ancient food and, like other fermented milk products, was a way of preserving milk for several days (cf. McGee 1984: 31–36). In Finland *viili* is still made in homes and is also produced commercially and sold in groceries and markets. Whereas milk itself is not advertised, commercials profile *viili* as a children's product with colourfully designed packaging to attract children. Mothers attest that *viili* quickly disappears when their children bring friends home. While considered especially appropriate for children, *viili* is also a popular adult snack food. Buttermilk commercials, on the other hand, target young single males and females who are independent, active, popular, and slim.

In the Finnish-American communities of northern Michigan, Wisconsin, and Minnesota – the largest density of Finnish Americans in the United States – milk and milk products seem to be more important in the diet than among non-Finnish Americans. Milk, sour cream, and butter are included in a great number of prepared dishes, e.g., in soups, stews, desserts, porridges, and breads (cf. Ojakangas 1964; Käkönen 1974; Kaplan, Hoover, Moore 1986: 144–162, 346–357; Larson 1976). Many prefer to drink milk (sweet or buttermilk) with their meals, whether eating roast beef, fish or soup, and in numerous homes, milk is served at every meal. However, unlike Finland, the choice of products treated for lactase deficiency is limited. Because this is a genetic deficiency, we can assume that the lactose intolerance rate among Finnish Americans is similar to that in Finland. *Viili*, as a cultured milk product, can be eaten freely even by those who are lactose intolerant.

Joyce Koskenmaki (1996) describes how her grandmother makes *viili*:

> Every morning she mixes one spoonful of *viili* with a cup of milk (wholemilk is best, she says, and no more than a spoonful per cup), puts it up in the cupboard in a warm place in a covered bowl or jar and lets it sit – about eight to 24 hours depending on the weather. It takes longer in the winter, shorter in the summer. After it is solid it can be refrigerated for up to a week.

It is widely believed that *viili* will separate in an electrical storm. When clabbered, it is chilled in the refrigerator before eating it with a spoon. *Viili* may also be whipped and drunk, which is more common in Finland than in the United States. When *viili* is made with unhomogenized milk, the cream rises to the top giving a dual texture of a creamy crust, which is a favourite part, and a sweet smooth filling. However, almost all milk offered in American markets is homogenized. It is available as nonfat, skimmed, 1 per cent cream, 2 per cent cream, and whole (4 per cent cream). It is illegal to sell raw milk, and milk that is only pasteurized and not homogenized is very difficult to find. In southeast Michigan, for example, the most densely populated region of the state, only one dairy sells a very limited amount of unhomogenized 'creamline' milk. In more remote areas, it is not available commercially.

Some cookbook writers refer to a 'long' and 'short' *viili* (Käkönen 1974: 187; Brown 1968: 139). Finnish nationals and Finnish Americans regard 'long' *viili* as the real thing. When a spoonful of long *viili* is lifted out of the bowl, then it stretches. 'At its best,' it is said, 'the...*viili* should be so elastic...[it] requires cutting with scissors' (Brown 1968: 139). One cookbook, in fact, describes it as 'elastic yoghurt' (Tanttu 1988: 24). It is this stringy, slippery texture that repulses some Finns (and many non-Finns) and which children compare with snot. The other type is 'short' *viili* – without any stretch and in texture, much like yoghurt. Ulla Käkönen, a Finnish-American cookbook writer, makes short *viili* by using buttermilk as her starter, but the result, she says, is not as tasty as 'long' viili (1974: 187). In most opinions, this is not *viili*. While there are varying degrees of stretch, 'short' *viili* in Finnish-American communities is not a recognized type. Some believe that the elastic quality increases the more the *viili* start is used. We suspect Käkönen's 'short' *viili* to be a substitute for those isolated Finnish Americans without access to a *viili* starter.

Among Finnish Americans, *viili* is widely known as 'feelia', an Americanization of the name for this most Finnish of Finnish foods.[2] Most first- to third-generation Finnish Americans love *viili*; they eat it as a snack or as a light meal, preferably with rye bread and butter and salted salmon. They usually eat it plain without sugar or fruit. Younger Finnish Americans (fourth and fifth generations) tend not to have this strong attachment to *viili*. Quite the contrary, many prefer yoghurt – the kind most available in the US is loaded with sugar and jam. While older generations talk about how much they like *viili* with descriptors such as enjoyable, flavourful, tasty – not flat tasting like yoghurt, tingly, yummy, healthy, and refreshing, the younger ones describe it as disgusting, the odour of vomit, slimy, snotty, and stringy. A third-generation woman provides a striking contrast to the fond descriptions of *viili*-lovers:

> I think *viili* is something you need to acquire a taste for. I'm not nuts about it, but can eat it. It seems quite bland, almost bitter to me. I remember being kind of intrigued by the thick 'skin' on the top. This was the cream portion of the milk that separated as it was forming; it kind of looked sick to me. The part underneath was much thinner, kind of stringy almost and would plop off your spoon if you didn't get enough mass on your spoon to cause it to break [from the rest of the *viili*]] (Szyszkoski 1999).

The origin of *viili* provides some of its mystique. One Finnish American asks, 'how do those bacteria know they're Finnish and make *viili* and not French yoghurt instead?' (*Finnish American Reporter*, January 1996: 3). Another explains that *viili* culture originated 'by letting the milk set in a freshly killed calf's stomach' (Lassila 1952). We have also heard that a special plant was used in making the first *viili* culture (cf. Ränk 1966). Origin of the original *viili*, however, is not important to the community; one always makes *viili* with a little of an earlier batch. What is important is that people who lose their starter, for whatever reason, know where to go for another or to one who can refer them to someone who should have it. Individuals take pride in their starters. Some are handed down from grandmothers and are highly prized and carefully tended.

This *viili* was brought from Finland by my grandmother at the turn of the century. To transport it over the long sea journey, she spread it on handkerchiefs and let them dry, then wrapped it up and carried it with her.

When she got to this country and again had access to cow's milk, she peeled the dried *viili* culture from her handkerchiefs and mixed it with the milk. This *viili* has been kept going for many years by her granddaughters. (Koskenmaki 1996)[3]

Others are valued because of their exceptionally good *viili* qualities, such as stretch and flavour. Because *viili* is not commercially available in the United States, people depend on each other for a starter when needed.

Viili is a very Finnish dish. Like the sauna, it is even regarded as having curative properties. A third-generation Finnish American attests that *viili*:

is known to be beneficial to digestion, especially to those whose genetic make-up is Finnish. It is especially good for gallstones (a common Finnish ailment) and can neutralize gastric distress. [My cousin says] daily consumption of *viili* can make a person live longer.' (Koskenmaki 1996)

Another testimonial affirms this as a shared belief:

My husband loves *viili*. About twenty-five years ago, my mother died and my dad brought all her *viili* ready made to my husband. He started eating it every day and cured his stomach. He was eating a few rolls of Tums every day. I did get raw milk for years and made it for him, but now we have no source of raw milk, so he just eats yogurt (store bought) and still has an o.k. stomach (B. Berg 1999).

Juusto is a sweet-milk cheese clotted with rennet, as opposed to the even older type of soft sour-milk cheese (cf. Ränk 1966). Traditionally *juusto* is made from colostrum, the rich first milk of a cow that has calved; although rare today this *juusto* is still regarded as the best. More commonly, commercial rennet is added to raw milk that is slightly warm. When the milk jells, it is sliced through, the whey poured off, and the milk placed into a colander or cheese cloth allowing the rest of the whey to drip away. On a baking sheet, the curd is formed into a large disk about one inch thick and put under a broiler or into a very hot oven for about fifteen minutes on each side until it is lightly browned.

In western Finland, where this cheese originates and whence came most of the Finnish Americans, this cheese is called *leipajuusto* (bread cheese) because it is baked. Like *viili, leipajuusto* is available commercially in Finland in groceries and markets. Most Finnish Americans know this cheese as 'squeaky cheese,' so called because it squeaks when you bite into it, or '*juustoa*,' derived from the word *juusto*, (cheese).[4]

In 1985 the Michigan Department of Agriculture in accordance with federal law made it illegal to produce *juusto* to sell. Under federal law, cheese must either be made from pasteurized milk or, if from raw milk, be aged at least 60 days. However, the fresh cheese *juusto* is made from raw milk and is best eaten warm from the oven. In 1989 a local Finnish American experimented with pasteurized milk, producing a cheese with the taste of cooked milk and a rubbery texture. Although clearly inferior, the producers distribute this *juusto* through a local dairy and markets in the Upper Peninsula, selling primarily to Finns who return for visits and tourists, and they ship to Finnish Americans around the country.

Despite the law, women with cows continue to make, give away, and sell *juusto*. Important social networks have developed about where to find it. With the steady decline of families with cows, however, fewer people are now making *juusto*. Nonetheless, everyone knows a cheese-maker or knows someone who knows one.

Juusto is central to traditional Finnish-American life. Whereas *viili* is commonplace and eaten daily, *juusto* is more often considered a treat and eaten on special occasions. Unlike *viili*, everyone – young, old, Finns, and non-Finns – like it. Cut into bite-sized pieces, it is commonly eaten as a snack and served with coffee. A traditional Finnish coffee table is laden with pastries, breads, sandwiches, cheeses, and *juusto*. Old timers put chunks of *juusto* 'to warm' in their coffee. It also plays a role in ceremonial and ritual tradition. At Finnish ethnic events, wedding receptions, graduations, anniversary parties, and funeral feasts, the presence of *juusto* is almost *de rigueur*. On the occasion of his birthday, one teen requested *juusto* in lieu of a cake, and his grandmother happily complied. A memory from the past emphasizes the role of *juusto* at another ritual:

> Once my mother and I visited my grand aunt in Chassell [a town in the Upper Peninsula of Michigan]. She was in the hospital, dying of cancer. She said, 'Ida, you will bake a cheese, won't you?' Mother said, 'yes.'
>
> When we left, I remember asking, 'does she think she will recover to enjoy the cheese?' Mother said, 'she means for her funeral.' (Lampi 1999)

Two aspects of these milk products have importance to Finnish Americans beyond the nutritional: one is cultural and the other social. Out of the large repertoire of foods brought by Finnish immigrants to Michigan, only a relatively small number have been maintained by their third-, fourth-, and fifth-generation descendants. Finnish Americans eat more hamburgers, pizza, and pasties than they do foods of Finnish origin. Yet the Finnish-American foods that survive have great symbolic value. To consume them, even to discuss them with family, friends or new acquaintances, is to affirm ones ethnic identity. Food, like few other cultural features, is a strong symbol of who we are. *Viili* and *juusto* serve this cultural function, just as do several other items in the contemporary Finnish-American menu.[5]

The social aspect of these milk products has to do with the network of interrelated people necessary to their distribution. Inevitably, the start necessary to make *viili* is lost: one forgets to save some of the last batch, someone eats it all, a bad batch is produced, or the starter goes bad from lack of use. No problem! One merely borrows some from a next door neighbour, a friend across town, or a cousin living on the family farm. The problem is more difficult for those Finnish Americans who have migrated beyond the Finnish-American settlements as many were forced to do by economic circumstances. For those still maintaining ties, it may just entail a *viili*-less wait until the next visit home. But for those who have lost sustained contact, it may mean also losing one more attribute of ones Finnishness. When an editor of a Finnish-American newspaper offered to send *viili* starts to readers who needed one, she was buried with requests stating the offer 'was a dream come true'. One recipient wrote (*Finnish American Reporter*, January 1997), 'Thank you for the *viili* culture you sent me. You have no idea of the joy in our home when we had a bowl of "feelia" – that wonderful flavor and "strings" [meaning the elastic stretch].'

Similarly, to know where to get the very best *juusto* is a demonstration of ones knowledge of and participation in the local Finnish-American community. Within easy range of any Finnish-American settlement are only a very limited number of *juusto*-makers. Sources become more and more scarce as the number of families keeping milk cows continues to decline and the enforcement of health department regulations make nearly all existing sources illegal. Sources remain, selling illegally or giving *juusto* without charge to relatives and chosen friends, but there has been a tremendous decrease of available sources for most of the Finnish-American community. Having a source of *juusto* has become an important indication of ones connections. If we were to devise a questionnaire designed to distinguish members of the Finnish-American community from those people merely of Finnish-American ancestry, one of the questions that would have to be asked (perhaps even as the only question) would be 'Do you know someone from whom to get a *viili* start or a source of *juusto*?'.

There is one further dimension. These cultural and social aspects are interrelated; to know the Finnish-American social network adequately enough to know sources of *juusto* and *viili* starter has symbolic importance. One has awareness of being a member of a larger social entity with its own specific cultural attributes, including foodways, that sets it apart from other Americans. This is a measure of just how Finnish you are.

ACKNOWLEDGEMENTS

We wish to acknowledge and thank those individuals who contributed information to this paper through questionnaires, interviews, letters, and conversation: Barbro Klein, Ellen Angman, Betty Berg, Karen Berg Douglas, Reino & Polly Hiipakka, Laina Lampi, Ruth Mannisto, Lorri Oikarinen, Bea Raisanen, Liisa Rantalaiho, Carol Saari, Judy Szyszkoski, Esther Tervo, Irene Vuorenmaa, and Toini Rajala Watkins.

BIBLIOGRAPHY

Alanen, Arnold. 'In Search of the Pioneer Finnish Homesteader in America'. *Finnish Americana. A Journal of Finnish American History and Culture*, Vol. IV (1981): 72–92.

Berg, Betty. Interview, June 1999.

Brown, Dale. 'The Vigorous Diet of Finland', pp. 135–159. *The Cooking of Scandinavia*. New York: Time-Life Books, 1968.

Fantastically Finnish. Iowa City, IA: Penfield Press, 1985.

Finnish American Reporter, November 1993: 7; From our Readers, January 1997.

Gough, Robert. *Farming the Cutover: A Social History of Northern Wisconsin, 1900–1940*. Lawrence: University Press of Kansas, 1997.

Hoglund, A. William. 'flight from Industry: Finns and Farming in America'. *Finnish Americana. A Journal of Finnish American History and Culture*, Vol. I (1978): 1–21.

Käkönen, Ulla. *Natural Cooking the Finnish Way*. New York: Quadrangle, 1974.

Kaplan, Anne R., Marjorie A. Hoover, Willard B. Moore. *The Finns*, pp. 144–162; 346–357. *The Minnesota Ethnic Food Book*. St. Paul, MN: Minnesota Historical Society, 1986.

Koskenmaki, Joyce. 'Sharing the Culture'. *The Finnish American Reporter*, January 1996: 3.

Laitala, Lynn Marie. 'Sharing the Culture'. *The Finnish American Reporter*, January 1996: 3.

Lampi, Laina. Interview, June 1999.

Lassila, Mrs. Edward, Detroit; Wayne State University Folklore Archive, accession no. 1952(42), Detroit, Michigan.

Larson, Amanda Wiljanen. *Finnish Heritage in America*. Marquette, MI: The Delta Kappa Gamma Society, 1976.

Lockwood, Yvonne R. and William G. Lockwood. 'The Cornish Pasty in Northern Michigan'. In *Food in Motion: The Migration of Foodstuffs and Cookery Techniques*, edited by Alan Davidson. Leeds: Prospect Books, and Oxford: Oxford University Press, 1983. (Reprinted in part in *On Fasting and Feasting: A Personal Collection of Favorite Writings on Food and Eating*, edited by Alan Davidson. London: Macdonald Orbis, 1983).

———. 'Pasties in Michigan's Upper Peninsula: Foodways, Interethnic Relations, and Regionalism', pp. 3–20. In *Creative Ethnicity: Symbols and Strategies of Contemporary Ethnic Life*, edited by Stephen Stern and John Allan Cicala. Logan: Utah State University Press, 1991.

McGee, Harold. *On Food and Cooking. The Science and Lore of the Kitchen*. New York: Charles Scribner's Sons, 1984.

Ojakangas, Beatrice A. *The Finnish Cookbook*. New York: Crown Publishers, Inc., 1964.

Rajanen, Aini. *Of Finnish Ways*. Minneapolis, MN: Dillon Press, Inc., 1981.

Ränk, Gustav. *Från mjölk till ost. Drag ur den äldre mjölkhushållningen i Sverige*. Nordiskamuseets handlingar 66, 1966.

Szszykoski, Judy. Interview, June, 1999.

Tanttu, Anna-Maria and Juha. *Food from Finland*. Helsinki: Kustannusosakeyhtiö Otava, 1988.

REFERENCES

1. According to Harold McGee (1984: 7–8), humans are exceptional among mammals because they drink milk after they begin to eat solid food. In fact, even among humans those who drink milk after infancy are the exception. This is not just a matter of taste or of custom. Most humans lose the lactose-breaking enzyme, lactase, by the age of 3 ½ years old. This inability to digest milk sugar became known only in the late 1960s, because of a strong Euro-centric bias and because most Westerners, particularly those of northern European background, are capable of digesting lactose in adulthood. Only about ten per cent of white Americans, as compared to 70 per cent of black Americans, are lactose intolerant. It has been suggested that a genetic trait of continuing lactase production arose in northern Europe people because it conferred the advantages of increased intake and improved absorption of calcium on a group whose dark, cold environment developed little vitamin D in the skin. Finns, on the other hand, seem to be an anomaly; many of them do not have this genetic trait, presumably because they migrated to Scandinavia more recently.
2. *Viilia* and its Americanization, *feelia*, are the genitive case, meaning 'some *viili*..'
3. Armenian immigrants describe bringing their starters for *madzoon* (yogurt) by the same method.
4. Like the explanation for 'viilia,' *juustoa* is the genitive case, meaning 'some cheese.'
5. Some of the more common foods are *nisu* (sweet cardamon bread), *mojakka* (stew of fish or beef), *riisipuuro* (rice porridge cooked in milk), *pannukakku* or *kropsu* (oven pancake), *korppu* (sweet crisp toast).

Names for Milk and Milk Products

Jenny Macarthur

Milk

Note: Milk may come from almost any domesticated animal: cow, goat, sheep, yak, camel, etc. Apparently no one has yet persuaded a pig to allow itself to be milked. I do not know why or whether the milk would be acceptable to human children. A topic for investigation?

France	Lait
Italy	Latte
Portugal	Leite
Spain	Leche
Netherlands	Melk
Germany	Milch
Norway	Melk
Sweden	Mjölk
Finland	Maito
Russia	Moloko
Poland	Mleko
Czech	Mléko
Serbo-Croat	Mlijeko
Romania	Lapte
Hungary	Tej
Greece	Gala, Ghala
Turkey	Süt
Israel	Halav
Arabic	Laban (haleeb)
Iran/Afghanistan	Shir
Hindi	Dudh, Doodh
Tamil	Pahl, Pal
Marathi	Doodh
Kashmir	Duad
Kannada	Halu
Telugu	Palu
Bengali	Doodh
Gujerati	Doodh
Burma	(Myanmar) Nwanou
Malaysia	Susu

Thailand	Nom
Laos	Namnom
Kampuchea	Tik-doh-go
Vietnam	Sua
Phillipines	Gata
Korea	Uyu
Japan	Gyu nyu, Miruku[1]
Pu tong hua (Mandarin)	Niu ju, Niu nai
Cantonese	Naaih, Ngau naoi
Swahili	Maziwa
Hausa	Madara
Xhosa	Ubisi
Maori	Waiu, Miraka
Wales	Llath

[1] Artificial 'cream' or 'milk' for coffee.

Cream

In English and French the name cream is used for any creamy preparation, for example face cream, cream of mushroom soup, cream of tartar, medical preparations, etc. In most European languages *crema* or *krem* means 'cream of' rather than true cream, although in tourist areas it may also be whipped cream. Menus may use the name *crema* for true cream even if the locals don't eat it much. It is of linguistic interest that all the countries which do not traditionally use milk products in their cooking seem have picked up the cream/*krem* names

Key: a: thick cream; b: soured cream.

France	Crème, Crème fraîche[1], Crème fleurette[2]
	a: Crème double; b: Smitane
Italy	Panna, Crema[3]
Portugal	Nata, Crema[3]
Spain	Nata, Crema[3]
Netherlands	Room, Crème[3]
Germany	Rahm, Sahne, Crème[3]
Norway	Fløte, Krem[3]
	b: Rømme
Sweden	Grädde, Kräm[3]
Finland	Kerma
	a: Paksu kerma, b: Happankerma
Russia	Slivki, b: Smetana[4]
Poland	Smietana
Czech	Smetana
Serbo-Croat	Krem, Kajmak[5]

Romania	Crem
Hungary	Tejszin, Tejföl, Krem[3]
Greece	Krema[3], Kaïmaki[5]
Turkey	Krema[3], Kaymak[5]
Israel	Krem[3], Shamenet
Arabic	Krem[3], Kreem[3], Eishta[5], Khishta[5]
Iran	Khame, Hahmeh, Sah sheer[5]
Afghanistan	'Qyma'q[5]
Hindi	Malai[5], Khoa[6]
Tamil	Paladai, Aadai, Thirattu pal[6]
Bengal	Sor
Sri Lanka	Yodaya
Burma	Malain[5]
Malaysia	Kepala susu
Indonesia	Rum, Sari susu
Thailand	Krihm
Laos	Ga'ti
Kampuchea	Kraem
Vietnam	Kem
Korea	Kurim
Japan	Kuriimu
Pu tong hua (Mandarin)	Ju lo, Ru luo, Nai-yo
Cantonese	Geih lim
Swahili	Kirimu, Mafuta ya maziwa
Hausa	Afarari
Xhosa	Icwambu
Maori	Kirimi
Wales	Hufen

[1] *Crème fraîche* is matured longer than British cream and is thus slightly sharper.

[2] *Fleurette* is roughly equivalent to British whipping cream.

[3] May be cream but usually 'cream of'. Some countries also use the name for whipped cream or artificial cream for coffee, etc. See introduction.

[4] The names smetana/smitane are used in Britain and France for an artificially soured cream, whereas in the Russian equivalents they mean fresh cream unless qualified. In Poland *smietana slodka* is fresh/sweet cream, *smietana kwasna* is soured cream and *smietana bita* is whipped and sweetened.

[5] These very thick creams are made in a similar way to clotted cream. Milk, often buffalo's, is simmered very slowly in a wide pan, sometimes aerated by taking ladlefuls and pouring them back slowly from a height. The cream is scooped off as it thickens and rises to the surface. This thick cream is usually pressed in wooden moulds but after that its treatment varies. It may be salted as a kind of cheese, especially in the Balkans. If it is aged it becomes extremely pungent. In India and the Middle East it is as often sweetened and used in milk desserts.

[6] This is milk boiled very slowly until it has dried. It is used mainly for sweetmeats.

Butter

France	Beurre
Italy	Burro[1]
Portugal	Manteiga
Spain	Mantequilla
Netherlands	Boter
Germany	Butter, Anken[2]
Norway	Smør
Sweden	Smör
Finland	Voi
Russia	Maslo[3,] Butyer
Poland	Maslo
Czech	Máslo
Serbo-Croat	Maslac
Romania	Unt
Hungary	Vaj
Greece	Voútiron
Turkey	Tereyag, Kaymakyagi[4]
Israel	Hem'a, Khem'a
Arabic	Zibda
Iran	Karey
Hindi	Makhan, Mukhan, Muska, Ghee[5]
Tamil	Vennai, Nayee[5], Ney[5]
Marathi	Loni, Thup[5]
Kashmir	Thany
Kannada/Tulu	Benne
Telugu	Venna
Bengali	Makhan
Afghanistan	Maska
Burma	Toba
Malaysia	Mentega
Thailand	Nuhi
Laos	Nammanbuh
Kampuchea	Boa
Vietnam	Bo, Buh
Korea	Boto
Japan	Batâ
Pu tong hua	Niu you
Cantonese	Ngau yau
Swahili	Siagi
Hausa	Manshanu
Xhosa	Ibhotolo
Hawaii	Waiupaka
Maori	Pata
Wales	Menyn

[1] Not to be confused with the Spanish *burro* which means donkey!
[2] South Germany.
[3] Also oil.
[4] See notes to cream.
[5] Clarified butter.

Buttermilk

Buttermilk is the rather sour by-product of butter churning. It has its fans among health food addicts but is mostly used in soda breads and scones (US biscuits). These days what is sold as 'buttermilk' is usually skimmed milk.

France	Petit lait, Babeurre
Italy	Siero di latte
Portugal	Leitelho
Spain	Leche de manteca
Netherlands	Karnemelk
Germany	Buttermilch, Molken
Norway	Kaenermelk
Sweden	Kärnmjölk
Finland	Kuoittu maito
Poland	Máslanka
Hindi	Matha lassi, Chaach
Tamil	Moru
Tulu	Ale

Cheese

Tom Stobart, (*The Cook's Encyclopaedia*, Cameron & Tayleur Books, 1980 and recently re-issued), describes cheese as having 'an incomparable array of tastes [which depend on] the combined effects of rennet and of the bacteria and [for some types of cheese] the moulds which attack it and cause chemical changes of unbelievable complexity, processes not understood until recently'. I do not know if anyone has even accurately counted the number or accurately recorded the history of cheeses, which date back to pre-history. There are dozens of books on cheese. I can find 151 currently available on the Internet.

Cheese from the Latin *caseus*: *Käse, kaas, queso*, etc. *Fromage* from the Latin *caseus formaticus* (Larousse), '*fromage fait dans une forme*'. The *forme* was usually some form of woven basket. Hence *fromage* and *formaggio*. The corruptions of the European names are one of the more amusing trails for the amateur etymologist.

France	Fromage, Fromage frais[1]
Italy	Formaggio
Portugal	Queijo, Queijinhos frescos[1]
Spain	Queso, Queso fresco[1]
Netherlands	Kaas, Maikaas[1]
Germany	Käse, Rahmkäse[1]
Norway	Ost, Fløteost[1]
Sweden	Ost, Flødeost[1]
Finland	Juusto, Raejuusto[1], Kermajuusto[1]
Russia	Ser
Poland	Ser
Czech	Syr
Serbo-Croat	Sir
Romania	Brínzâ[2]
Hungary	Sajt
Greece	Tiri, Tyri
Turkey	Peynir
Israel	Gevina, Gvinah
Arabic	Jebna, Djeub'n, Gibna
Syria	Joban, Ka-ree-shee
Iran	Panir
Hindi	Paneer, Channa[1]
Tamil	Palkatti, Uraibaledu
Marathi	Paneer, Khava,
Kashmir	Tsaama
Kannada	Ginnu
Telugu	Junnu
Bengali	Paneer, Ponir
Gujerati	Paneer
Afghanistan	Panir
Sri Lanka	Keju
Burma	Deub-ghe
Malaysia	Kéju
Indonesia	Kédju
Thailand	Nuhi kaang
Laos	Fomah[3]
Kampuchea	Promaat[3]
Vietnam	Pho mat[3]
Phillipines	Kesong
Korea	Chissu, Chi-ju[3]
Japan	Chiizu[3]
Pu tong hua (Mandarin)	Kan lo, Gan lë,
Cantonese	Chisi, Foo yee[4], Nom yee[4]
Swahili	Jibini[5]
Hausa	Cuku
Xhosa	Isonka sobisi

Hawaii	Waiupala paakiki
Maori	Tiihi
Papua New Guinea	Sis[3]
Fiji	Jisi[3]
Tahiti	Pata pa'aru
Old English	Quesy, Cese, Cyse
Wales	Caws
Ireland	Cais

[1] Fresh cheese or cream cheese.
[2] *Brínzâ* is the general name for cheese but usually a soft sheep's milk cheese, not the same as the Swiss *sbrinz* which is a hard cheese.
[3] Corruptions of English or French.
[4] Respectively red and white cheese.
[5] A corruption of the Arabic.

The Milk-tie

Jeremy MacClancy

Milk is a unique universal. It is the only common food produced in humans, for humans. It is the only food every mother makes, the only food every person consumes, at the beginning of their life. Since breast-feeding is so central a part of infant-rearing it is unsurprising that a host of food historians (e.g. Hardyment 1983; Fildes 1986), medical researchers (e.g. Wickes 1953; Hamosh and Goldman 1986), anthropologists (e.g. Raphael 1973; Raphael and Davis 1985; Durham 1991) and feminists (e.g. Coward 1989; Carter 1995; Draper 1996) have studied its nutritional pros and cons, its cross-cultural variability, its patriarchal politics, the evolution of Western attitudes towards it, and the polemic about supplying powdered milk to mothers in the Third World. Yet in the midst of all this work about milk in the West and beyond, what is surprising is that one distinctive and historically significant topic appears to have been neglected: the establishment of a indissoluble 'milk-tie' between infants of different parents who suck at the same breast, and the important social, political, and economic consequences which follow from it.

The aims of this paper are (1) to carry out the first (to my knowledge) systematic survey of the literature on the milk-tie, and (2) to try to derive some generalizations about this practice.

******

The milk-tie is not a universal phenomenon: in some cultures women never give suck to others' infants, regarding the very idea as distasteful, unnatural or even dangerous to the life of the woman who attempted it (Fildes 1988: 265). Among the Baganda of East Africa, for instance, mothers refuse to allow surplus, expressed milk to be fed to other babies: those Baganda infants whose mothers are unable to feed them die (Jelliffe 1962: 22).

But the milk-tie *is* a strikingly widespread phenomenon, practised by peoples from the Balkans to Bengal, from Marrakech to Mandalay. It is all the more disappointing then that the vast majority of data available on this custom consists of mere snippets of information collected by travellers, doctors and ethnographers. All that these here-gathered gobbets give us is some idea of the historical or contemporary extent of the practice.

Starting with the westernmost examples first, Dunn writes of the campaign waged in the later eighteenth century by one Moroccan tribe against another.

When the latter finally surrendered, the former did not subordinate them but:

> united with them in a pact known in the region as *tafargant* (interdiction). The initiating ritual involved the exchange of milk from lactating mothers. ... *Tafargant* stipulated not only peaceful relations and mutual aid but also strict prohibition on marriage between the two tribes. This taboo implied symbolic brotherhood between them, but it may also have had the practical function of eliminating one major cause of tension (Dunn 1973: 97.)

Maher (1984: 107) states that, in the Middle Atlas region of Morocco, a mother must have the permission of her husband before she nurses another's child. All the examples of milk-kin that Maher came across obtained between a mother and relatives of her husband. None obtained between a mother and her relatives. This suggests that each husband was careful to ensure that his wife's wet nursing only served to strengthen the links between his immediate family and his kin.

Information on milk-kinship in the Near East is tantalizingly brief. Hammel (1968: 30) and Filipovic (1960) both mention that, in the Balkans, siblingship can be created between two children if they are suckled by the same mother, but neither author gives further details. In traditional Turkey, a mother's child and her nurseling become 'milk-siblings' and thus cannot intermarry (Davis 1977: 237—38).

There are more reports about milk-kinship in the Middle East, all of which state that while the practice is still maintained today, the number of children so nursed has declined greatly in recent decades. On the island of Socotra, 550 miles east of Aden, infants are nursed by women from outside the father's clan; if both parents are from different tribes, 'they are placed with the tribes or clans that stand closer to that of the mother' (Naumkin 1993: 282). In the southern Egyptian village studied by Ammar (1954: 102), people remembered who had wet nursed whom so that the local marriage taboos would not subsequently be broken. Wet nursing rarely occurred among close relatives, where future marriage of the children was anticipated. Ammar underlines the importance of the milk-tie by reporting the locally-repeated tale of Harun al-Rashid, an eighth-century caliph of Baghdad, who decided to kill one of his viziers and to punish the man's family because he believed they were plotting to depose him. Though the vizier's mother appealed to Harun, whom she had nursed as a child, he went ahead with his decision anyway. One chronicler considered this, 'An example of tyranny that overrides one's duties towards his breast-feeder who is like mother' (Ammar ibid.: 99).

Among the urban elite of Saudi Arabia, Altorki (1980: 240—41) argues, milk-kinship was created for two primary reasons:

> *(1) Domestic Convenience.* Since a woman was (and still to a great extent is) compelled to veil before any man other than a close kinsman, whom she could not marry, a man might ask his slave woman to nurse his daughters, so that they would not later have to veil to her son(s).
>
> *(2) Forestalling Potential Marriages.* Given the local preference for the intermarriage of cousins and the prevalence of extended households composed of the families of adult brothers, a jealous man could prevent an undesirable marriage by having his wife nurse the children of the envied brother.

The members of the Iranian landowning elite with whom Khatib-Chahidi (1992: 119) spoke, thought both of these reasons highly bizarre. For the sake of domestic convenience, these Iranians did not have to establish milk-kinship with their servants. Instead they got around the problem by means of temporary fictive marriage contracts (a traditional device) between the servant and one of the children of the household head. If they *did* have to call on the services of a wet nurse, their only concern was to avoid a choice of milk mother which *could* affect future marriage arrangements.

According to Granqvist (1947: 252), who lived among Palestinian Bedouin in the 1920s, local Muslim infants could be nursed by Muslim, Christian or Jewish mothers. By the same token, a Muslim could nurse a Christian, and the consequences would be understood in local terms. In 1881 in Damascus, the milk of the British Consul's wife failed. As her son records:

> A girl was duly produced *(by an astute sheikh of the 'Anizah group)* and according to my mother's testimony I drank her milk for several weeks. This is in the eyes of the Badawin entitles me to a certain 'blood affinity' with the 'Anizah; for to drink a woman's milk in the desert is to become a child of the foster mother. This fact has been of assistance to me in my dealings with the Badawin' (Dickson 1952: 7).

Going beyond the Middle East, Lyall (1882: 221) stated that among the Rajputs of India chiefs chose wet nurses for their children from a well-known pastoral tribe. The nurse's family held a recognized hereditary status of 'kinship by milk' and when the once-nursed man finally assumed the chieftaincy, his milk-brothers often attained much influence and position at his court. Also, in the Muslim Thai village studied by Hanks (1963: 128), if a wet nurse was employed, her child had to be of the same sex as the child to be nursed. This was not only because the milk intended for a child of the opposite sex was believed to cause disease and even death in the nurseling, but because incest might occur otherwise when the milk-siblings became adults.

Almost all the reports about milk-kinship cited so far are overly pithy and fail to site the phenomenon adequately in its various contexts. To gain a more exact idea of the ends to which milk-kinship was put, it is necessary for us to look at two further, more intricate examples, from the Caucasus and northern Pakistan.

For a farsighted father, marrying off his children strategically was one way to extend, in the right direction, his family's network of kin and relatives; using his children to create milk-ties with people they *could not* marry was another way. In pre-Revolutionary Russia, Christian villagers of northern Georgia and their Muslim counterparts over the Caucasian mountains used to sell their produce to one another by making hazardous treks through the high passes (Dragadze 1987). Since cross-creed marriages were prohibited, the best way for a pair of already close trading partners to strengthen their relationship was for one of them to nurse and bring up a child of the other. In order not to appear suspicious, the family of the 'adopted' child rarely made the trek over the mountains to visit. The child returned home before entering adolescence so that it could not be thought the 'adoptee' family were exploiting his, or her, labour. After all, the ultimate aim of this tie was not for the partners to make a balanced exchange but to ensure the survival of the partnership across generations, for the milk-siblings would speak the same language, know the same traditions, and help each other with transport, trade and hospitality.

In Georgia today, milk-siblingship is seen as an important, intimate relationship. The parents exchange favours, gifts, and visits, and it is expected that the milk-siblings will be on close terms with each other throughout life. Neither they nor their children can intermarry.

The most radical, elaborate use of the milk-tie for political purposes occurred in the tiny kingdom of Chitral, northern Pakistan (Biddulph 1880: 82–3; Jettmar n.d.) In this markedly hierarchical state of royalty, aristocrats, commoners and slaves, which survived as a functioning polity until the 1950s, the rulers used to give their children away at birth to be fostered by noblewomen. These women would share the nursing of the children with as many of their female dependents as possible. In this way the welfare of a royal child involved a large number of people. Foster-parents of a princess received land and other gifts on presenting her at court when she reached seven; on her marriage, one of her milk-siblings would accompany her to her new home. Those who had reared a prince would not only receive similar gifts on his handing-over but stood to gain even more: if at an early age he were made the governor of a province, his milk-brothers became his main advisors and ran his executive. A prince was usually made governor of the area where he had been fostered so that he would be fully acquainted with the locals and would consider himself, to a certain extent, one of them.

Nobles, who headed tribes or clans, followed the royal example by farming out all their infants to families of lower status, who in return enjoyed particular privileges and were excused the payment of certain tithes. As a contemporary visitor to the region observed, 'the foster parents continually show great devotion and abnegation to this cuckoo in their nest, and their own children suffer. I have known cases where the foster fathers have spent all their subsistence on some useless brat of the aristocratic class.' (Schomberg 1938: 225) Again like their superiors, nobles might have their babies passed around a whole village or local tribe. Thus an infantile aristocrat suckled by several dozen different women might grow up to have 50 milk-mothers, 50 milk-fathers and hundreds of milk-siblings, who would support and protect him if need be.

It seems like a cunningly simple political system, a state ruled at all levels by massively extended families linked by blood, marriage and milk. The trouble is, the system had a fatal flaw: no regulated accession to the throne, whether by seniority or achievement, leading to very bloody conflicts between the parties of different pretenders. When a king died, aspiring princes rapidly mobilized all their kin. Pretenders reminded nobles of their milk-ties and nobles did the same to their relatives of lower status. During these interregna, aristocrats acted as the bodyguard of their pretender; commoners and slaves formed the rank and file in the forces of their noble milk-brothers. Those groups who failed their prince, as did sometimes happen, were stigmatized inedibly. Generally, however, almost all groups gave unswerving loyalty and effective protection to their candidate for the crown.

The stakes were high. If a pretender was successful, he gave high offices, land and other gifts to his aristocratic milk-brothers, who passed on some of this largesse to their own supportive milk-kin. But if his bid for the throne failed and he was not killed in the process, he and all his noble backers were banished immediately. This enforced exile was unavoidable for it was feared, rightly, that otherwise a pretender's milk-brothers would strive to assassinate the new incumbent. They fled to the mountainous areas on the margins of the monarchy, which the king's men could never properly bring under control and which acted traditionally as places of refuge for his defeated brothers and their milk-kin. There the vanquished prince and his band made a living by obtaining help from the local peoples, or violently subjecting them to his own will and establishing his own mini-state. Those aristocrats who could not easily accompany them on their flight (old people and mothers with infants) lost their homes and land and were forced to live in misery. Commoners of the defeated party might be enslaved to victorious nobles or sold to traders.

It is easy to see these conflicts from above, as primarily the bloody consequences of competition at court. But they may also be viewed from below, as battles between tribes for power, with the princes as valued assets or risky liabilities. Thus powerful tribes, headed by ambitious nobles, far from being the puppets of milk-brother princes, could become the driving force of the system, persuading their royal kin, when the moment arrived, to make a bid for the throne.

There were also other, less risky ways for tribes and clans to exploit the system to their own advantage. Once, when the Roshte tribe incurred the wrath of the king, he confiscated their house and lands, which he handed over to others. To save their own lives, the Roshte had to flee. Only one of them, a woman with an infant, decided to remain. A friend of hers had just borne the king his first son, of which he was very proud. The friend, clutching her baby, stole out of the palace and gave him to her to suckle. She immediately took the child to her former house, burst in, and proclaimed that she was now the milk-mother of the prince. When its inhabitants realized what she was shouting, they had no option but to leave. The king, though enraged, could not object to the demands of his new kinswoman. He restored all rights to the Roshte.

So far, I have given a list of instances of milk-siblingship, founded on the sucking at the same breast by genealogically-unrelated infants. Yet it is important to note that, in areas where milk-kinship is already established, the symbolism of the practice may be exploited performatively in order to create structurally similar relations between adults. In these cases, the milk-kinship based on infantile breast-feeding is used as a model for the creation of a lifelong tie. Granqvist (1947: 114) states that a Palestinian woman who wished to adopt a stranger boy or man, could do so by publicly putting her nipple into his mouth, saying, 'Thou art my son in God's Book, thou hast sucked from my breast'. According to Granqvist (1931: 65), 'One reason for adoption is often the fact that a woman has to be alone with a strange man for some time as on a journey, and to protect her reputation she adopts him'.

The creation of a milk-tie may also be exploited in order to shore up a reputation that is on the point of being ruined. In Georgia, if a husband thought his wife unfaithful, he called the suspected paramour to his house, bared his wife's right breast, put salt on it, and asked the man to kiss it. The suspect had no option: if he kissed it, he would be milk-tied to the woman for life and so could not, under threat of punitive retribution, have sex with her; if he did not kiss it, he incriminated himself and faced punitive retribution. Once the deed was done, the husband would address the couple: 'Man, behold your mother. Woman, behold your son'. He could now rest assured; his wife and new milk son-in-law could meet openly without fear of raising any suspicion, for incest was out of the question.

A similar use of the milk-tie was employed among the Afghans of the Hindu Kush, as noted by a British colonial officer:

> In cases where conclusive proof *(of adultery)* is wanting, and which are brought for settlement before the ruler, guarantee is taken for the future by the accused placing his lips to the woman's breast. She thenceforth is regarded as his foster-mother, and no other relations but those of mother and son can exist between them. So sacred is the tie thus established esteemed, that it has never been known to be broken. (Biddulph 1880: 77)

One British colonial officer, working in the same general area, who had agreed to participate in a ceremony of elective kinship with a local youth, found that the ritual entailed more than he had expected:

> A goat was procured, quickly killed and its kidneys were removed. These were cooked at a fire and cut into morsels by an officiating (local) who fed us both…on the point of a knife. At short intervals we had to turn our heads to one another and go through the (sickly) motion of kissing with our lips a foot or so apart. But – the surprise was in reserve…My coat and shirt were suddenly torn open and some butter was placed on my left breast, to which the youth applied his lips with the greatest energy and earnestness! I jumped up as if shot – but the thing was over! (Robertson 1896: 30–31, cited in Parkes n.d.: 2).

What the official had not anticipated was that the ceremony would turn into a rite of submission, whereby the subordinate places oneself in a state of symbolic dependency, like that of a child to a mother.

******

Our list done, we can attempt to generalize about the nature of the milk-tie.

The first generalization we can make is that the extent of the tie is greatly variable between cultures, and even at times within cultures. Some authors speak of milk-kinship as though it were exclusively a relation between a pair of milk-siblings (e.g. Simoons 1976: 316; Farb and Armelagos 1980: 96). But these representations of the practice are far too restrictive, for milk-kinship, even in the most restrictive cases, always involves more people than just the pair of siblings. In the Moroccan case studied by Maher, the milk-kin are a close circle of the husband's relatives. In contrast, in the eighteenth-century Moroccan example provided by Dunn, two whole tribes entered into a single milk-tie. In Georgia, a pair of families are united by each tie while in Chitral a consciously created network of such ties could unite whole regions of the former kingdom.

In Islam, even though the Koran and the *hadith* (pronouncements of the Prophet) are explicit and fulsome about whom milk-kin may not marry, Muslim jurisprudentialists have debated, in an at times tortuous manner, the definition of what exactly constituted a milk-tie (Altorki 1980). To them, the tie was not knotted by a single feeding but determined by a complex formula. Learned scholars crossed pens on such matters as exactly how many feeds, and of what volume, were necessary to knot the milk-tie. Estimates ranged from one to ten feedings to a few drops on separate occasions. Some jurists contended that an infant's involuntary pause while sucking marked the end of one feed and the beginning of another; others would only recognize this break if the mother had deliberately interrupted her suckling. Some argued it was sufficient for a child to drink five times from a pot containing a woman's milk; their opponents counter-argued a child had only to drink from a pot containing a woman's milk collected on five separate occasions. Further issues in this arcane debate included the nature of testimony required for proof of the relationship, and whether cheese made from a woman's milk was an acceptable substitute.

These debates might, at first glance, seem to be a culturally particular example of the legal pedantry which jurisprudentialists of any culture might indulge in. But these debates were not exercises in Islamic hairsplitting. Exactly how long, and when, an infant sucked at a woman's breasts were relevant legal questions, as the belated discovery of a distant milk relationship between husband and wife led to mandatory divorce. Further, as Khatib-Chahidi argues (1992: 123), practising Muslims need to know to whom they are related and in what ways if they are to perform their devotional duties effectively, because, for some of them, merely looking at a potential marriage partner 'while saying their prayers would mean their devotions were nullified and had to be repeated'. Thus some of the rulings relating to milk-kinship may:

> reflect the behavioural implications of the forbidden degrees of kinship for marriage: they represent not so much an indication of whom a person may or may not marry, but those of the opposite sex with whom one may or may not act in a relaxed manner (*ibid.*)

The second (and key) generalization we must make is about what the milk-tie *does*. Khatib-Chahidi (ibid.: 124) characterizes it as 'a means to get friends or allies'. This is far too restrictive. As a well-grounded generalization, all we can state is that the milk-tie is a means of establishing a lasting connection between two groups of people. It is a tie that in many ways is meant to be as significant, and as lasting, as those based on blood. Like human blood, human milk is an essential life-enabling substance. And by using milk (whether a nutritionally significant amount or not) to create deliberately a link of kinship, people try to naturalize the cultural. In other words, they exploit the parallels between blood and milk in order to make this fictive form of kinship appear as natural as that based on genealogy.

Perhaps what is most interesting about the milk-tie is the variety of purposes to which that connection can be put, and the associations we can make between certain purposes and certain forms of social organization. It is evident that, in societies where rules of intermarriage consolidate the coherence of certain sub-groups (and so isolate them in the process), the establishment of milk-ties is a means of creating lasting links with groups of people whom one is not allowed to marry into. This is clearly the case for caste-divided societies such as the Chitrali; for groups who need to maintain cooperative relations with their neighbours, such as the Rajputs or the tribes of eighteenth-century Morocco; and for religiously-defined groups who wish to uphold links with members of other faiths, such as the Christian villagers of northern Georgia or the Palestinian Arabs studied by Granqvist. (The prompt provision by a sheikh of a local wet-nurse for the British Consul's son may be interpreted in the same vein.)

The establishment of a milk-tie may be used, not to bring people together in a mutually beneficial manner, but to *increase* the social distance of one group from others to the ultimate benefit of only one party. Thus, in the example of the jealous Saudi brother, he was able to exploit the consequences of creating a milk-tie in order to forestall marriage between his children and those of his envied sibling.

The examples of the suspicious husbands and, in eighteenth-century Morocco, of the imposition of peace on a subordinate tribe illustrate a further end to which the milk-tie can be put: making others one's milk-kin as a way to control their behaviour.

******

As far as the evidence will allow us to state, the practice of milk-kinship is in decline, thanks to the increasingly widespread distribution of formula milk, to the general shift from extended rural families to nuclear urban ones, and to the gradual decline of kinship as the central model of cooperative interaction. This present decline is no reason for us to continue to ignore the practice. For milk-kinship is not just a historical oddity (though knowledge of its existence and types is important for the historical record); it is not just an ethnographic curiosity (though information about its nature and functioning is important for our understanding of the varieties of kinship). Perhaps above all, knowledge of the milk-tie and its diversities helps to illuminate the social uses to which food may be put, and the ways that the uniqueness of one food may be put to unique ends.

ACKNOWLEDGEMENT
I thank Peter Parkes for supplying the Chitrali material.

REFERENCES
Altorki, S. 1980, 'Milk-Kinship in Arab Society: An Unexplored Problem in the Ethnography of Marriage', *Ethnology XIX (2)*, pp. 233–44
Ammar, Hamed 1954, *Growing up in an Egyptian Village: Silwa, Province of Aswan,* London: Routledge and Kegan Paul
Biddulph, J. 1880, *Tribes of the Hindoo Koosh,* Calcutta: Office of the Superintendent of Printing
Carter, P. 1995, *Feminism, Breasts and Breast-Feeding,* London: Macmillan
Coward, R. 1989, *The Whole Truth,* London: Faber and Faber
Davis, J. 1978, *Peoples of the Mediterranean. An essay in comparative social anthropology,* London: Routledge and Kegan Paul
Dickson, H.R.P. 1949, *The Arab of the Desert. A glimpse into Badawin Life in Kuwait and Saudi Arabia,* London: George Allen and Unwin
Dragadze, T. 1987, *The Domestic Unit in a Rural Area of Soviet Georgia,* unpublished D.Phil. thesis, University of Oxford
Draper, S.B. 1996, 'Breast-Feeding as a Sustainable Resource System', *American Anthropologist,* 98 (2), pp. 258–66
Dunn, R.E. 1973, 'Berber imperialism. The Ait'Atta expansion in southeast Morocco', in E. Gellner and C. Micaud (eds.), *Arabs and Berbers. From Tribe to Nation in North Africa,* London: Duckworth, pp. 85–108
Durham, W.H. 1991, *Coevolution. Genes, culture, and human diversity,* Stanford: Stanford University Press
Farb, P. and Armelagos, G. 1980, *Consuming Passions. The Anthropology of Eating,* Boston: Houghton Mifflin
Fildes, V. 1986, *Breasts, Bottles and Babies. A History of Infant Feeding,* Edinburgh: Edinburgh University Press
—— 1988, *Wet Nursing. A History from Antiquity to the Present,* Oxford: Basil Blackwell
Filipovic, M.S. 1963, 'Forms and functions of ritual kinship among south Slavs', *V Congrès international des sciences anthropologiques et etnologiques,* Tom.II, vol.1, Paris, pp. 77–80
Granqvist, H. 1931, *Marriage Conditions in a Palestinian Village* Helsingfors: Societas Scientiarum Fennica
—— 1947, *Birth and Childhood among the Arabs. Studies in a Muhammadan Village in Palestine,* Helsingfors: Söderström & C:O Förlagsaktiebolag
Hammel, E.A. 1968, *Alternative Social Structures and Ritual Relations in the Balkans,* Englewood Cliffs, NJ: Prentice-Hall
Hamosh, M. and Goldman, A.S. 1986, *Human Lactation 2. Maternal and Environmental Factors,* New York: Plenum
Hanks, J.R. 1963, *Maternity and its Rituals in Bang Chan,* Cornell Thailand Project. Interim Reports Series No. 6. Ithaca, NY: Southeast Asia Program, Department of Asian Studies, Cornell University
Hardyment, C. 1983, *Dream Babies,* Oxford: Oxford University Press
Jelliffe, D.B. 1962, 'Culture, social change and infant feeding: current trends in tropical regions', *American Journal of Clinical Nutrition 10,* pp. 19–45
Jettmar, K. n.d., 'Introduction', in K.Jettmar (ed.), *Milk Kinship in Chitral,* by Shahzada Hussam-ul-Mulk. Unpublished Ms.
Khatib-Chahidi, J. 1992, 'Milk Kinship in Shi'ite Islamic Iran', in V.Maher (ed.), *The Anthropology of Breast-Feeding. Natural Law or Social Construct,* Oxford: Berg, pp. 109–33

Lyall, Sir Arthur 1882, *Asiatic Studies,* London: John Murray
Maher, V. 1984, 'Possession and dispossession: maternity and mortality in Morocco', in H. Medick and D.W. Sabean (eds), *Interest and Emotion. Essays on the Subject of Family and Kinship,* Cambridge: Cambridge University Press, pp. 103–28
Naumkin, V. 1993, *Island of the Phoenix. An Ethnographic Study of the People of Socotra,* Reading: Ithaca Press
Parkes, P. n.d., *The Gift of Milk: Pastoral Reciprocity in the Hindu Kush* paper given at the 1993 International Commission of the Anthropology of Food conference, Oxford Brookes University (attended by the author)
Raphael, D. (ed.) 1973, *The Tender Gift – Breast Feeding,* Englewood Cliffs, NJ: Prentice Hall
Raphael, D. and Davis, F. 1985, *Only Mothers Know. Patterns of Infant Feeding in Traditional Cultures,* Westwood, CT: Greenwood Press
Robertson, Sir George Scott 1896, *The Kafirs of the Hindu Kush,* London: Lawrence and Bullen
Schomberg, R.C.F. 1938, *Kafirs and Glaciers. Travels in Chitral,* London: John Murray
Simoons, F.J. 1976, 'Food Habits as Influenced by Human Culture: Approaches in Anthropology and Geography', in T. Silverstone (ed.), *Appetite and Food Intake,*Berlin: Dahlem Konferenzen, pp. 313–29
Wickes, I.G. 1953, 'A History of Infant Feeding' Parts I and II, *Archives of Disease in Childhood* 28, pp. 150–58, 237–56

The Health Hazards of Milk

H. Morrow Brown

The public are constantly assured of the nutritional value of milk, which was given free to school children until Margaret Thatcher put a stop to it, and the Milk Marketing Board used to exhort us to 'drink a Pinta Milka Aday' to help dispose of the surplus. The introduction of pasteurization and tuberculin testing of cows has prevented the spread of tuberculosis, abortus fever, and E Coli infections, except for those who prefer raw milk, where there is still some risk. While milk is an excellent food for the great majority, it can be very bad for the minority who are allergic or intolerant to it. Unfortunately doctors and health visitors are usually so brainwashed regarding the health-giving aspects of milk that they often have great difficulty in accepting that milk can be a health hazard, apart from the possibility of infections, which have been eliminated by pasteurization..

The dairy revolution

That milk can cause illness is by no means a recent discovery. In 460 BC Hippocrates recorded that milk could cause gastric upsets and hives, Galen described allergy to goat's milk in the second century AD, and Prince Charles Edward Stuart, the pretender to the English Throne in 1745, is reputed to have had the 'bloody flux' due to milk.

Human milk from mother or wet-nurse was the usual infant feed until about 100 years ago, when advances in animal husbandry directed at maximizing milk yield, along with pasteurization, sterilization, freeze-drying, and other advances in food technology, brought about enormous expansion of the Dairy Industry, which marketed its products as a dietary essential and a healthy food. As a result, by the beginning of this century infant formulae based on cow's milk were being mass-produced as an alternative to breast-feeding, which became unpopular and unfashionable until the comparatively recent realization of the many great advantages of feeding babies on the custom-made human product. Yet, as cow's milk was meant for baby cows, not baby or adult humans, it should be no surprise that milk can cause disease. Bovine somatotrophin may now be given to cows to increase milk yield even more, in spite of misgivings from the USA, where it has been used for some time. Nowadays enormous quantities of cow's milk products are consumed compared with 100 years ago, hence we have lifelong exposure to much larger amounts of cow's milk proteins than our great-grandparents. It should be no surprise that such a fundamental change in the national diet can have adverse effects for some, yet awareness that cow's milk can actually be harmful varies greatly amongst health professionals trained to regard milk as an essential nutritious food.

By 1905 the first reports of allergic reactions to milk were published in American medical journals, but not until 1944 in the UK. Estimates of the prevalence of milk allergy and intolerance vary from 0.5 per cent to 7.5 per cent, no doubt depending to some extent on the enthusiasm of the investigator and the strictness of the diagnostic criteria, which tend to be vague. Prevalence is least in infants who have been breast-fed, even for a short time, and the many general benefits of breast-feeding to the baby are now fully recognized. In adults many diverse illnesses can be due to milk at any age, even over 70.

Lactase deficiency

It is hard to believe that milk, usually regarded in the West as an essential food, cannot be tolerated by about half of the world's population. Certain ethnic groups, particularly Asiatics and Africans, are congenitally unable to digest lactose, the sugar of milk, because after infancy they normally become deficient in lactase, the enzyme which converts lactose into glucose. Because northern races continue to produce lactase lifelong, they can continue to take milk without problems. Lactase deficiency in infancy causes watery diarrhoea, colic, bloating, and vomiting, often confused with other causes of these symptoms. In adults a severe gut infection can sometimes cause temporary lactase deficiency, with the same symptoms. Simple lactase deficiency is the most straightforward problem caused by milk. If diagnosed it is easily dealt with by taking lactase tablets, but the main subject of this paper is allergy or intolerance of milk proteins – a much more complicated matter.

Allergy and intolerance are different

It is important to define the meaning of these terms at the outset. *Allergy* to milk, or any other food, is present when taking a *very small* amount results in a reaction within a *short* time, even a few seconds when very sensitive indeed. A reaction may take many forms, and can cause anaphylactic shock, which can cause death if not properly treated without delay. Specific IgE Antibodies can be found in the blood, and pricking an extract into the skin produces a reaction within minutes. Allergy to peanuts, which has received much publicity of late, is a good example of a severe food reaction, but it is very rare for milk to cause an extreme reaction, intolerance being much more common.

Intolerance to milk or milk products is quite different, as there are no antibodies in the blood, no positive skin test reactions, and *large* quantities of milk are required to trigger a reaction which occurs after several hours, sometimes even the next day. It is not usually realized that formula-fed babies are being given the daily equivalent of ten litres of milk for a ten-stone [70 Kg] adult! Symptoms tend to be vague, and are often very difficult to attribute to milk without manipulating the diet to demonstrate a cause and effect relationship.

The allergens of cow's milk are 25 separate proteins, which are also found in most mammalian milks. The differences between cow's, goat's and sheep's milk are so slight that it is uncommon for a change to milk from another animal to be beneficial. I have heard that camel's milk is really different. Heat distorts these protein molecules, so that some patients can tolerate sterilized milk. Many infant formulae for allergic infants have been specially treated so that they are unlikely to react, but it can be difficult to find the formula which suits one particular infant.

Problems caused by milk allergy or intolerance

Any part of the body can be affected by food allergy, depending on which part has become sensitized. For example the food can affect the lips, mouth, tongue and throat by direct contact causing itching and swelling. If swallowed it may be rejected by vomiting, and if nearly all is got rid of in this way the patient will quickly recover. If nothing happens until the food has passed from the stomach into the small intestine colic and acute diarrhoea will result, and again recovery will follow if it has all gone.

If the lining of the gut does not react the food may be absorbed into the blood stream, and pass via the circulation of the blood to all parts of the body. The result depends on which organ is sensitized, the allergen being delivered by the blood to the tissue fluids and then it accesses the cells. For example allergy in the skin will manifest as eczema, the bronchi asthma, the nose rhinitis, the brain hyperactivity, the ears fluid, and so on. If both bronchial tubes and skin are sensitized we will have the common association of asthma plus eczema. The picture may be complicated by allergy to other foods, or to inhalants such as dust mite and cats, so that an underlying food allergy may not be suspected. Other causes in the environment such as pets, dust mites, and pollens in the summer cause symptoms by contacting the nose or the bronchi or the skin from outside.

Clinical presentations of milk allergy or intolerance

The case history is the most important investigation, but it takes time, hence is often neglected. Milk allergy and intolerance can affect any organ of the body. The best way to illustrate what can happen is by a series of short case histories, or 'anecdotes'. Current medical thinking rejects anecdotal reports, accepting only 'statistically significant evidence based medicine', but this is counterproductive when applied to food allergy problems.

Allergy on the breast

Breast-feeding is not always ideal, because the allergens of milk or other foods pass through in mother's milk and can cause symptoms in the baby. Double blind trials have proved that colic and screaming in breast-fed babies can be due to milk in the maternal diet, and also other foods such as egg. This possibility should be considered when a breast-fed infant keeps everyone awake. It is not widely enough realized that mother's diet can affect an allergic baby, and it has been advised recently that if there is allergy in the family mother should avoid peanuts during pregnancy because the foetus can become sensitized. This may explain why some children react dangerously the very first time they have a peanut. When consulted recently regarding peanut allergy in two children, I discovered that their mothers had become friends in the antenatal clinic, and that both had had a passion for peanuts when pregnant!!

Thus it is possible for the baby to be sensitized to cow's milk before birth, but until the practice was stopped, the commonest way that babies became sensitized to cow's milk was when well-intentioned nursing staff gave them a bottle in the night rather than waken the mother. This practice may still persist in some hospitals.

A good example was a breast-feeding mother who noted that whenever she took wheat or milk the baby got a rash and diarrhoea. The GP referred her to a paediatrician who told her to stop reading magazines, and that skin tests could not be done until age six! When seen one drop of milk on the tongue caused alarming swelling, skin and blood tests were very positive, and the child was in danger of anaphylaxis if given milk, so the mother's diagnosis was correct. Milk could have caused apparent cot death.

Milk allergy and intolerance runs in families

A mother I had seen with milk allergy as a child recovered in her teens, and could take milk without problems. After her first baby she took extra milk to help breast-feeding, but the baby had colic and kept her parents awake for three months. Within 24 hours of mother stopping cow's milk the baby slept through the night, and so did his parents for the first time since he was born! This child was the fourth generation of milk allergics I had seen in that family, emphasizing how milk allergy can afflict successive generations. Problems due to milk often disappear spontaneously in a few years, but can recur in much later life. This is why a history of infant feeding difficulties may suggest milk as a cause of allergy problems in adult life, even in the sixties!

Milk is a beef product
Most dieticians do not appear to realize that milk is from cows, so it is a beef product, so beef as well as milk must be prohibited to begin with. In my long experience about a quarter of those who react to milk also react to beef, and severe problems with the gastro-intestinal system can result, as in the following examples.

A bottle-fed baby was a screamer from birth, with diarrhoea which smelt so badly as to make the home almost uninhabitable. The hospital dietitian had put him on a milk-free diet, but without improvement. Remarkably, the parents had taken him with them on a bus tour of Europe for two weeks just before I saw him, and he had been perfectly well until after returning home, when he promptly relapsed and was referred for investigation. On enquiring what he was fed on during the trip I found that they had taken a suitcase full of tinned baby foods with them which by a sheer chance were from a manufacturer which does not use a beef-broth base. On return home they had reverted to the usual brand containing beef, with immediate relapse. Avoiding all trace of beef as well as milk cleared his problems in a week, and reintroduction again caused relapse.

Another patient aged 45 had developed ulcerative colitis severe enough for removal of the colon to be considered. He had heard of milk as a possible cause and was improved on avoiding it, but made a complete recovery when beef was completely avoided. Then he gave himself a large steak as a treat, and suffered a severe relapse lasting ten days.

I was recently consulted regarding an eighteen year-old girl who had had ulcerative colitis for about three years, only slightly controlled with drugs, having daily blood in the stools causing severe anaemia, and many malodorous motions. Her prognosis was one of indefinite chronic illness, probably ending in surgery. She was often having restaurant meals, and always chose steak. The clue was that her mother had had feeding problems as an infant, as had the patient, so I suggested a milk-free diet. She began to improve slightly in a week, definitely in a month, and by two months she had no bleeding, no anaemia, and no malodorous stools. She resented dietary restrictions and took some pizza with cheese topping, with resultant relapse, then tried sheep's milk on cereal with a much more obvious result. She may finally have been convinced of the importance of avoidance of all trace of milk products by a severe relapse following having as little as some milk in two cups of coffee.

The complete opposite, chronic constipation, is a rare presentation of milk intolerance, three cases being seen in twenty years. The most curious example is a girl aged nine with a history of infant feeding problems followed by chronic asthma. A paediatrician had insisted that all her problems were due to marital discord which led to the mother's divorce, but avoidance of milk not only resolved all her respiratory problems within a week but she also passed normal daily stools for the first time since infancy. It then transpired that her previous bowel habit was to pass an enormous motion every two weeks, which blocked the loo. Reintroduction of milk reproduced both respiratory and bowel problems.

Another lady had had acute diarrhoea every Saturday morning for twenty years for which she had had detailed and expensive investigations at a teaching hospital. But the answer was in the history, which was that after stopping work at noon on Saturday she passed a baker's shop, where she would buy herself her weekly treat of a cream bun. She would eat this while walking home, where she would arrive just in time for the explosion. No Buns = no cream = no diarrhoea!

The possible role of dietary factors, especially milk and beef, is seldom considered today, reliance on suppressive drugs being the usual approach to treatment. It is a matter for conjecture how frequently food is the cause of many of the illnesses mentioned in this paper because manipulation of the diet is so very seldom practised. It is 50 years since Truelove, in Oxford, first showed that colitis could be caused by milk, and John Hunter at Addenbrookes has continued this work by showing that dietary manipulation is very effective treatment not only for colitis but also Crohn's disease and irritable bowel syndrome – but the profession as a whole still ignore this research.

Asthma can also be caused by milk products and other foods and sometimes it appears when gut problems or eczema clear up. Perhaps the best example is of the 28 year-old lady who had a history of severe infant feeding problems which changed to chronic asthma. When first seen she had discharged herself from a psychiatric hospital to which she had been admitted for a severe emotional crisis, thought to be caused by giving her steroids for her asthma. The fact that she had just had her first baby, that her husband had broken his leg and lost his job, and the bronchodilator spray on which she relied to control her asthma had been taken away from her, had been ignored completely! The acute asthma was treated with high doses of steroids, milk was eliminated, and she became free from asthma without any treatment whatever for the first time since infancy. Provocation tests with small amounts of milk produced alarming repeatable reactions, and she has led a normal life since. All tests were negative, the only evidence pointing to milk as the cause being the history.

Eczema is a common manifestation of allergy to milk but the clinical picture is often confused by the unsuspected role of inhalant allergens such as dust mite or pets. One boy aged two with very severe eczema improved on avoiding milk, but recovered completely on holiday abroad. On the way home from the airport they picked up the dog from the boarding kennels and by the time they reached home he was scratching himself to pieces. A girl of five had severe eczema which cleared completely on a milk-free diet, but relapsed severely on visiting her grandmother who had a dog, so had she lived with granny excluding milk would have been ineffective. Each eczema case is individual with a unique pattern of response, but steroid creams are so effective that most skin specialists no longer search for causative factors, and often refuse to test the skin!

Emotional difficulties due to milk
The emotional effects of milk are so variable and bizarre that the main difficulty is selecting the best illustrative cases

Rosemary was nineteen when she was referred because of intermittent allergic conjunctivitis, swelling round the eyes, and being very withdrawn, stupid, and unable to concentrate for several years. Her mother could not drink milk as it made her sick, and cheese gave her sister diarrhoea. The patient had been a screaming baby with diarrhoea, and she had continued to have loose bowels. On avoidance of milk all her emotional and physical difficulties vanished within two weeks, she was able to stop her antidepressant tablets, and her appearance completely changed, presenting as an attractive lively girl. She could think clearly for the first time in two years, and took a weekend job as a waitress in a busy restaurant, remembering orders more efficiently than the others and not becoming stressed. Reintroduction of various milk products was shown to reproduce the problems, and after about two years she could again tolerate milk. Subsequently she married, and had feeding difficulties with the baby which were, as usual, dismissed by the doctor and health visitor, but resolved on avoiding milk. She and her husband have run a very successful Health Food Store ever since, having been inspired to this enterprise by her personal experiences.

Jonathan was a large baby said by the midwife to be too big to breast-feed, but he had vomiting and colic on formula feeds. He eventually settled down, as many do, but at thirteen months he began incessant screaming and banging his head on the cot, followed by uncontrollable temper tantrums lasting for hours at any time of day or night. He was also very clumsy and uncoordinated, and became destructive and would inflict pain on himself by pinching arms and legs to produce large bruises. His mother described him as 'a manipulating destructive monster' who made family life a nightmare and threatened to break up the marriage. To preserve her sanity she placed him with a registered child minder when aged two and went back to teaching as head of a local school, but when he cut his head and required stitching at the local hospital it was noted that he was covered in bruises. Mother was accused of battering him, a dreadful accusation for a headmistress, but eventually the paediatrician believed her and apologized. At nursery and infant school he was aggressive, antisocial, and had frequent infections finally diagnosed as asthma when aged seven. His behaviour, asthma, aggressiveness and violence became worse, he would jump or run on the spot for long periods, make silly noises, and throw himself down the stairs to hurt himself. His behaviour became vindictive and destructive, and he kicked holes in his bedroom wall.

A new GP referred him to a child psychiatrist who blamed his condition on parental mismanagement and their inability to communicate with him, and suggested he was being provoked by his sister, who was behaving normally. Finally his asthma got so bad that he was admitted to hospital, followed by referral for allergy investigation, which found that he was very allergic to the family cat, which was removed, but the main problem was milk. Within a week of prohibiting milk products his behaviour became normal, and before long all his teachers could recognize when he had been cheating by taking milk chocolate or ice-cream because his behaviour pattern relapsed within a few hours.

A similar case misbehaved only in the summer, because he had many other summer allergies which, when added to the hidden milk allergy, caused him to misbehave. When milk was excluded he was as good as gold, except when he cheated. He became head of the class with a totally different personality.

Behaviour problems due to foods may be contributing to delinquency and even crime, and dietary manipulation has been tried experimentally in prisons here and in the USA with very encouraging results. Unfortunately these results have failed to attract any support, and the Police Superintendent involved has retired a disappointed man.

Milk as a cause of arthritis

Many cases where milk was associated with arthritis have been seen over the years, but the most remarkable was the 53 year-old lady who had arthritis for about four years which was severe enough to prevent her driving. She had noticed that when she was unable to eat for a few days because of a gastric upset her joints were much better. This story often suggests to the patient that food may have some relationship to the arthritis, but unfortunately is often scoffed at by medical advisers. On a milk-free diet there was rapid improvement in her joint swelling, pain, and stiffness, and taking milk caused a relapse after six hours for two days. Her husband then noted that since this remarkable improvement had occurred her joints were always painful the morning after intercourse, suggesting that enough milk protein could be delivered in his semen to react on her joints.

It was then defined that this effect did not occur when a condom was used, or when he also abstained from milk. To prove this by double-blind or blindfold techniques with other partners was clearly unacceptable, so skin tests using the husband's semen were arranged when he was taking milk and when he was not. Positive reactions occurred only when he had been having milk, and since the milkman was sacked the improvement has been maintained. As an unexpected bonus, husband's chronic eczema of the ears of 30 years duration cleared up completely.

A consultant rheumatologist who became interested in the role of foods in causing arthritis conducted a carefully controlled double-blind trial which was published in the *Lancet*, a very prestigious journal. Unfortunately her colleagues remain unconvinced, and very few have followed her example, so that we will never know how many cases of arthritis are caused by food who might actually be cured by avoiding the causative foods, which is most commonly milk.

Conclusion

We are what we eat, yet neither patients nor their medical advisers appear to realize that our food, especially milk, can disagree with our immune systems and cause illness. If only the medical profession could be persuaded to consider the diet, especially milk, as a possible cause of illness instead of prescribing the latest drug to suppress the symptoms, many patients might benefit. For example, a new drug developed at vast expense is vigorously promoted to every doctor in the country by an army of reps, the drug is being prescribed nationwide within a year, and the development costs are soon replaced. In contrast, a new concept published in a specialist journal may take 50 years to be generally accepted, or, more likely, never become part of mainstream medical thought. In 1962 I met the late Dr Albert Roe, the father of modern food allergy studies, at a conference and spent two whole days discussing food allergy. In this country there was no significant change until the Anaphylaxis Campaign – formed in 1993 by a father who had lost his daughter to peanut allergy, not by a medical man – demonstrated that very large numbers of people have dangerous food allergies. The speciality of allergic diseases was at long last officially recognized in this country in April 1999.

BIBLIOGRAPHY

Bahna, S.L. & Heiner, D.C. (1980), *Allergies to Milk.*, New York, Grune & Stratton.
Rowe, A. (1972), *Food Allergy*, Springfield, Il, Charles C Thomas.
Hunter, J.O., & Jones, V.A. (1985), *Food and the Gut*, Bailliere Tindall.
Freed, D.L.J. (1980s), *Health Hazards of Milk*, Bailliere Tindall.

The Art of Making Brie de Meaux Fermier

Lizabeth Nicol

To quote Pierre Androuet, the famed French cheese-master: 'Cheese is the soul of the soil. It is the romantic link between humans and the earth'. From my little corner of the world, in the Brie where I live and work, come some of the finest cheeses you will ever taste. Some are known throughout the world and you will recognize their names immediately – the *brie de Meaux, brie de Melun* and *brie de Coulommiers*. Others you may never have heard of before today: the *brie de Montereau, brie de Nangis, brie de Provins, Explorateur, Gratte-paille, Jehan de Brie*, the *triple-crème Duquesne*, the *Fontainebleau*. Each one has its own special taste and character and all are part of the rich heritage of the art of cheese-making which is traditional in the Brie.

I have fallen in love with one of the most special and most rare of Brie cheeses, even in France: the *brie de Meaux fermier* – a farm-made raw-milk rather than pasteurized, industrially-produced cheese. If you have ever had the chance to savour the subtle yet penetrating flavour and felt the velvety, almost honey-like texture of a *brie de Meaux fermier*, then you will understand what I mean when I call cheese-making an art. It is only through experience and tasting that one can understand and appreciate the difference between the pleasant but somewhat bland bries that usually grace our tables and the exquisitely rich, creamy, mellowness of a *brie de Meaux fermier*.

'To taste once again the wonderful flavours of brie as it was in the olden days. What a supreme pleasure, what a joy!' exclaims Stephane Ganot, *affineur* (cheese-ager) in the town of Jouarre in the Brie. He explains, 'to make this happen you need good cheeses, those which are made twice a day, immediately after milking the cows, all from the same herd, of course. This is when the milk is at just the right temperature to begin the curdling process. To be the best, the milk must come from cows in the same pasture, eating the same things. This will ensure that the cheese will have the true *goût du terroir* (flavour of the land), which is at the heart of the unique taste of the old-fashioned bries.'

There are those who purport that there is little difference between a cheese made with raw milk and one made with pasteurized milk. I beg to differ as *brie de Meaux fermier* is made only with raw milk. Milk, of course, is the most important element in the making of cheese as it gives it all its flavour and texture. The taste and quality of a raw-milk cheese will vary with the seasons, depending on what the cows are eating (fresh green grass in the pasture, or hay in the barn). The knowledge, handed down through generations, and the experience of the cheese-maker and *affineur* ensure consistent taste and quality. This is the true art of making cheese.

The artisans who create the wonderful *brie de Meaux fermier* that I have described come from two Briard families. The cheese-maker, Madame Madeleine Clain of Glandons-de-Glandelu and Monsieur André Ganot, owner and *maître affineur* of the ageing cellars Fromagerie Ganot along with his wife, Madame Marie-France Ganot and their son Stephane, who will continue the tradition. Now over 200 years old, Fromagerie Ganot is the oldest and one of the last surviving ageing cellars for the *bries fermiers* of Meaux, Melun and Coulommiers.

In order to ensure the highest quality, milk should not be transported any great distance and Madame Clain makes her cheeses, twice a day, immediately after milking her cows. They are *Bleu-blanc Belges* and especially bred for giving milk for cheese-making. The milk must be used while still naturally warm as then it is not necessary to heat it. This is most important, as this gives the cheese its special character and is why a pasteurized-milk cheese cannot compare to a raw-milk cheese. One of the reasons for pasteurizing milk is because it has to be transported and during that time the milk will cool and germs and bacteria will start to proliferate. Heating the milk to 72°C will destroy all the harmful germs but will also destroy the natural 'flore' and lactic acid bacteria which give the cheese its special flavour and contribute to the curdling process. When pasteurized milk is used, a commercially-produced lactic acid bacteria will have to be reintroduced into the milk and the resulting cheese will be blander.

It will take 28–32 days to create a *brie de Meaux fermier*. The first step in making the cheese is to coagulate the milk and create the curds and whey. This is carried out in rooms that are maintained at a temperature of 16–18°C, which has always been, in fact, the normal temperature of the dairy areas of the old farms of the Brie. Now the artistry of making the cheese begins. The cheese-maker must add just the right dose of rennet. If she adds too much the milk will curdle too quickly and the resulting cheese will be dry and hard; if she adds too little it will be runny. Coagulation will take two and a half to three hours. The cheese-maker will judge the curd ready by using the *test de la boutonnière*. This is done by pressing a finger (done nowadays, according to strict rules of hygiene, with a gloved hand) into the curd, extracting it immediately and making sure that the resulting hole is as tight as a buttonhole If this is the case, she will move on to moulding the cheese.

The cheese-maker puts the curd into moulds, which are 35–37 cm in diameter and 2.5 cm thick. She does this by hand using a *saucerette,* which is a sort of large circular paddle with serrated edges. She carefully cuts the curds into horizontal slices and fills the moulds. This is a delicate operation and in the Brie it is accomplished in just one go. The moulds contain a series of *hausses* or rings and as the whey progressively drains from the curd the rings are removed. After twelve hours of draining the curd is dense enough for all rings to be removed and after 24 hours the cheese-maker gently removes the cheese and places it into an adjustable sort of spring-form mould fixed with a clip. She will wait another twelve hours and then turn the cheese by placing it between a wooden crate (*cageot*) and a tray. Twelve hours later she will turn the cheese a second time. The resulting cheese will weigh 2.0–2.5 kilos. Because of the large diameter and thinness of the cheese all of these manual operations demand a great deal of dexterity and skill.

The cheese is now ready for the next step, *le salage* or salting. The salt is very fine and dry with a low degree of humidity (8–10 per cent) and she will use about 125 grams (3 grams per litre of milk). Again, the skill of the cheese-maker comes into play. If salting occurs before the cheese is sufficiently drained the rind will become red and slightly greasy; if too much salt is used the cheese will be too dry and not able to age properly. The salted cheeses are turned twice during the following 48 hours. The adjustable mould is then removed and they are placed on a *clayette,* which is a straw mat. These mats, which are woven from rye straw grown in the area, are very important in the evolution of the cheese as they serve for further draining and for transmitting '*la fleur*' (*Penicillium candidum*) which creates the flowered rind of the Brie. They are then stacked on wooden racks in the *haloir à sechage* or drying room, which is a ventilated cellar.

Although they are called drying rooms, the humidity is maintained at 70–75 per cent and the temperature is carefully controlled at 12°–15° C. Looking after the cheeses at this stage demands a lot of attention. The mats must be changed if they become too damp. The temperature must not be too high or the cheeses will become too soft and 'run' too quickly. If it is too low the ageing process will be slowed along with risk of contamination. The cheese will stay about a week in the drying room and it is here that the fine, white mould or flowered rind will form.

Now the cheese is ready to be put into the capable and skilled hands of the *affineur* or cheese-ager. Cheese like wine is a living thing and needs special handling. An *affineur* carefully takes the cheese through every stage of ageing using his talents and experience to develop the very best taste and texture. His goal is to manage the ageing process so that the cheese will be delivered to the consumer when it is at its peak of savour both in taste and texture. The terrain of the Brie is largely clay and it is difficult to dig cellars, so many individual farms simply do not have cellars for ageing. The Brie has a long tradition of ageing cheeses and for centuries the cheese-makers of the Brie have been taking their cheeses to *affineurs* to be aged.

The art of ageing cheeses is a meticulous job requiring all the senses of a skilled professional. The cheese-ager must watch over the cheeses carefully observing the changes in colour that denote the moment the cheese is ready to be sold. This is especially delicate with soft cheeses of the flowered rind varieties. The rind will progressively lose its velvet-like pure white mould on the outside and will take on a slightly reddish cast. Inside, the initial chalky white *pâte* will soften and ripen into a soft, lovely golden to dark ivory colour. He must also be sensitive to the scent of the cheeses. At the beginning of their lives they are subject to acidifying bacteria. As they ferment and grow more alkaline, they become suppler, tender, almost melting, and give off the mellow scent of a perfectly ripened cheese. If left too long they will turn ammoniac giving off that rather disagreeable scent. As he turns the cheeses each day, his sense of touch will tell him how far the cheese has advanced in the ageing process. He will be careful to change any mats that are too damp. He must also be sensitive to fluctuations in temperature as this is an important part of the fermentation process. The cellars at Fromagerie Ganot are maintained at a temperature of 12°C.

In true artesian fashion, the Ganots sell their cheeses at the fourteen weekly regional markets of the Brie. They also prepare a shipment for the Rungis Wholesale Market each day, where their cheeses will then find their way to cheese platters and trolleys of some of the greatest restaurants in Paris.

As we have seen, the art of making fine cheeses is as much a labour of love as it is labour intensive and demands a high level of expertise and experience. Production remains relatively small and thus the cheeses command a high price in the market.

Of course it is necessary to protect such a wonderful artisan product to ensure that it will continue to be produced. One of the means of protection is at the local level through organizations and associations. The *Confrérie du Brie de Meaux* (literally meaning 'brotherhood' but more like a guild of cheese-makers) was created in 1990 and the *Confrérie du Brie de Melun* in 1994. They worked together with the more administrative and officially recognized *Union Syndicale Interprofessionnelle de Défense du Brie de Meaux* and the *Syndicat Interprofessionnel de Défense du Brie de Melun* to obtain one of more major means of protection for these cheeses, the AOC or *Appellation d'Origine Contrôlée*.

The AOC was first created in 1935 for wines and spirits and was extended in 1990 to cover all agricultural products – cheese representing fifteen per cent. The AOC plays a fundamental role in preserving the true qualities of regional cheeses. It has often been quoted that there are over 365 cheeses in France and there are only 34 which hold the AOC title today: eighteen are cow's milk cheeses, five are goat's milk cheeses and two are ewe's milk cheeses. There are also two butters and one cream.

The four requirements and conditions for receiving the AOC are quite severe:

1) The area of production must be well-defined and be the original area where the product was first produced;
2) The product must meet strict and exact standards of production;
3) The product must be established as one which is well-known and esteemed;
4) The product must have the agreement of the commission.

It can take as long as eight years from the date of application for an AOC to actually receiving it. First the producers must create a 'dossier' or application. This will contain information on the history of the cheese, the area of production, which will include not only the geographical area but also take into account the climate, the type of earth, the moulds to be found in the area and the type of cow's milk used in making the cheese, as well as methods of production and ageing. This dossier is submitted to the INAO *(Institut National des Appellations d'Origines)*, a commission is named to investigate and discuss the merits of the request and finally, if approval is given, the Ministry of Agriculture publishes an official announcement of the AOC in the *Journal officiel*.

Milk, however, remains the most important element in the production of cheese. And how that milk is handled will have a direct influence on the quality of the product. Some feel that a *fermier* cheese guarantees a higher quality of product than just the AOC because *fermier* cheeses are made from the raw milk of one herd of cows rather than a blend of milk collected from different farms, which is then possibly pasteurized.

With our new European norms the AOC are also gradually becoming AOP (*appellations d'origine protégée*), which was recognized by the EU in 1996. It is recognized generally that if a product has been awarded the AOC it will almost automatically receive the AOP.

Over the last few years, raw-milk cheese production has diminished. The makers of some of our finest unpasteurized cheeses have been forced out of business for various reasons: health scares which create bad publicity and cause a loss of sales, the need to buy new equipment to modernize their operations because of stricter health rules, etc. But the few small- and medium-sized producers who have mastered the technical aspects and adhere to the strict rules of hygiene that are now in place *must* be allowed to thrive and develop their activity.

Most of this hinges on the issue of raw vs. pasteurized milk for cheese-making. It is definitely the richness of the raw milk's biological life that produces the full, rich and varied flavours that make each cheese unique. No one has made me more aware of this than Sister Noëlla Marcellino, a friend and Fulbright scholar, who makes raw-milk *Saint Nectaire* at the Convent of Regina Laudis in Connecticut. She has been working with the INRA *(Institut National pour la Recherche Agronome)* and has made me realize the ultimate importance of raw milk and the role of *geotrichum candidum* (GC), yeast similar to a fungus that can be found in the cheeses of the Brie, Normandy, Franche-Comté and Auvergne. She has been researching the bio-diversity of fungus strains found on the rind of traditional French cheeses. She contends that pasteurization adversely alters the cheese by removing that essential element, GC. She has sampled and tasted cheeses from all over France and has gathered a collection of 175 strains of GC. She has shown that this yeast, which naturally occurs in the environment, has a biological diversity depending on the region where the cheese is produced. It appears on the *robe* (rind or crust) and affects the flavour of the whole cheese. It can reduce bitterness as well as any rancid, mouldy, cardboard-like or plastic notes in the cheese. Although it is difficult to relate the scientific complexity of her research, she has established the environmental bio-diversity of GC through its microscopic morphology, i.e. DNA. This bio-diversity is so unique that differences can be traced in cheeses produced within the same region yet at a distance of only two and a half kilometres. Because pasteurization requires use of a uniform commercial strain of GC the cheese will lose this unique element of its make-up. Her 'archive' of freeze-dried cultures should last for at least 50 years and thus we will be able to safeguard a collection of native strains that might otherwise become extinct if pasteurization should become mandatory.

She has also made me aware that we are in danger of losing the heritage and whole art of cheese-making and ageing that has been handed down to us through the centuries. We do not want to become a planet of antiseptic food products that have no substance, taste or character. These local, artesian producers need our help. As lovers of fine cheeses and consumers, we can, of course, support them by buying their products. We must also give our support to national organizations such as The American Cheese Society (ACS) which has a platform to commend and uphold the use of both pasteurized and unpasteurized milk in cheese production. We can also petition the Codex Alimentarius Commission (CAC), which defines the technical criteria and norms for food products to allow the continuation of the trade and importation of raw milk cheeses. This body was created in 1961 by the Food and Agriculture Organization (FAO) and the World Health Organization (WHO) of the United Nations for the World Trade Organization to establish standards and regulate the trade of foodstuffs between member countries. The current proposal includes a rule that requires member countries to make only pasteurized-milk cheeses available for world trade. You can take a stand and protest the prohibition of importation of raw-milk cheeses by contacting them directly over the Internet at codex@fao.org or through the site www.fromage.com.

Vive le brie de Meaux fermier!

ACKNOWLEDGEMENTS
I would like to thank M and Mme Andre Ganot, M Stephane Ganot, Mme Madeleine Clain, M Jean Garsuault of the Institut International de Fromage and Sister Noella Marcellino for their invaluable help in my research.

SOURCES
L'Association Nationale des Appellations d'Origine Laitières Françaises (L'ANAOF)
Le Centre Nationale des Arts Culinaires (CNAC)
L'Institut National des Appellations d'Origine (INAO)

BIBLIOGRAPHY
Androuet Pierre, Chabot Yves, Bernini, Gerard, *Le Brie,* Presses du Village, 1997
Girard, Sylvie, *Le Monde des Fromages,* Hatier, 1994
Vialard, Catherine, *Fromages des Terroirs de France,* Editions Solar, 1998

Medieval Arab Dairy Products

Charles Perry

In addition to the usual cream, butter, cheese and yoghurt, the medieval Arab world knew several less familiar dairy products. The most detailed descriptions are given in the tenth-century book *Kitâb al-Tabîkh* and fourteenth-century *Kitâb Zahr al-Hadîqa fi al-At'ima al-Anîqa*.

Bîrâf may be the least exotic; it seems to have been sour cream. The instruction in *K. Zahr* is to leave milk outdoors overnight in hot weather with a sieve over it to keep things from falling in, and then to skim it. The recipe recommends eating it by itself or with syrup, honey or sugar, and adds, 'The doctors say to take a drink of *sikanjabîn safarjal* (syrup flavoured with vinegar and quinces) after it, or to suck a quince or pear.'

Libâ', like the ancient Greek *pyriate*, was cooked beestings (colostrum), the extra-rich milk given immediately after birth, or rather a mixture of regular milk and beestings in proportions of 1:1 or 3:1, depending on the recipe. The mixture was cooked and left overnight to solidify. The author of *K. Zahr* warns, 'If there is too much beestings, it dissolves the milk and overpowers it, and it becomes like stone with no good flavour in it, and it is not good. The peasants in a village of mine used to make it and ripen it, and it never tasted good to me until I showed them how to mix it with milk.' For the thrifty, there were also recipes for *faux* beestings consisting of a pound of milk mixed with one egg yolk and four whites and cooked until solid, making a sort of unsweetened custard.

Hâlûm (from the Coptic *halôm*) is still the name of a well-known cheese. Today's *hâlûm* is a goat's milk or sheep's milk cheese made layering fresh cheese with mint, and often it is stiffened by the mozzarella method of boiling and pulling the curd. *K. Zahr's* recipe was simpler: the milk was boiled with thyme until reduced by a third, cooled, thickened with rennet and layered in a container with fresh thyme and peeled citrus fruits – bitter oranges, large citrons (*kubbâd*), small citrons (*turunj*) and lemons. Finally, boiled milk was poured over everything and it was sealed from the air with a layer of olive oil.

Qarîsha sounds like a sort of processed cheese, made by cooking six parts cheese with one part fresh milk, but lexicographers describe it as sharp and sour. Perhaps it was then cultured like yoghurt; in parts of the Fertile Crescent, the word is today a synonym for *qanbarîs* (see below).

Yoghurt was the basis of a number of products. *Qanbarîs* was yoghurt drained to thicken it. This product is still made in Lebanon (the name now pronounced *'anbarîs*); it's softer than the generally similar Arab product *labneh*, or the Greek *yaourti tou poungiou* or Turkish *süzme*.

K. Zahr's recipe begins by adding one part yoghurt to twenty parts fresh milk, which suggests that old yoghurt was being used as a culture. But perhaps middle-eastern cooks were not in the habit of keeping a starter for their yoghurt; some yoghurt recipes simply say to leave milk out overnight, which probably worked well enough in a kitchen where yoghurt was regularly made and the bacterial spores would have been in the air. It may be that this didn't work in cold weather, however, because *K. Zahr* gives recipes for making both 'yoghurt' and '*qanbarîs*' in winter by curdling milk with an acid ingredient, sour grape juice for yoghurt and vinegar for *qanbarîs*.

Jâjaq was a sort of flavoured 'yoghurt cheese,' something like Boursin. You mixed yoghurt with salt, thickened it by draining the whey, as for *qanbarîs*, and mixed it with herbs. *K. Zahr* uses only wild mustard as a flavouring, but the tenth-century *K. Tabîkh* calls for garlic, celery leaves, mint, tarragon, rue, both smooth and ridged cucumbers (*khiyâr* and *quththâ*), lettuce stems, cardoons and ground almonds.

K. Tabîkh also describes a rather similar product called *khilât* (mixture). It was one part salt to about three parts (unspecified) herbs to ten parts yoghurt.

The Arabic word *laban* can mean fresh milk, but it usually means yoghurt. Oddly, these medieval books call for products with names which ostensibly mean yoghurt in Persian (*mâst*) and Turkish (*yâghûrt*), but which are thickened with rennet, like cheese.

Laban yâghûrt was made from a mixture of cow's and buffalo's milk, according to *K. Zahr*. The recipe describes adding some yoghurt as a starter, and then mixing half an ounce of rennet with water to thicken a *qintâr* (about 45 kilos) of the cultured milk (the recipe does not make clear whether the yoghurt should have thickened from bacterial action before adding the rennet). 'If you make it in the evening, it is ready in the morning,' it promises, 'and if you make it in the morning, it is ready in the evening.'

In the tenth-century book, *laban mâst* was much the same, except that the milk was simply left out until it soured, rather than being cultured with a starter, and then thickened with rennet.

Shîrâz was this sort of rennet-thickened yoghurt which had been mixed with salt and drained in a sack, like the thickened yoghurt products. This would have had more the consistency of a real cheese, but it was not pressed in a mould as was cheese.

And then there were the relishes called *kâmakh*. *Kâmakh ahmar* (brown *kâmakh*) was five parts fresh milk to one part salt and one part rotted barley.

In my experience, moulds (*Penicillium* spp.) quickly turn it into a thick, brownish paste tasting like salty blue cheese.

The author of *K. Zahr* observes that he doesn't like *kâmakh ahmar* and will not give a recipe for it, but he does recommend *kâmakh min qamh* (*kâmakh* of wheat). It is a somewhat obscure recipe, but it sounds as if it would give something of the same result as *kâmakh ahmar*. You hull wheat, fry it and grind it coarsely, then separate the flour from the coarser parts of the grain and knead it into small balls which are left in the sun 'until they dry and become like elixir' – but we never hear of these intriguing elixir-like dough balls again. You put the coarse remainder of the grain in a jug with a little milk and enough water to cover and insert three fig twigs, which may well bear *Penicillium* mould spores. Everything is sprinkled with salt and olive oil and left for three days.

Kâmakh rîjâl (or *rîjâr;* the name comes from the Persian word *rîchâr*, meaning an electuary, or medicine which is designed to be licked up), also known as *kâmakh abyad* (white *kâmakh*), is one of the recipes found in nearly all medieval Arab cookbooks. A typical recipe is 25 parts yoghurt and 25 parts fresh milk mixed with one part salt, put into a large hollowed-out gourd (*K. Zahr* observes that a pottery vessel could be used instead) which is set in the sun from the beginning of June to the end of September, with more milk being added daily as necessary. Nothing much happens for the first four or five weeks, but then it develops the flavour of cheese – the ordinary Cheddar-like cheese, not a blue cheese. This is the effect of slow-acting bacteria in the milk; to achieve it, the milk must be preserved in some way from faster-acting microbes. In ordinary cheese, this is done by reducing the moisture level through curdling and pressing. Here the milk is preserved by salt alone. The result, says the recipe, could be flavoured with mint, garlic, nigella or even rose petals.

I would like to thank Jane Levi and Anissa El Helou for testing some of these recipes for me. They reported some difficulty in thickening yoghurt with rennet, but in principle the products in question were palatable. They were enthusiastic about *jâjaq*, particularly for the unexpectedly pleasant almond element, but described the *faux* beestings as nasty and nothing, to El Helou's certain knowledge, like real beestings.

Images of Progress: Milk Advertisements in Greece

Elia Petridou

Today Greece holds a European record: evaporated milk represents around 35 per cent of total milk consumption, in a country where the local dairy industry of pasteurized milk is among the most developed sectors of the economy. Intensive advertising is one main reason identified by marketing executives in the dairy industry to explain the popularity of evaporated milk. Especially in the last two decades, due to the rapid development of the Greek dairy industry and the increasing competition from fresh pasteurized milk, the advertising of evaporated milk became more intensive. In the last two decades, television advertising has been an important means of communication between the dairy companies and the Greek consumer, to the implications of which I now turn.

Advertisements tend to establish associations between the products they promote to wider issues in society. They function as a material expression of discourses surrounding the issues, i.e. as an objectification of culture at a specific historical moment. At the same time, as an objectified interpretation of cultural concerns, advertisements play a crucial part in the shaping of culture.

In the present paper, I focus on milk advertisements and the way they have functioned as an objectification of the cultural tensions brought about by modernization. From the early years of Independence in the nineteenth century the project of bringing Greece up to western standards has been a main concern, which has informed perceptions of Greek identity throughout the twentieth century. In the last fifteen years the language of progress and development has been extensively used in milk advertising, shaping at the same time the symbolic meaning of milk in Greek society.

Western originals and Greek realities

After the independence of the modern Greek state in the nineteenth century, the modernization of Greece and its categorical inclusion to the countries of the West became top priority. The modernization of the Greek society involved 'the enterprise launched by the intellectual and mercantile élites in the late eighteenth century to designate the Greek-speaking Orthodox a national community, free the Greek territories from Ottoman control, and define the

Greeks as western'.[1] The ideology accompanying that enterprise was called 'Greek Enlightenment', and like the Western Enlightenment advocated its longing for progress. In Greece, however, progress was attributed a new meaning, as it also denoted the upward course that would bring the Greeks closer to Europe. Progress became equivalent to Europeanization.[2]

The modernization of Greece, however, did not follow the western pattern that the élites had envisaged. The structural dependence on the West culminated in a formation of economic and social imbalances that were not present in western societies. The belated modernization of Greece was 'imperfect' in the sense that it could not 'culminate in a faithful duplication of western prototypes'.[3]

It is frequently the case that peripheral societies view the discrepancy between western originals and their own realities as a structural failure. In Greece, the adoption of imported models of modernization and the tendency to imitate 'the West' became a dominant social ideal throughout the twentieth century. Though in different forms that varied according to the historical phase, the project of modernizing Greece and bringing her up to western standards has been unceasingly present in everyday discourse. Milk advertisements constitute one field where these cultural tensions become materialized.

Milk advertisements: an imagery of progress and modernization

Evaporated milk

Evaporated milk was introduced in Greece after World War II by two foreign companies. Nestlé introduced the evaporated milk *GALA VLACHAS*,[4] which derived its name from the Vlachs, the pastoral nomadic people of central Greece. In 1951, Friesland Dairy Foods entered the Greek market with *NOUNOU*,[5] which became the leading brand that shaped the development of the market of evaporated milk in Greece.

Television advertisements in the 1980s and 1990s presented evaporated milk as the guaranteed milk for the development of a strong body and of children well equipped to face a demanding future. Evaporated milk was promoted as safe milk of good quality that Greek mothers could *trust* for their children. For example, most of Nestlé's adverts ended in their Greek version with the phrase: 'With Nestlé, I'm confident.'

In a 1989 advertisement of Nestlé, a boy opens a huge door to space and a friendly spaceship approaches him with a tin of evaporated milk. In the voice-over the 'mother' says:

Open the door of the coming century, my baby. You will live it and win it. Take knowledge as compass. Then the future will hoist the sails. And I will give you strength for shield, and care for company'.
[Singing] *GALA VLACHAS* by Nestlé: it brings up strong children. With Nestlé, I'm confident.

The same concept, that of building a strong body for a demanding future, was also used by Nestlé's competitor, the evaporated milk brand *NOUNOU*. A 1986 *NOUNOU* advertisement presents children today as the creators of the world of the twenty-first century. They are presented among toys that range from computers and plane models to astronaut costumes and robots. According to the advert, *NOUNOU* contains all the nutritional elements that help children develop strong bodies and minds, and make them strong to face the future:

> These are the children that will make tomorrow the world of the twenty-first century. *NOUNOU* gives them today all the necessary strength and nutritional elements that they need to build a new better world.

The language of parental responsibility and rational decision-making for the child's future is widely used in *NOUNOU* advertisements, not only in the 1980s but even more explicitly in the 1990s. The parent is invited into a process of rational thinking in order to come up with the 'right decision' for the child's future:

> When you know that everything depends on ... your right choice, would you let your child grow up without you being sure of its milk? ... Would you trust anything less for your child? (*NOUNOU*, 1992)

In this discourse the notion of *trust* plays an important part, especially when it is associated with scientific achievements. The parent is invited to trust *NOUNOU* milk because it is the outcome of advanced scientific research. This claim derives its validity from the belief that the path to healthy eating goes through science. The role of science as the guarantee for healthy and happy children is incisively illustrated in a 1994 *NOUNOU* advert, which is structured on the divide between the rational world of scientists and the romantic world of children. The advert is part of a wider campaign which was based on the slogan 'the child's milk – science for us'. On the one hand, scientists in laboratories are shown in white coats working in front of computers; on the other, a little girl is shown peacefully asleep in her fairy-tale-decorated bedroom, while her mother is lovingly offering her a glass of milk. The voice-over emphasizes the developments of technology and science, and involves the parent in a discourse of *trust* and *responsibility*.

When it comes to the child's milk, every moment in the day or the night is valuable. Valuable for the scientists, who ensure the quality *NOUNOU*, and valuable for the children's world. Inside the model laboratories of children's nutrition, there, where the valuable base of life is chosen, milk, and where the strictest quality control is carried out, there, where scientific developments become a matter of responsibility towards the child and the parents, safe, nutritious milk is created: your beloved *NOUNOU*. *NOUNOU*: the child's milk, science.

At the end of the 1980s the profitable market of evaporated milk in Greece attracted many foreign firms. Together with an increase in competition, there was an increase in television advertising. In order to associate the product with images of 'good quality', advertisements of imported evaporated milk widely used the cultural category 'Europe'. Sometimes they diverted from the language of science (as in 'European products are scientifically more advanced and therefore you should trust them') to the language of nature (as in 'grasslands in Europe are richer and more peaceful and therefore the milk is of better quality'). For example, the advert of the imported brand *AGROKTIMATA AROZA* (Aroza Farms) showed idyllic grasslands with cows, accompanied by classical music, while in the voice-over emphasis was put on the fact that the milk comes from northern Europe and is, therefore, pure and healthy:

> In the green-clad grasslands of northern Europe, in a peaceful and fertile nature, the evaporated milk *AGROKTIMATA AROZA* is produced with love. Pure, healthy, from selected stout cows. Evaporated milk *AGROKTIMATA AROZA*: with all its basic vitamins. Weekly new imports.

Still, the advertisements which celebrated the 'natural' advantage of 'Europe' rather than its scientific achievements, were rejected as romanticized and silly by other advertisements which insisted on the dominant scientific language. An advertisement by *BEBELAC* employs the argument of *scientific care* and rejects the image of 'grasslands, clever children and happy cows' put forward by the competitor. A man in black suit wearing thick glasses presents the evaporated milk PANDALAC with an air of responsibility inviting parents to trust it:

> If you have children the name *BEBELAC* is well known to you. Here is a new *BEBELAC* product: the evaporated milk *PANDALAC*. We gave to it the same scientific care as to our baby products. And it has a very nice taste. We are not telling you about green fields, clever children and happy cows.[6] But we are telling you with a lot of confidence and

responsibility that *PANDALAC* is very good milk. Trust it. It is a *BEBELAC* product. It will become your milk forever.

Images of 'Europe' as a way of promoting evaporated milk have been widely used in advertising in Greece during the last fifteen years. In most cases the language of technological and scientific advances is employed to provide an incentive to consumers, especially parents, to regard evaporated milk as safe milk of very good quality. In the second part of the paper, there will be a shift of focus from evaporated milk to fresh pasteurized milk and to the Greek dairy industry. My intention is to show how through advertising fresh milk answered back to the argument of progress and modernization. Rather than promoting fresh milk as a fresh product of the Greek nature, its promotion was based on the argument of technological development.

Pasteurized milk

Until 1987 evaporated milk dominated in the milk market with 53.6 per cent market share while pasteurized milk had less than 36.9 per cent.[7] In 1987, however, the positions started to reverse. *DELTA*, a Greek dairy company situated in Athens which had shown rapid development during the 1980s, 'went two steps ahead' in pasteurized milk technology. Homogenization and carton packaging brought for the first time pasteurized milk in the first position, and *DELTA* leader in the milk market.

In 1987, *DELTA*'s first advertisement on television aimed at informing about the meaning of homogenization as well as explain the new opening device of the milk carton. The advertisement was in the form of dialogue between Madame Sousou, a female character drawn from the writings of a Greek satirical writer, and her maid. Madame Sousou, full of airs and graces, strives to make up for her low class origin and be accepted in the circles of the Greek bourgeoisie by despising everything Greek and imitating French manners. Madame Sousou is sitting comfortably on her bed when her maid enters the bedroom with a tray of a *DELTA* milk carton and a glass:

Sousou: Entrez!
Maid: Your milk, Madame.
S: What is this, poor?[8] I drink only fresh milk.
M: But it is fresh!
S: And how do you comprehend that it is fresh? Illiterate!
M: There! By the 'tent'[9] up here! Fresh, and pasteurized it is, by *DELTA*.
S: Ah, French.
M: But...

S: Silence, poor! And how does it open, my child?
M: There you are, madam! Like this, and this, and this…
(The camera focuses on the maid's fingers opening the carton)
S: What they do, the Europeans!
M: But …
S: And, there no crust on top.
M: It's because of…ecogenization.
S: Homogenization, poor. This is how the French give to milk all its taste.
M: But Madam! *DELTA* is Greek.
S: Greek? I knew it, poor. That poor Greece has made progress!

In *DELTA*'s first campaign, the choice of Madam Sousou is not coincidental. In a satirical way, it is a critique on the Greek tendency to imitate foreign manners and the belief that foreign goods are superior. Adopting in effect the competitors' argument of 'European' superiority, *DELTA* chose to promote fresh milk based on a language of modernization. As it was stated in the advert, through its technological development the Greek company brought Greece 'two steps ahead'.

In the same year *DELTA* launched their second campaign to promote the new half-litre carton. The marketing strategy was to promote the new product as lighter and easier to handle. For the new, 'modern' packaging, *DELTA* needed a 'modern' image to associate it with. In a set of four commercials that came out in 1987 and 1988, *DELTA* fresh milk was associated to young, dynamic men and women, who know how to use computer technology to achieve their goals. All adverts contained electronic music with computer sound effects and three phrases:

> Fresh *DELTA* milk: the milk of a new/young (*néas*), distinct generation.
> When you know where you are going.
> Fresh *DELTA* milk: the milk of a new age.

Each advertisement tells a different story: a young man sets up a broadcasting station, another coordinates a group of instruments with use of computer technology, another designs with her robot-assistant a glass chamber that brings a dead rose back to life. The fourth advert presents a young woman assembling a turbo-powered car.[10] The girl, blond with blue eyes, is dressed in jeans-uniform and white sport shoes, and has an air of confidence while doing what is usually considered to be a man's job. When she succeeds in starting up the car, she exclaims 'yeah', and the music from computerized sound effects changes into country music.

This advert, like the previous one, celebrates progress in the form of technological development. Although the image of what is 'modern' changes from a 'European' to an 'American' ideal, in both adverts the good command of technology is valued and, therefore, used in the promotion of fresh milk. Fresh milk is promoted as 'modern' and is associated with images of technology.

In the campaign that followed, the promotion of *DELTA* milk is followed by a strong assertion of national identity. While in the first advertisement *DELTA* brought Greece 'two steps ahead', in this advertisement *DELTA* becomes identical to the geographical boundaries of the country with the slogan 'Here we drink *DELTA*'. In slow motion a group of young people, men and women, witness a milk explosion caused by a huge milk carton that falls and explodes creating waves of milk. The white liquid takes shape and forms the map of Greece. The music, powerful and dominating, is used to emphasize the awesome event that is taking place. In Greece 'we drink *DELTA*': by emphasizing that point, *DELTA* once again bring up the issue of foreign values and Greek identity. Here feelings of foreign superiority are replaced with a strong sense of patriotism, which was expected to find response within the younger generation. When I played the advert to a group of Greek students, a young man commented:

> the slogan 'here we drink *DELTA*' establishes that WE are here and that *DELTA* is ours. It is like saying to a foreigner: take away from here your nice European products, here we drink *DELTA*. It's very patriotic! To hell with the multinationals. We had enough being always the losers.

In 1993 *DELTA* launched a new campaign with the slogan 'The care of *DELTA*'s milk starts from the moment milk is born and never stops'. The campaign was built on the idea that *DELTA cares*. It cares for the raw material, its immediate transportation and processing, its preservation and safe delivery, and, consequently, for providing consumers with the best quality product on an everyday basis. The campaign consists of a main advert accompanied by a set of 'testimonials' in which *DELTA*'s employees give a 'testimony' about their work.

The main advert is a poetic image of rural Greece, of the place where '[Greek] milk is born'. The voice-over is a poem by the distinguished Greek poet George Seferis. The image is a symbolic adaptation of the poem: the whiteness of a milk silo is paralleled to the whiteness of a chapel, a milk container travels through the Greek landscape, signs of Greek localities appear on the road, and a boy offers a glass of milk to an old wrinkled man, undoubtedly a Greek farmer. Towards the end of the poem, a man clad in *DELTA*'s uniform is shown to converse with a priest on a stone bench.

In the advertisement Greek tradition is merged with modern technology. The emphasis is not only on modernization, but on Greek modernization achieved by a Greek company. At the same time, the message is that the company is responsible and *cares* for every stage in the production of milk. Consumers should, therefore, trust *DELTA* fresh milk.

The 'testimonies' given by *DELTA*'s employees aim at providing further information on how *DELTA cares* about the raw material and its transportation. A chemist in charge of one of *DELTA*'s milk collection stations guides the camera around the installations while commenting that:

> every drop of fresh *DELTA* milk that is produced in this area reaches you intact and wholly fresh. And I say this with certainty because I organize both the daily collection from the farms, and the strict quality control. With such an equipment and such a perfectly organized system, I can take full responsibility.

The language used by the chemist refers to the notions of *trust* and *responsibility*, which were widely used in the advertisements of evaporated milk. In another 'testimonial' a driver of a milk-container addresses the camera apologetically:

> Yes, I was driving fast. Sometimes you have to. Now, for example, the road was blocked. And I transport fresh milk that has always got to be on time. *DELTA* has organized here an enormous system; and they take care that it works everyday like a clock.

The campaign entitled 'care' lasted from 1993 to 1996. Its aim was to prove to the Greek consumer that *DELTA* is an organized, reliable company that has both the means and the will to produce milk of superior quality. Drawing once more on the discourse on progress, the association of good, nutritious milk goes through the path of technological development. Only this time, technological development is Greek and is presented as a merging of Greek traditional values with modern technology.

Conclusion

Advertisements tend to interpret and present in a materialized form cultural discourses that are present in society and inform everyday experience. Throughout the twentieth century there has been an urgency in Greek society to bring itself closer to western standards. In the last fifteen years both evaporated and fresh milk brands have been advertised intensively on Greek television. Evaporated milk advertisements stress as competitive advantage the European provenance of the milk. In a language that celebrates the achievements of technology and scientific research, the advertisements invite Greek consumers to trust evaporated milk as a product of high quality standards.

The Greek dairy industry visualized themselves as the torchbearers in the forefront of industrial development and modernization. This vision acquired objectified form in advertising. In the end of the 1980s the Greek industry started a dynamic campaign for the promotion of fresh milk. The message to the Greek consumer was that Greek companies are now capable of competing with their European competitors and of producing high quality milk. So, instead of advertising fresh milk on the basis of nature and freshness, quite to the contrary, fresh milk became associated with technological and industrial development.

In this paper I tried to highlight the role of advertising in the construction of meaning. My aim was to show that in order to gain a deeper insight into the symbolic meaning of milk in Greek society, it is important to consider the politico-economic framework within which advertisements are produced and the values of which they celebrate.

REFERENCES
[1] Jusdanis, G. (1991) *Belated Modernity and Aesthetic Culture,* Minneapolis: University of Minnesota Press (Introduction xiii).
[2] cf. Fillipidis, D. (1984) *Modern Greek Architecture,* Athens: Melissa
[3] Jusdanis, ibid.
[4] lit. the milk of the Vlach-woman.
[5] from the French word *la nounou* (nanny).
[6] On this phrase he removes his glasses and puts them back again when he mentions the words 'confidence and responsibility'.
[7] Source: *OIKONOMIKOS* 23/4/92. In 1988 the market share of evaporated milk went down to 44.5 per cent while by 1990 fresh milk reached 48 per cent.
[8] In this context the term 'poor' (*ptokhí*), which has a material as well as spiritual content, is used to express a feeling of superiority.
[9] The opening device of the carton reminds one of a tent/envelope.
[10] The advertisement was given the first award of the General Secretariat of Equality (Ministry of the Interior) and the First Channel of the National Television.

Cheese in Art

Gillian Riley

Things that are good to eat and look good as well are indeed a feast for the eyes, but we have a problem with delicious food that has little or no visual appeal. Take haggis, *brandade de morue*, hummous... and certain kinds of cheese. A fruit and vegetable stall is a wonderful subject for a still-life painter who can exploit his technique to depict glowing combinations of everyday and exotic products, perhaps a dash of human interest, and the raw materials for anyone who wishes to indulge in tedious moralizing using the heavy symbolic baggage attached to many of the subjects. The laden table in a banquet scene displays all the luxury and wealth to which the painter and his clients aspired, a savoury mass of pattern, colour and costly accessories. Cheese might be part of the pattern, but by its very nature cannot be considered as a prime subject for visual display. The texture and taste of a fine cheese often lurk beneath an uncouth outward appearance; unlike a glowing nectarine or a freshly picked ripe peach. Adrian Coorte never attempted to immortalize a cheese the way he gave luminous life to an inert bunch of asparagus.

Perhaps some of the most detailed early pictures of cheese are in the illustrations to health handbooks, the *Tacuina Sanitatis*, illuminated manuscripts produced in the Po valley within a few decades of each other during the late fourteenth and early fifteenth centuries. Based on earlier works by the eleventh-century Arab doctor Ibn Botlân, they describe all foodstuffs, states of mind and weather conditions which affect human health, including different cheeses and their properties, according to the system of humours, where both the food and whoever eats it are analysed in terms of their temperaments – hot, cold, dry or wet – making it possible to diagnose and treat imbalances in health by making adjustments to diet. This is my somewhat soft-centred interpretation, along the lines of modern 'holistic' medicine, but in fact the realities of medieval and renaissance practice were bogged down in academic theories which usually did the hapless patient more harm than good. As Fioravanti, the late sixteenth-century pragmatic medical man, part natural healer and part charlatan, asked in one of his many attacks on the awfulness of academic doctors, 'who has ever seen these humours, what proof have we that they exist?'.

Most of these health handbooks were illustrated with representations of each item discussed, alongside a brief account of their properties, often providing a charming view of their preparation and consumption. They describe different milk products: fresh milk, which is temperate and sweet when warm, is good for the chest and lungs and bad for fevers, but can be neutralized with seedless raisins; butter is warm and moist, it is good for the lungs upset by cold and dryness, but renders the stomach apathetic and can be counteracted with acidic substances; fresh cheese is warm and moist, it softens and fattens the body, but causes occlusions, although walnuts, almonds and honey will counteract this; old cheese is dry and of moderate warmth, roasted it placates dysentery, but can form kidney stones and to prevent this is best eaten between courses; ricotta is cold and moist, it nourishes and fattens but can cause colic and indigestion which can be warded off by eating with butter and honey.

Cheese seems to have a manichean role in both art and life. The mysteries of its production, the sometimes gruesome appearance of the finished product and the not always pleasing smell prompt feelings of love and hate, of overweening desire and profound loathing. As we have seen, medical opinion reflected this dichotomy – cheese was good for you, it opened the stomach, it aided digestion, it helped if taken at the beginning of a meal, alternatively it closed the stomach and helped things settle down eaten at the end. But it was also said that cheese provoked foul and noxious humours, was impossible of digestion, disturbed repose and caused sorrowful and unhealthy imaginings. The ripeness of a serious mature cheese combined both the quintessential flavours of a high gastronomic experience, and the hideous outward appearance and aromas of corruption and putrefaction. Moralists were not slow to exploit this. The maggots inside a ripe cheese could signify corruption from within, a timely warning of the way pure and innocent milk could be changed into an oozing mass of rank and fetid slime. Thus the innocent soul, exposed to both the influences of the wicked world and the fatal presence within of original sin might deteriorate and perish. 'Too sharp makes jagged' or 'early ripe early rot', ripeness produces decay, warped great minds produce bad ideas…and so on and so forth.

On the other hand Dutch cheeses were celebrated in scenes of plenty, many a laden table had as the centre of its composition a cheese-stack of perhaps two or three hard cheeses with a plate of butter on top, and fraught though many such still lifes and kitchen scenes were with symbolic interpretations of every object on view, from apples and grapes to walnuts and monkeys, it is hard to see in these ziggurats of wealth anything more than a complacent display of the conspicuous consumption that only the rich could afford. Although over the centuries bread and cheese may have become shorthand for a cheap snack, the amount of capital that went into the purchase of such a cheese stack implied wealth, not frugality.

When Jan Steen painted a smug virtuous family saying grace before a meal he demonstrates the virtue of thrift and sobriety which enabled a humble artisan to buy in a whole ham and a great big cheese. These people are not deserving poor, they are the middling sort of citizen, showing considerably more respect for their hard-won protein than the dissolute family Jan Steen modelled so closely on his own, who scatter cheese, pies, fruit and expensive worldly goods in a riot of tipsy merriment.

Early painters of laden tables made fine mature cheeses part of their displays of fine food, bestowing on them a dignity that *pronk* still-life artists a generation later gave to the brightly coloured lobster or richly decorated pie. One of these, Clara Peeters, was painting in Antwerp mysterious compositions of everyday objects, luxury goods and food, many with eucharistic and symbolic undertones but valued by food historians today as precise descriptions of things like bride pies, celebration cakes, sugared comfits and exquisite professional patisserie. Her cheeses had the same numinous quality as her confectionery. Very little is known about Peeters, we have no idea who her patrons were and whether she painted to a brief from them or supplied her own themes and props to create them. She put a lot of herself into her work, from initials on a knife that may have had some ritual significance, to ghostly appearances in the reflections of the interior of her studio on the shining bosses of a gilt jug. Perhaps the cheeses were a reflection of her tastes or those of her patrons, but we may never know the reason why she painted them.

Simon Schama writes of the levelling nature of cheese, good 'Fatherland Food' in a striving young republic, the universal enjoyment of which 'dissolved rank within national community'. He quotes a Spanish emissary's curiosity about a group of men devouring bread and cheese on a barge; he was told they were 'Their High and Mightinesses the States of Holland'.

Perhaps the most eloquent comment on the rich dairy produce of the Netherlands is to be found in the work of the painter Aelbert Cuyp, whose sleek, well-fed cattle in a golden landscape seem to personify the buttercup-coloured floods of cholesterol which brought wealth to his native town of Dordrecht. To the mass of consumers the problems of interpreting the role of cheese in art simply did not exist; they ate it to live and ate it for enjoyment, and some of the enjoyment is here in these pastoral evocations of a prosperous dairy industry.

The food historian who gets closest to the agonies and ecstasies of our attitudes to food, Piero Camporesi, writes perceptively of our ambivalent attitudes to cheese in *Le officine dei sensi* (Milan, 1985). He quotes the learned doctor J.P. Lotichius's *De casei nequitia tractatus medico-philologicus novus*, published in Frankfurt in 1643. Lotichius seems to have had a serious cheese problem. His denunciations have a hysterical ring to them; if the moon is indeed made of cheese then Dr Lotichius is alive and having a bad time on the dark side of it. His denunciation of the 'foul and slovenly sluts' whose unclean hands shape the monstrous fetid forms imply a problem with women as well. Camporesi reminds us of the unchronicled role of peasant women in the history of food culture, of the deft, sensitive touch of generations of clean and caring hands, of a sensuous intuitive handling of this mysterious, incomprehensible process, of turning a perishable elemental fluid, akin to water, blood and sperm, into a portable, long-lasting, life-enhancing product: the whole process of handling and forming the cheese very much like giving birth and nursing a newborn infant. He explains the atavistic loathing of Lotichius and his contemporaries as the reaction of civilized urban man to the barbaric lifestyle of the Mongol hordes of mountain and steppe, roaring around, semi-intoxicated on dire potions of fermented milk products, in orgies of pillage, rape and democratic tendencies.

Camporesi does not go on to make the connection between the rigid Calvinist disparagement of a luxurious papistical indulgence which might have been behind this loathing of cheese, but Schama and Josua Bruyn's observations about the condemnation of cheese by Dutch moralists does have bearing on this possibility. Bruyn makes a strong case against cheese in an article in *Simiolus* vol. 24, 1996, no. 1 'Dutch cheese: a problem of interpretation'. Here he seeks to dismiss the idea that cheese in still lifes might refer to successful items in the national economy and prefers the interpretations which see it as a symbol of the eucharist, equating Christ with heavenly milk, or contrasting the purity of butter (Christ) with the iniquities of cheese (mortal corruption and sin). Bruyn considers that after the scientific investigation of the processes of coagulation and fermentation in cheese by Johann Joachim Becker in 1669 the vulgar dread of putrefaction and decay no longer reinforced the strictures of the moralists and cheese was thus of less use to them in paintings.

Both Camporesi and his bilious Germans are perhaps over-impressed by the dark elemental forces at work in the preparation of cheese. Those much-maligned female hands were just getting on with a practical down-to-earth job, doing something they knew had to be done in a certain way and making sure it was done properly, for the survival of their families. This practical approach was a long way from the mystical wrestling with semi-occult powers that repelled and fascinated male academic doctors. Fortunately the Italian authors Giulio Landi and Pantaleone da Confienza had a more prosaic approach and their works are an invaluable source of information and insight into the status of cheese in Italian society.

Pantaleone da Confienza published his *Summa lacticiniorum* in 1477, a time when medical men were constrained to refer in both theory and practice to ancient authorities, relying more on partially understood classical texts than modern empirical ways. In spite of this to us blinkered not to say perverse approach, Pantaleone, clearly a lover of cheese, gives vivid first-hand descriptions of all the typical cheeses of Italy and many in Europe, particularly France, with sensible suggestions about when to eat them, depending on the food they precede or follow, and the health of the person enjoying them.

A century later Bartolomeo Scappi included in his monumental *Opera* a list of menus devised and prepared by him for various grand occasions. A meal in August, the hottest month of the year in Rome, for sixteen guests, was a delicate balance between festive food and delicate refreshing morsels. The first course began with white and red melons and included fresh cheeses from Romagna and clotted cream; the second course had some fine dishes with poultry but no cheese; the third course offered pastries, cooked and raw fruit, fresh young almonds, raw artichokes and chunks of Parmesan, fresh cream cheese and a *neve di latte*, a kind of syllabub, strewn with sugar. The final course was a delicate collection of comfits and sweetmeats, with the usual offering of perfumed toothpicks and posies of fresh flowers. Papal etiquette was in line there with medical theory – cheese served at the points when its digestive properties were most efficacious. Another much grander banquet for the coronation of Pope Pius V was of five main courses of 125 different dishes before the final dessert of comfits and so forth and after the sumptuous fare on offer it is not surprising that the cheeses in the penultimate course were so varied, from the usual chunks of Parmesan to fresh and dry *raviggioli fiorentini*, cheese from Majorca, goat's ricotta mixed with butter, fresh *marzolini* and *provature* and a *neve di latte*. All described with such affection by Pantaleone almost a century earlier. The *marzolini* in particular were appreciated by Michelangelo; these delicate white sheep or cows milk matured cheeses were a Tuscan speciality and he had consignments of them sent to him in Rome several times a year, though they were too prosaic ever to appear in his paintings.

Giulio Landi wrote his *Formaggiata di sere stentato* in 1652 as a dedicatory offering to present alongside one of the famous hard cheeses of his native Piacenza to his patron Cardinal Hippolito de' Medici. The text is loaded with convoluted turns of phrase and tedious displays of wit and smut that almost distract us from the nuggets of common sense to be found within. The usefulness of gifts of Parmesan as sweeteners cannot be underestimated, and the gastronomic and nutritional qualities of the cheese are described with loving accuracy. Landi is lyrical in his description of the comely young women who make the cheese, and equally lyrical in his praise of Parmesan as a condiment, without which soups, stews, lasagne, ravioli, annolini and macaroni, omelettes and fritters, not to mention stuffing for roast meat and chicken, would be quite without savour, while as accompaniment to good bread Parmesan knows no equal. He goes on to berate the strictures of medical men, who bedevilled by their own purely academic theories , malign this beneficent and delicious food of which they have no practical experience or understanding. He scorns the sententious proverb *caseus est sanus quem dat avara manus* and claims as Oscar Wilde did later that if enough is as good as a feast, more than enough is even better than a feast, a sentiment which modern nutritional analyses bear out, indicating that the high protein content of Parmesan is present in an easily digestible form, with a good balance of useful minerals, a low cholesterol content and a significant amount of calcium. I have seen with my own eyes a fine healthy child weaned at a tender age on vast quantities of Parmesan grated into vegetable soup, as recommended by one of Italy's foremost paediatricians.

That Parmesan, and in particular that produced around the city of Piacenza, is the crowning glory of Italian, and indeed European cheese, was held to be a universal truth, and its appearance in Italian still-life paintings of the period is unlikely to have any other significance than the celebration of huge mountains of this majestic delicious king of cheeses. It is depicted among displays of good things to eat, shown being grated by buxom kitchen maids, seen towering above items on a kitchen shelf and is the heroic craggy centre of a fine composition by Meléndez.

When Landi claimed that it was best offered at table whole, in all its splendour, rather than cut into chunks, he is perhaps inadvertently giving an explanation for the cheese stacks in Netherlandish paintings. These gastronomic giants are shown in their full majesty to demonstrate the munificence of the spread and hence the wealth of the host, a contrast to the insignificant hunks eaten by the less well-off as and when they could afford it. The amount of capital locked up in two or three forms of fine old cheese must have been considerable, compared with a few stuivers from a workman's daily wage for a modest portion, so cheese in all its forms, in spite of all the wise and erudite explanations for its appearance in art, might be seen as a benign boast rather than a symbol of sin and corruption, an image of pleasure rather than a cause of retribution, and a great help to historians in their humble task of finding out about what people had to eat and what it might have looked like.

Animal Husbandry and Other Issues in the Dairy Industry at the End of the Twentieth Century

Cherry Ripe

Cow's milk has a lot to answer for. Our demand this century for ever greater quantities of this no-longer-seasonal fluid has had profound effects. Not just on reducing domestic animal bio-diversity,[1] but arguably it was responsible for the greatest food contamination scandal this century – also 'Britain's most expensive peacetime catastrophe'[2] – the cost of which at last count was four billion pounds and caused the slaughter of eight million cattle. Bovine Spongiform Encephalopathy (BSE or 'mad cow disease') began in – and spread through – dairy herds.

Where once cattle were multipurpose animals (draft, dairy, meat and sometimes fuel) today in industrialized societies, production has become specialized. Most steers and fat cattle which provide beef or veal are a by-product of the dairy industry. Dairy cows are impregnated and give birth to ensure continuous lactation, and their resulting offspring are fattened for meat, despite not necessarily being a breed genetically predisposed to, or most suitable for, eating.

As a result of the desire to produce ever cheaper food, what we have done to the animals which produce this now ubiquitous liquid not only raises questions of ethics in animal husbandry, but points to the increasing dominance of industrial agriculture and of mono-cultures.

Just the one black and white breed, the 'universal dairy cow', the Holstein-Friesian – dubbed the 'milk machine' – has come to dominate the dairy industry in the industrialized world, particularly in Western Europe. It has almost eradicated from the English landscape traditional British dairy breeds such as Jersey, Guernsey, Ayrshire and Dairy Shorthorn.

Yet paradoxically, the Holstein-Friesian is not the world's most productive dairy breed. That distinction goes to a small cow off the Horn of Africa, a dairy cow from the island of Socotra (off Ethiopia and Somalia) which produces more per unit of feed input than any other dairy cow in the world.[3]

However high input/high output production systems in industrialized countries favour the black and white dairy cow. As a result of the adoption of intensive methods and breeds which respond to them to the exclusion of other, locally-adapted types, the Food and Agriculture Organization of the United Nations (FAO) estimates that worldwide two breeds of domestic animal become extinct every month. In the five years since the Symposium's theme was *Going Today, Gone Tomorrow? – Endangered Foods and Cuisines*, the situation hasn't improved. FAO now estimates that half of the remaining 117 breeds of domestic animals in America are at risk of extinction, and as reported in 1994, in Germany only five of at least 35 indigenous breeds of cattle remain and a similar trend is happening in the former Soviet Union, largely as a result of Western aid.

According to Keith Hammond, Senior Officer (Animal Breeding and Genetic Resources) with FAO in Rome, 'the major reason there is a large number currently at risk is that they have been pushed out by this massive expansion over the last half of this century of the few high input/high output types, and their acceptance by farmers'.

This expansion into industrial agriculture has been accompanied in the last five decades by an increasing disregard for the welfare of the animals which provide our food. Little more than a generation ago many farmers knew their animals individually by name. Even today in the Haute Savoie in the French Alps, a farmer with six cows is considered well off; one with nine is considered wealthy by local standards. Hearing that my English cousin, a former dairy farmer, once owned and milked more than 200 cows, a Savoyard farmer pronounced him a '*milliardaire*'. Now dairy herds in industrialized northern Europe can run to thousands. They are confined, kept indoors much of the year, fed high-protein diets and frequently turned into carnivores.

The most plausible explanation of the cause of BSE ('mad cow disease') is that it was the result of turning ruminants (herbivores) into omnivores and even cannibals. The justification for this is that dairy cattle fed on protein produce greater quantities of milk rather than on their natural diet, grass. Between five and fifteen per cent of an intensively-reared (factory-farmed or feed-lot) animal's diet can be animal protein – bone meal, or even processed chicken manure, which may have chicken carcasses in it.

Although BSE first showed up in dairy cattle in England in 1985 (some claim earlier, but it was first diagnosed and named in 1986) the cause of BSE is widely attributed to feeding dairy cattle abattoir waste from cattle, and sheep infected with scrapie – a similar spongiform disease of the brain peculiar to sheep. (Since 1988, the British government has insisted that animals acquired BSE through their feed. But despite a ban on feeding ruminants to ruminants which was introduced in the UK in July 1988, calves born eight years later were still found to be infected, presumably meaning that it had been transmitted in utero.)

Foremost in the spread of BSE – and its ineffectual initial eradication program – was, in retrospect, political expediency. Because the (then) ruling Conservative party was so heavily dependent on its rural constituency, to avoid upsetting this rural vote it failed to respond as quickly and effectively as it might have done.

It would have been considerably easier to contain BSE in its early stages by adopting a pragmatic slaughter ('zero tolerance') policy – however politically unpopular – similar to the French: one strike (case of BSE) and the whole herd is out (killed). Certainly in the UK provision for such measures exist in the case of an outbreak of Foot and Mouth disease, a highly contagious virus which is fatal neither to cattle nor humans: one case in a herd and there's mandatory slaughter of the entire herd. However a similar strategy was not implemented in the case of BSE.

Certainly more stringent early containment measures – by preventing its spread – would ultimately have meant the slaughter of many fewer cattle, but at the expense of a loss of rural votes. Here, political expediency triumphed over issues of public health. Lack of early containment measures also meant that many more affected beasts found their way into hamburgers, pies and sausages.

This was further exacerbated by inadequate compensation to farmers. Initially, even if a farmer thought he might have an infected cow, there was an actual financial disincentive to report it. When in 1988 the government did get round to offering farmers compensation for infected animals – those actually showing symptoms – there was a shortfall of around £300 per beast between the compensation payment and the price the farmer could get by sending that same cow off to the abattoir and into the food chain. (This situation existed for two years before even offal was banned from human consumption.) And while the government was insisting there was no threat to human health, there was not even a moral obligation – beyond their own conscience – on farmers to ensure that suspected beasts were kept out of the food chain.

Furthermore, even after the compensation to farmers for infected beasts was upped to their market value, because BSE could take up to ten years to incubate in a beast and the average lifespan of a British dairy cow is around six years (by when she is past her productive prime, and is abattoir-bound), infected animals who were asymptomatic (without the staggers) also found their way into the food chain.

There was also a widespread – if, in retrospect, misguided – public assumption that if a foodstuff had government approval – endorsement even, as beef then had in Britain – it posed zero health risk. There is the still poignant image of then Agriculture Minister John Gummer in May 1990 feeding his daughter Cordelia a hamburger to prove how safe ground beef was.

For nearly a decade despite scientific evidence pointing to the possibility of BSE leaping the species barrier – first cats fed on pet meat, then zoo animals such as cheetah and puma fed on affected carcasses, and monkeys injected with it, all succumbed[4] – the government played down the possibility of human infection until March 1996 and the announcement in Parliament of the identification of a new 'floral' strain of Creutzfeldt-Jakob disease (nvCJD). (That the lessons from BSE haven't been absorbed fully came in an admission by the Department of Agriculture in Ireland in June 1999 that dog and cat flesh – carcasses of abandoned pets destroyed by local authorities – was being processed into bone meal and used in feed for pigs and poultry.)[5]

Despite the rush to produce greater quantities of ever-cheaper food by more unnatural (factory farming) methods, Animal Liberationists have yet to turn their attention to the dairy industry. Their focus has been on veal calves and battery chickens. With the latter they can rightly claim some victory: battery cages for chickens in the European Union will be abolished by 2012. (No new cages will be allowed to be built after 2003.) And while increasingly in British supermarkets you can find 'cruelty free meat', the treatment of the mammals the results of whose lactation down the centuries has provided us with so much nutritious food (cheese, butter) and gastronomic pleasure – has yet to come under closer scrutiny.

In the United States dairy cattle are even worse off. They have become disposable, discarded while still in what was once considered their youth.

Cows naturally have a lifespan of 15–20 years. As the daughter of a farmer, in my childhood in the 1950s we had cows which lived to the ripe old age of 29 or 30, one of whom from memory bore her last calf aged 27. The world's oldest cow, Big Bertha, an Irish specimen, lived to age 49 – a month short of her fiftieth birthday. (She also held the world lifetime breeding record for the greatest number of calves: 39.)[6]

However these days in America, where they have also turned ruminants into omnivores, many dairy cows are despatched to the abattoir at a mere 48 months of age, deemed to be at the end of their productive lives. Many are given the genetically engineered bovine growth hormone (rBGH) or Bovine Somatotrophin (rBST) to boost milk production. BST is claimed to cause an increase in mastitis (udder infection) to counter which the cows are given more antibiotics which can end up as residues in the milk. rBST also increases levels of Insulin-like Growth Factor (IGF1) in the milk. IGF1 is thought to increase the risk of certain cancers.

But in America, milk ain't just milk anymore. That cows work hard to produce it is not sufficient. Nor is it enough that they lay down years of their lives to offer up greater quantities of it more quickly than ever before in history. Instead of fifteen years of productive life, they're now clapped out by age four, and off to the knackers, quite possibly to be recycled into cattle feed.

Nor are cows so overworked that they can't be pushed to give more of it: around half the American dairy herd is given rBST – dubbed 'crack for cows' by former Greenpeace advocate Jonathon Porritt – to make them even more productive. Unable to achieve similar levels of milk production as afforded by the mammoth dairying operations with their industrial methods and economies of scale, many small American family-run dairy farms have been forced out of business, unable to compete. Worse still, when some farmers – perceiving a marketing advantage – wanted to label their milk as 'BST-free', Monsanto (from memory) took them to court to prevent them!

But after all this, will Americans drink it the way it comes out of the cow? No, it has to be fixed. Pasteurized. Homogenized (so the cream doesn't float). Boosted with added vitamins. Or have its fat or its lactose removed. De-toxed. Pity the poor cow: she hasn't done the job properly. In Australia, we do many similar things: taking out the dairy fat and adding it back again so that it is at a consistent level all year round – or replacing it with canola oil, and adding calcium. But in America tampering is epidemic. There, milk comes in a dizzying array of types. Whether it's a trendy Chicago supermarket specializing in healthy produce, or a relatively small neighbourhood foodstore on New York's Upper East Side, you can find in their cool cabinets between 30 and 40 versions, from organic and 'BST-free' to vitamin D- and vitamin A-enriched.

There are whole shelves of Lactaid, a lactose-reduced milk. It alone comes in at least half a dozen types from '100 per cent lactose-reduced' to '2 per cent fat' or '1 per cent fat'. (What's left? Opaque water?) Presumably this is so that consumers can drink it guiltlessly while munching on a bag of French fries. There are fat-free flavoured milks like chocolate, and bizarrely, even something (as if it wasn't an oxymoron) called 'fat-free half and half'. (Traditionally 'half and half' is half cream, half milk.)

Paradoxically, many of the people who deliberately buy low fat milk have no hesitation in sitting down to a double serving of scoop after scoop of full-fat ice-cream, apparently oblivious to the inherent contradiction.

Finally on the New York supermarket shelf, in a nod towards tradition, there's Meadowbrook Farm Dairy milk. It comes in old-fashioned looking squarish glass bottles, with nothing to indicate that it's been manipulated in any way. For some Americans at least, milk in glass bottles nostalgically harks back to less complicated times.

As if it's not bad enough tampering with milk, Americans cannot leave butter alone either. It's solidified cream, right? Not anymore. 'I Can't Believe It's Not Butter' is actually the name of a product. It comes in at least three types, including 'sweet, unsalted'.

Another implies intelligence. 'Smart Beat' claims to be 'smarter than butter' and 'smarter than fat free'. It's also *trans* fat-free and lactose-free. Then there are 'Butter Buds'. They are powdered granules for sprinkling on – or stirring through – hot food, or mixed with water to make a spread. They're made from maltodextrin (derived from maize), dried butter, salt, guar gum and baking soda. (On landing back in Australia with packets of these, they were deemed an illegal import: airport quarantine officers delicately slit my individual sachets and emptied their contents of yellow, salty powder into dump bins for incineration: probably the best thing for them.)[7]

Milk is a highly perishable commodity. Until refrigeration it was the ultimate local product, unable to be transported any distance – unless preserved as cheese or butter. It is therefore an example of one of the simplest of all agricultural products to have become industrialized.

Bizarrely, in an example of 'nutritional colonization', these days, in previously lactose-intolerant cultures in Asia, locals are being encouraged by television advertising campaigns by multinational food companies to mix up a suspension of water and white powder (derived from Europe, and no doubt subsidized surpluses) from 2.5 kilo tins to pour over some ground-up, reformed, pre-cooked dried and extruded cereal – a puritanical idea of what was right and proper to break one's fast on – as an appropriate first meal of the day for the modern middle class child from Yogyakarta to Kuala Lumpur. It is another way of the multinationals such as Nestlé getting their tentacles into Asia at the same time as off-loading Europe's milk mountain.

In the meantime those Savoyards are being told by the 'Bacteriological Police' (as Prince Charles labels them) that the way they have made cheese, such as Reblochon, for centuries from the raw (unpasteurized) milk of their red and white Abondance cows – whose personalities they know as well as those of their children, above whose stalls they traditionally sleep in winter for warmth – is unhygienic.

Just as battery chooks have become a political issue, eventually how we treat all the animals which supply our food will also become part of the ethical debate. But the broader conclusion to be drawn from this is the consequence of going against nature. Perhaps those feed companies which encouraged the feeding of animal remains to ruminants should have heeded the prophetic words of German philosopher Rudolph Steiner back in the 1920s: 'if the ox were to eat meat directly instead of plants....the ox would go crazy'.

REFERENCES

[1] Cherry Ripe. 'Dying of Starvation in the Supermarket – Bio-diversity and the Industrialization of the Food Supply', *Disappearing Foods. The Oxford Symposium on Food & Cookery 1994*, Totnes 1995.
[2] *The Guardian*, July 8, 1998.
[3] Interview with Keith Hammond, Senior Officer, Animal Genetic Resources, the Food and Agriculture Organization of the United Nations, Rome. September 1998.
[4] R.W. Lacey, *Mad Cow Disease: The History of BSE in Britain*. Ipsela Press, Jersey. 1995.
[5] *The Food Magazine* 46, July–Sept. 1999, p. 5.
[6] *Guinness Book of Records*. Big Bertha, born March 1944, died February 1994.
[7] Cherry Ripe, *Ripe Enough?* Allen and Unwin, Australia, 1999.

Sandesh:
An Emblem of Bengaliness

Colleen Taylor Sen

Throughout the Indian subcontinent, Bengalis – inhabitants of the Indian state of West Bengal and the Republic of Bangladesh – are famous for their love of *mishti*, or sweets. Most Bengali sweets are made from milk products and sugar, sometimes flavoured with spices, fruit, nuts, coconut, and other ingredients.[1]

In a Bengali household, no meal is considered complete without some kind of sweet dish. Incomes permitting, Bengalis eat sweets throughout the day: as desserts at the end of meals (a custom not followed in many other parts of India); at afternoon tea, an important meal in a Bengali household; and as snacks at the countless sweet shops found in every town in West Bengal and Bangladesh, including uncounted thousands in Calcutta alone. Stories are told about wealthy Bengali *zamindars* (rich landowners) of past centuries who ate nothing but sweets from morning to night. This was seen as evidence not only of their wealth but their refinement, for an appreciation of the finer points of sweets has been viewed as a mark of a cultivated person.

Sweets are an essential component of Bengali hospitality. Bengalis send sweets to friends and superiors as gifts, eat them to celebrate passing an examination or getting a new job, and offer them to the gods at *puja*s. Sweets are a marker of rites of passages in a Bengali's life: the birth of a child, pregnancy, marriage, even death.

The apogee of the Bengali sweet-makers' art and the sweet most emblematic of Bengaliness is *sandesh*. *Sandesh* (the word is both singular and plural) are small sweetmeats made from *chhana* – casein made by cutting boiling cow's milk with lemon juice – and sugar, perhaps flavoured with nuts, fruits, rose water, even chocolate and vanilla extract. The mixture is cooked in a little clarified butter to make a smooth paste that is pressed into moulds shaped like flowers, fruit, or shells. *Sandesh* are dry (i.e., not served in syrup) and range in texture from soft and spongy to very hard. The word *sandesh* means 'news' or 'message', probably from the old practice of sending food, particularly sweets, with a messenger as a gift.

Sandesh comes in more than a hundred varieties. Their names reflect their flavouring, texture, shape, size, design, ingredients, and the poetic fancies – or advertising flare – of their creators. Examples are *Desh gorob* (glory of the nation), *Manoranjan* (heart's delight), *Monohara* (captivator of the heart), *Pranahara* (captivator of the soul), *Abak* (wonder), *Nayantara* (star of the eye), *Bagh* (tiger), and *Abar Khabo* (I'll have another). Top quality *sandesh* have a delicate, complex flavour that must be carefully savoured. Connoisseurs debate the virtues of their favourite variety and manufacturer with the passion and expertise of a French oenophile.

This paper will review the origins of *sandesh* and other Indian sweets in the context of Indian and Bengali culture and society; describe its manufacture; and provide a list of *sandesh* varieties.

History of sweets: the Indian subcontinent

The Indo-Aryans who migrated into what is now Pakistan and Northwest India starting around 1500 BC were a semi-nomadic cattle-breeding people who settled in small villages and eventually took to agriculture. 'The cow was the measure of value and was a very precious commodity,' writes a leading historian.[2] Cows are referred to no less than 700 times in the *Rig Veda*, the great collection of Indo-Aryan hymns.

The Indo-Aryans' staple diet consisted of dairy products, vegetables, fruit, barley and other cereals, and, for feasts, the flesh of oxen, goat, and sheep. Cow's milk was drunk fresh, boiled, or as cream. Yoghurt (*dahi*) was eaten with rice, cereals or folded into milk. Butter was produced by churning, and *ghrta* (today's ghee) or clarified butter was made by heating butter and skimming off the solids to produce a clear substance used for frying and as flavouring. By the third century the dairy industry was well organized: commercial milking and churning were in the charge of a state official who sold curd and other products.[3]

Sugar cane, a giant grass that needs rich moist soil and great heat to thrive, and sugar refining are both native to India. *The Natural History of Su-King* written in the seventh century AD noted that the emperor Tai-Hung sent people to learn the art of making sugar in Lyu (India) and especially in Mo-Ki-To (Bengal). In the early fifteenth century, another Chinese writer about Bengal mentions its 'white sugar, granulated sugar, candied or preserved fruits'.[4]

Indians have used sugar in many forms: raw, granulated (Sanskrit *Sarkara*), as a thick juice, as sugar candy, and as *gur* or *jaggery* – raw sugar juice boiled down to make a hard fudge-like product. Culinary texts from the twelfth to the sixteenth centuries describe a number of sweet dishes based on sugar and milk products, including an item called *sandesh*. However, no information is given about its content or manufacture, so that it is unclear whether it is the same as the modern version. Sweets were made by castes of professional sweet-makers, called *halwai* in North India and *mayuras* or *moiras* in Bengal[5] (from the Sanskrit *Modaka-kara*, confection-maker.) Sweets made of milk and sugar, even those purchased from outside, were eaten by all social groups. [6]

Khoa or kheer, a thick paste made by boiling down milk, appears to have been historically the main ingredient in Indian milk-based sweets. It is still the basis of most of the sweets indigenous to Northern India, Pakistan, Western and Southern India (where the sweet repertoire is relatively limited compared with that in Bengal). Ancient texts describe the curdling of milk with an acidic product to produce curds or casein, and even the manufacture of a kind of cheese called *dhandhawat*. But the texts are ambivalent and details sparse. Several authors cite a religious prohibition by orthodox Hindus against 'cutting' milk and note that Krishna, who grew up among cowherds, is depicted as a child stealing butter and cream but never *chhana*.[7] The reason for such a taboo is not clear, and it no longer seems to be in effect. An informal survey of orthodox Hindus and Jains found they do not have any inhibitions about eating sweets made from *chhana*.

History of sweets: Bengal

In Bengal, the majority of commercial sweets are made with *chhana*. What is the reason for its prevalence? One explanation is that sweet-makers adopted its use from Portuguese confectioners. In the mid- to late seventeenth century, some 20,000 Portuguese traders and their descendants had settled around Hooghly (near the site of present day Calcutta), near Dacca, and elsewhere in the region. They lived in great luxury, with many domestics and professional bakers who specialized in the preparation of sweetmeats, breads, cakes, cheese, and other delicacies. The French traveller François Bernier, who lived in India from 1659 to 1666, wrote: 'Bengal is celebrated for its sweetmeats, especially in places inhabited by the Portuguese, who are skilful in the art of preparing them and with whom they are an article of considerable trade.' [8]

The Portuguese produced curds by breaking milk with acidic materials. By the mid-nineteenth century, Bengali sweet-makers in Calcutta and elsewhere were using *chhana* as their main ingredient. [9] They also began to expand their repertoire by inventing new varieties, such as *rasgolla*, a light spongy white ball of *chhana* served in sugar syrup; *ledikeni*, a dark-coloured fried *rasgolla*; *cham-cham*, patties dipped in thickened milk and sprinkled with grated *khoa*; *pantua*, sausage-shaped spheres fried golden brown and dropped in sugar syrup; and *sandesh* in all its incarnations.

A critical factor in the development of the Bengali sweet industry was the emergence in the mid- to late nineteenth century of an affluent urban middle class. From 1833 on, Calcutta was the capital of India and the second city of the British empire, a thriving centre of commerce where ambitious Bengalis got rich as traders and middlemen. Drawn by the city's glitter, wealthy landowners moved here and built palaces. Educated Bengalis became government clerks, doctors, barristers, and other professionals.

With their disposable income, penchant for hospitality, and love of sweets (fortified by a cultural aversion to alcohol), these people became avid patrons of the sweet shops, who vied for their custom by inventing and promoting new varieties. A shop opened by Bhola Moira in 1851 seems to have been the first to sell *sandesh*. Other prominent sweet-makers of the mid-nineteenth century include Makhlan Lal, S.K. Modak, Bhim Chandra Nag, Nabin Chandra Das, and Sen Mohashir, the last three of which still operate today. *Rasgolla* is said to have been invented in 1868 by N.C. Das, who wanted to invent a new sweet to compete with *sandesh*. After several attempts, he came up with *rasgolla*. However, it only caught on when a wealthy timber merchant purchased large quantities and gave it to his friend and relatives. N.C. Das's son, K.C. Das, promoted its sale outside of Bengal and today *rasgolla* is available everywhere in India and abroad. [10]

The most famous story concerns the invention of *ledikeni*, dark-brown balls of semolina, *chhana*, and sugar filled with raisins, fried, soaked in sugar syrup, and rolled in powdered sugar. According to one version, B.C. Nag was challenged by Lady Canning, wife of the first Viceroy of India, to create a new sweet for her birthday. In another version, she tasted a new sweet and liked it so much that he named it after her. The latter is more likely: With an eye to marketing, sweet-makers started to give catchy names to existing or new products. A *sandesh* was named after Lord Ripon, Viceroy from 1880 to 1884. In the 1960s, a *sandesh* was called *Bulganiner Bishmoy* (Bulganin's wonder) in honour of the visiting Soviet premier.[11] *Sandesh* makers add chocolate, apples, ice-cream, and other non-traditional flavourings, and make *sandesh* shaped like slices of toast, sandwiches, cakes, chops, pastries, biscuits, and other Western food items.

Sandesh: its manufacture and varieties

To get a first-hand look at *sandesh* manufacture, I visited the firm of Sen Mohashir ('The Respected Mr. Sen'), which was founded by the grandfather of the present owner in the 1870s. The company owns a factory in North Calcutta, six retail outlets, and a farm with 80 cows and twenty water buffaloes where it produces all its own milk. Sen Mohashir makes 70–80 items, including 50–60 types of *sandesh*. The factory produces soft sweets in the morning and sends them to the shops to be consumed the same day. Hard sweets are made in the afternoon, since they last longer and can be sold over the next day or two.

Each season has its specialities. In the spring, soft *sandesh* is made with the season's new jaggery (molasses); in summer, *sandesh* are flavoured with mango, jackfruit and other fruits of the season and moulded in the shape of these fruits; in winter, hard *sandesh* are flavoured with sugar made from the sap of the date palm (*kejur gur*). *Sandesh* cost between three and five rupees each (between eight and twelve US cents). The main determinant of price is size.

Making *sandesh* sounds easy but is difficult, especially for the home cook. The first step is to make the *chhana*. Milk is heated slowly in a large pot. When it starts to boil, a little lime or lemon juice or citric acid diluted in water is gradually added. (For the home cook, the ratio is three to four tablespoons of lemon juice to two litres (eight cups) of whole milk). The milk is again brought to a boil and simmered until it curdles. It is immediately removed from the fire, then poured through a muslin cloth or cheese cloth. The cloth is gathered from all sides, tied on top, squeezed, and then put under a heavy wooden block until almost all the water is drained out. The *chhana* is then kneaded until it becomes smooth and pliable.

The next step is to mix the *chhana* with sugar or sugar syrup, stirring it gently in a large conical shaped pan in a little clarified butter over low to medium heat until the mixture thickens and becomes smooth. This is the critical stage in the process, and the man who does it is the key person in a sweet-making operation. He watches the mixture carefully, constantly adjusting the burner with iron tongs.

Small variations in the length of cooking are said to change the taste of the final product. Some aficionados believe *chhana* prepared over a wood-fuelled fire has more flavour than that made over coal. Another important variable is the quality of the ingredients, especially the milk and the *ghee*. Connoisseurs believe that milk produced from cows raised in certain areas is sweeter and more flavourful than that from other areas and that *chhana* made by the sweet-maker himself is superior to that purchased in the market. Although cow's milk is considered superior to buffalo's milk, some sweet-makers, including Sen Mohashir, add a little buffalo's milk to add body and texture.

Flavouring, spices and colouring are added to parts of the *chhana* and distributed among the *sandesh* makers. They sit on the floor, clad in *lungis* (a piece of cloth wrapped around the waist), and with great dexterity mould the individual pieces of *sandesh*, hand forming them into ovals or balls or pressing them into the moulds. Some are made of *chhana* flavoured in different ways; for example, the popular *sandwich sandesh* is made of alternating layers of green-coloured pistachio-flavoured *chhana* and plain white *chhana*. Some have a filling: a small piece of date, a little piece of flavoured *chhana*, thickened cream, or sugar syrup. The table below lists some well-known varieties of *sandesh*.

The role of sandesh *and sweets in Bengali culture*
In Bengal, hospitality and the offering of sweets are inseparable. '*Mukha mishti korao*' – 'Sweeten your mouth' – is the phrase used when offering them to guests. The number, quality, and provenance of the sweets are an index of a guest's status. In remote villages, where sweets are not available or people are too poor to buy them, visitors may be offered a glass of sugar-flavoured water. Friends and family members who travel bring back sweets from famous shops. Some towns are associated with certain items; for example, *jal bhara sandesh* with Chananagar and *sitabhog* and *mihidana* with Barddhaman district. Every festival is celebrated with sweets, and sweets are offered to the deities during *puja*.

Sweets play a role in the lifecycle rites.[12] When a woman becomes pregnant, her in-laws treat their friends to sweets and give her all the sweets she craves. The news of the birth of a baby is circulated with sweets on the sixth day after the birth. At a child's *mukhe bhat*, or first rice-eating ceremony, relatives bring sweets along with their gifts.

Before marriage, when the groom's relatives come to see the bride and perhaps finalize the marriage, the bride's party gives them sweets. If they accept them, the marriage is on. Before the wedding, the gifts sent by the groom's party to the bride's home include clothes, fish, and sweets, including fish-shaped *sandesh*. At the wedding dinner, *rasgolla*, *sandesh*, and other sweets are ordered for the occasion from a sweet shop. Conspicuous consumption is paramount on such occasions: a barrister is said to have presented his bride with a *sandesh* model of Calcutta's Gothic High Court at the wedding![13] After a death, the people who carry the body to the cremation grounds are given *neem*, a bitter leaf, followed by sweets.

Today Calcuttans are heard to complain that the quality and variety of sweets is deteriorating. Here, as elsewhere, concerns about health and appearance are driving people to try to limit their intake of sweets. Some of the old sweet shops have passed into the hands of non-Bengalis, and the *moiras* in their *lungis* are being replaced by machines. A thorough investigation of the Bengali sweet industry, including a survey and classification of the sweets in West Bengal and Bangladesh, would be an interesting and valuable study.

ACKNOWLEDGEMENTS

I gratefully acknowledge the assistance and contributions of Professor Dipesh Chakrabarty, University of Chicago; Research Professor Ajit K. Danda, Asiatic Society, Calcutta; Surajit Dasgupta, Calcutta; Bapa Ray, New Delhi; and Professor Ashish Sen, University of Illinois at Chicago.

BIBLIOGRAPHY

Achaya, K.T. *Indian Food: A Historical Companion.* Delhi: Oxford University Press, 1994
Banerjee, Satarupa. *Book of Indian Sweets.* New Delhi: Rupa & Co., 1994.
Banerji, Chitrita. *Life and Food in Bengal.* New Delhi: Rupa & Co., 1993
Bhattacharya, J.N. *Hindu Castes and Sects.* Calcutta: Thacker and Spinks & Co., 1898
Campos, Joaquim Joseph. *A History of the Portuguese in Bengal.* Calcutta, Lond: Butterworth, 1919.
Chaudhuri, Nirad C. 'Gastronomy in India,' in Amiya Chaudhuri, *Traditional Indian Cooking,* Delhi: Orient Paperbacks, 1990, pp. 25-26.
Das Gupta, Minakshie, Bunny Gupta and Jaya Chaliha. *The Calcutta Cookbook.* New Delhi, Penguin, 1995
Kkare, R.S., ed. *Culture and Reality: Essays on the Hindu System of Managing Foods.* Simla: Indian Institute of Advanced Study, 1976.
Kirchner, Bharati. *The Flavors of India.* New York: Galahad Books 1992.
Prakash, Om. *Food and Drinks in Ancient India,* New Delhi: Munshi Ram Manohar, La., 1961
Roy, Pratap Kumar. 'The Food and Sweets of Calcutta,' in *Calcutta: The Living City,* Volume II. Ed. Sukanta Chaudhuri. Calcutta: Oxford University Press, 1990. pp. 337-340.
Sen, Colleen Taylor, 'The Portuguese Influence on Bengali Cuisine,' *Food on the Move: Proceedings of the Oxford Symposium on Food and Cookery, 1996,* Prospect Books 1997, pp. 288-298
Sinha, A.K., 'The Sweetmeat Areas of West Bengal and its Tradition: An Anthropological Perspective', *Man in India,* September 1988, pp. 268-277.
Thapar, Romilla. *A History of India:* Volume I. Baltimore: Penguin Books, 1966
Toussaint-Samat, Maguelonne. *History of Food.* Trans. Anthea Bell. Cambridge, Ma.: Blackwell, 1992.

REFERENCES

[1] Most commercial sweets are made from *chhana,* casein or curds that are produced by 'cutting' boiling milk with lemon juice or some other acid. Home-made sweets are often made from *khoa,* a thick paste made by boiling down milk, wheat and rice flour, coconut, lentils, and other ingredients.

[2] Thapar, p. 34.

[3] For historical background, see Prakash; Achaya; and Toussaint-Samat.

[4] Ironically, the Bengali word for sugar is *chini* from the word for China. Today the state of West Bengal is a net importer of sugar, mainly from Uttar Pradesh.

[5] Typical Bengali *moira* surnames are Modak, Nag, Nandi, Rakshit, and Saha; see Bhattacharya. Whereas Hindu sweets are milk-based, Muslim sweets usually contain some farinaceous product, such as semolina, wheat flour, or chickpeas, as the main ingredient. A good example is *halwa,* bars of semolina or flour, sugar, and ghee that may contain nuts, fruits, carrots, etc. The word '*halwa*' comes from the Arabic word for 'sweet.'

[6] In the complex web of Hindu dietary rules and customs, food is supposed to suit a person's age, stage of life, caste, and the season. Both milk products and sugar are considered *sattvic,* or pure, and thus conducive to spirituality and serenity. Dishes made from these products may be may be offered to the deities and eaten by orthodox Brahmins and widows. The

method of preparing sweets by cooking with ghee over fire renders them *pucca*, which means they are considered fully cooked, less liable to pollution, and can be consumed outside the family. See Khare; and Achaya.

[7] Banerji, p. 14.

[8] Quoted in Achaya, p. 175. For a discussion of the Portuguese in Bengal, see Sen.

[9] Bengali Brahmins, vegetarian in theory, have been less rigid in their dietary habits than those in other parts of the subcontinent. They eat fish, for example, and thus would not have an aversion to *chhana*-based products or even cheese. At least until the late 1950s, a kind of soft cheese called 'Dacca cheese' was on the market. Another product still available in Calcutta is a salted smoked cheese called Bandel cheese after a Portuguese settlement.

[10] For background on the origins of Bengali sweets, see Sinha, pp. 268-277; and Roy, pp. 337-340.

[11] Das Gupta, Gupta and Chaliha, p. 352.

[12] Personal communication of Professor A.K. Danda, Asiatic Society of India.

[13] Das Gupta, Gupta, and Chaliha, p. 352.

Table: Description of Sandesh

Bengali name	English trans.	Ingredients	Remarks
Desh gorob	Glory of the nation	Saffron, pistachio, skim from the top of cream	Four layers
Sandwich	–	Three layers of white/green/white (flavoured with pistachio)	Triangular shaped
Rose cream	–	Flavoured with rosewater	–
Boishaki	Spring	Cashew nuts, dates, vanilla ice-cream flavour	Green colour. Decorated with silver foil
Malai chop	Cream chop	Round, with cream inside	Hard texture, covered with silver foil
Cake	–	Vanilla flavoured	Cake-like texture
Monohara	Captivator of the heart	*Khoa* filling, pistachio and cardamom dust	Very soft
Babu	Sir	Cashew filling, saffron flavoured	
Vanco	Abbr. 'vanilla/cocoa'	Chocolate filling	
Dilkhus	Happy Heart	Saffron, *keer*, and *sandesh*	
Peshawar	City in Pakistan	Layers of unflavoured and saffron-flavoured *chhana*	Four layers
Parijat	Fairy-like	Rose-water	Soft texture
Jol bhora	Filled with water	Sugar-syrup centre	Hard texture
Kanchagolla	Fresh *gullas*	Cardamom	Soft and granular texture
Amrita kumbha	Nectar of the Gods	Powdered sugar, ice-cream flavour	Bowl-shaped
Sarpuria	Cream-coated	Cardamom flavoured, coated with cream	Cut into squares
Batabi	Lime	Pistachios	Coloured green
Khejurer	Date	Chopped fresh dates and cardamom powder	
Chocolate roll	–	White *sandesh* filled with cocoa-flavored *chhana*	Resembles a swiss roll
Pranahara	Captivator of the soul	*Chhana* mixed with ground *khoa*; rose-water flavour	Very soft
Karapak	Hard cooked	No flavouring; cooked a long time until dry	Very hard
Dim	Egg	Part of *chhana* is dyed yellow and used as 'yolk' in oval *sandesh*	Cut in half to resemble eggs
Kaju	Cashew	*Chhana* is mixed ground cashews, saffron, and rosewater	Frozen as squares
Ice cream	–	coarse *chhana* flavoured with vanilla essence	
Notun gurer	New date-palm	Jaggery flavour	Light brown colour

Yoghurt in Iran

Margaret Shaida

Although yoghurt is a familiar item in our supermarkets today, it is a relative newcomer to the West. It has become popular here only in the latter half of this century. The word 'yoghurt' is Turkish, and many people think that the product itself comes from Turkey. This belief is supported by the oft-repeated assertion that it came with the Turkish and/or Mongol invasions from Asia in the twelfth and thirteenth centuries.

As with so much food production, the origins of making yoghurt go back to the earliest times and cannot be ascribed to one particular tribe or people. However, certain generalizations can be made. It is a product of the prehistory dairy economies that thrived in the nomadic regions of Central Asia, an area that covered more than half the known world, from the Middle East to the Gobi desert. Most of the early peoples living in this vast region were nomadic, living off their flocks of sheep and goats. Their prime source of protein was milk. From milk to yoghurt is a short step. As Harold McGee says, 'if left alone, fresh milk quickly teems with lactic acid bacteria that sour it and eventually cause the casein micelles to aggregate'.[1] Yoghurt, in fact, is one of the oldest food products. It is an effective way of preserving milk in a hot climate. The fact that it can be dried makes it even more suitable for long-term preservation.

A long history

Yoghurt has existed in Iran (where nomadic tribes also survived well into this century) for thousands of years, and its existence certainly preceded the Turkic invasions. In Persian, the word for yoghurt is *māst*. This is an Old Persian word, dating back to the Achaemenian era (over 500 BC), indicating a very early knowledge of yoghurt in Iran.

Yoghurt has been an established part of life in Iran for so long that many Persian proverbs and homilies refer to yoghurt and its by-products. For instance, in relating the story of a very frightening situation, somebody will be sure to say *Māst-roo sefid shod* (the yoghurt went white) i.e. the situation was so scary that even the yoghurt went white with fear. There is also a proverb that to take infinite care over a delicate job is the equivalent of removing a hair from a bowl of yoghurt.

There is a range of yoghurt dishes in Iran, known for centuries, as *Boorani*. These are often served as an accompaniment to a main meal or as a light luncheon on a hot summer's day. Some still retain their old names – *boorani-ye esfenaj* (yoghurt with spinach), *boorani-ye kangar* (yoghurt with cardoons) and so on, while others are known simply as *māst-o kheeyar* (yoghurt with cucumber), *māst-o mooseer* (yoghurt with shallots) or *māst-o laboo* (yoghurt with beetroot).

Such dishes are usually made with *māst-e keesé'-i* (strained yoghurt). As the name *keesé* (bag) implies, yoghurt is strained through a muslin bag to a thick, creamy consistency. This is vital when mixing it with other ingredients, as fresh yoghurt has a tendency to separate. As another popular saying goes, 'to strain the yoghurt' is like 'cutting a man down to size' especially a man who boasts without justification.

The original health food

Today, yoghurt is known in the West as the ultimate 'health food'. Stories of Bulgarian peasants living to well over a century on a diet of yoghurt were rife in the 1960s and 1970s. This came as no surprise to the Iranians, who have long recognized the benefits of consuming yoghurt. When an American, Justice William Douglas, was travelling in Iran in the 1960s, he related how, during a restless night, he had been given a bowl of yoghurt to help ease his stomach upset. 'After I had eaten most of it, I went to sleep at once; and I woke up well.'[2] He explained that '*māst* (yoghurt) harbors no bacteria hostile to man and has some that kill many unfriendly ones'. I, too, can personally vouch for the efficacy of yoghurt in curing attacks of 'Tehran Tummy', the equivalent of Delhi Belly!

While many of these stories have been largely dismissed, there is still a strong belief in Iran that the consumption of yoghurt is a cure for certain ailments. Indeed, it was stated on the BBC Radio 4 in April 1999 that the *Lactobacillus bulgaricus* in yoghurt significantly reduces the symptoms of diarrhoea and helps to speed recovery.[3]

According to Forough Hekmat, writing in *The Art of Persian Cooking* in 1961, yoghurt 'mixed with a good quantity of finely chopped fresh garlic, is one of the oldest cures for malaria'.[4] In 1974, Nesta Ramazani wrote in *Persian Cooking* that 'a number of nutritionists today recommend yogurt as a general health food and [which is] of particular value when one is taking antibiotics or sulfa drugs'.[5]

Such beliefs are hard to reject out of hand, especially when they are based on another ancient belief, that of the judicious blending of 'hot' and 'cold' foods. This is the Iranian equivalent of the more widely known Chinese 'yin' and 'yang' and was the subject of a paper given at this symposium in 1985.[6]

Cookery uses

Yoghurt is an indispensable part of Iranian cooking today. Cream is rarely eaten and even milk is consumed infrequently. Stews and soups are made with yoghurt, while others have yoghurt decoratively swirled into them at the table. It is added to many dishes to give a creamy, tart flavour. It is mixed with egg and saffron to make a number of rich rice dishes, and contributes to a wonderful *tahdeeg* (the celebrated crispy bottom-of-the-rice). It is even used in making pastry, most notably in the preparation of *ghotāb*, deep-fried almond-filled pastries. It also makes an appearance in the recipe for *zoloobiyā* (a sweet syrupy confection known as *jelebi* in India).

There is nothing new in any of this. According to the seventeenth-century cookery book written by the chef of the Safavid court in Isfahan,[7] yoghurt was much used in soups, stews, and rice dishes, and was frequently used as a garnish, including a particularly delicious-sounding 'lasagne' made of fried aubergine layered with cooked meat and yoghurt mixed with garlic and mint, and topped with yoghurt and saffron swirls.

More recently, at the end of the last century, the English doctor C.J. Wills, working in Iran, wrote that 'bread, eggs, *māst* and cheese form the staple food of the labouring classes in Persia'. He added that yoghurt, 'for the first twenty-four hours…is sweet and delicious, tasting like a Devonshire junket, but as a rule the Persian does not care for it until it has become slightly acid.'[8]

Wills went on to say that when the yoghurt has become a little acid, about half a pint of water can be mixed with it. Then 'a little cut mint is added and a few lumps of ice, and a cooling drink (or *doogh*) is made, which is supposed by the Persians to be a powerful diuretic. It is without question a capital thirst-quencher in hot weather.'

An ancient beverage

This brings us very neatly to the subject of *doogh*. Here again, it has been suggested that 'the Mongols brought it (*doogh*) with them'.[9] However, Frederick Shoberl, who visited Iran in the early nineteenth century, bears witness to the fact that *doogh* has a long and noble history in Iran, existing there long before the arrival of the Mongols. 'The antiquity of this beverage is so great,' wrote Shoberl, 'that Plutarch mentions it as part of the ceremony at the consecration of the Persian kings, to quaff a large goblet of this acidulated mixture.'[10]

Doogh originally meant 'milk' in Old Persian. By Sassanian times (third and fourth centuries AD), it had come to mean sour milk. Finally, in Middle Persian, it came to mean yoghurt diluted with water, a meaning that it has retained to the present day. In this respect, I should add that the verb 'to milk' is *dooshidand* in Old Persian. The past tense is *dukhtidand* and the name of the young maids who did the milking is derived from this: *dukhtar* in Persian (and 'daughter' in English). It is interesting to note that today, any white liquid is called *doogh* or *doogh-āb*. For instance, in the construction industry, diluted plaster is called *dooghab-e ahak*, while diluted cement is called *dooghab-seemon*.

The fact that *doogh* is known throughout the Middle East and India by different names (ay-ran, laban, lassi, etc.) would suggest that it developed simultaneously in various parts of the ancient world. *Lassi*, familiar to Indian restaurant diners in Britain today, is served with either sugar or salt. In Iran, *doogh* is served only with salt and mint.

Doogh has also made an appearance in Persian literature. Sa'adi commented that if a stranger offers you *doogh*, it may well consist of only a spoonful of *māst* and a double measure of water, meaning that the further a person travels from his home, the greater his lies are likely to be.

In more recent years, plain water has been replaced with the pure, aerated spring waters of the famous Cheshmeh Ab-e Ali (Ali's Spring) in the Elburz Mountains near Tehran. This spring of water was said to have appeared in the barren foothills at Ali's command[11] and the *doogh* made from this miraculous water was specially cherished. In the 1960s and 1970s, it was bottled and sold under the name of *Doogh Abe-Ali*. So vast were the numbers of bottles available that it would have taken only the most determined believer in miracles to accept that all these millions of bottles of *doogh* were made with water from Cheshmeh Ab-e Ali. They are still produced today, and can be found on sale not only in Iran but also in Iranian shops from London to Los Angeles.

The ubiquitous kashk

Another by-product of yoghurt is *kashk*. In Iran today, *kashk* means dried yoghurt. This is initially dried into small hard balls that can then also be ground into a powder. Both are reconstituted by the addition of water, and used to enhance soups and stews.

The word *kashk* can also be found in early Persian literature, and there is little doubt that it *is* a Persian word. However, it appears that *kashk* may well have meant something quite different in Iran in earlier times. Today, kashk (or *keshk, keshik*, etc.) means a number of different products in various countries of the Middle East. Here, I venture into waters almost as cloudy as *doogh* itself.

There is much confusion about the meaning of *kashk*. Charles Perry touched upon this 'vast confused history' in a note written to *Petits Propos Culinaires*,[12] while Françoise Aubaile-Sallenave investigated the whole question meticulously in 1994. Her paper, presented at SOAS in 1994,[13] went a long way to clear the air. As she said in her introduction, 'the complexity of today reflects the larger complexity of the past'.

She discussed the derivatives and the different meanings of *kashk* in different countries over the centuries. As far as Iran is concerned, Françoise Aubaile-Sallenave concluded that the original Persian word, *kashk,* meant a barley preparation or gruel, to which was added leaven and water or fermented milk. She goes on to suggest that this Persian word passed on to the rest of the Middle East and eventually came to designate various and complex preserved foods that were made from cereal and a ferment.[14] However, this does not really explain how this preserved barley product came to mean dried yoghurt in contemporary Iran. Françoise Aubaile-Sallenave suggests that, 'Iranian-speaking pastoralists, for whom dried sour milk was a staple, and who had no easy access to barley, applied the word *kashk* by analogy to dry sour milk.' But Charles Perry seems to think that dried yoghurt actually came with the Turkoman tribes. In support, he cites the thirteenth-century Arabic cookery book *Kitab Wasf al-At'imah al-Mu'tadah* which says that dried yoghurt was a 'Turkoman style *kashk*'.[15]

It seems impossible to come to any absolute conclusion. As with so many foods in the Middle East, the shifting borders and tribes over the centuries make it impossible to attribute the origins of one dish or product to any one modern nation. Suffice it to say that in Iran today, *kashk* means dried yoghurt. It is sold in solid balls, in powder form, or as a reconstituted liquid, for use as a convenience food to add richness to soups and baked dishes.

The Iranian panacea

The benefits of yoghurt are not confined to the kitchen alone. In Iran, it is recognized as an excellent balm for sunburn as well as a beneficial skin treatment. In the medicine cupboard, it is used as a cure for diarrhoea, a relief for fever, and an aid to long life. In the Iranian kitchen, it is irreplaceable. Despite the arrival of modern chilling and freezing technology, yoghurt remains an integral part of Iran's cuisine and culture. It thickens and enriches soups and stews; it is used as a tenderizing marinade for meat; it replaces milk and eggs in pastry-making; it is the base of numerous snacks and dips; and is the accompaniment for many rice dishes. It will always be found on the table, at breakfast, lunch and dinner. No Iranian home or kitchen would be complete without a daily supply of home-made yoghurt.

ACKNOWLEDGEMENTS

I am indebted to my husband Hassan and also to our good friend, Nasser Engheta of Los Angeles, for their assistance in the preparation of this article.

REFERENCES

[1] McGee, Harold, *On Food and Cooking*, Allen & Unwin, 1986, p.31.
[2] Douglas, William, *Strange Lands and Friendly People*, 1960. Quot. in Ramazani, see note 5.
[3] Hart, Tony, 'Leading Edge', BBC Radio 4 Programme about the Edinburgh Science Conference, 8 April 1999.
[4] Hekmat, Forough, *The Art of Persian Cooking*, Ebn-e-Sina, Tehran, 1961, p. 175.
[5] Ramazani, Nesta, *Persian Cooking*, University of Virginia Press, 1974, p. 69.
[6] Tilsley-Benham, Jill, 'A Look at Sardi/Garmi in Iran' in *Oxford Symposium Proceedings, 1984 & '85*, Prospect Books, 1986.
[7] Afshar, Iraj (ed.), *Cookery in the Safavid Period*, (in Persian), Iran Radio and Television Publications, Teheran, 1st edition, 1981.
[8] Wills, C.J., *The Land of the Lion and the Sun*, Ward, Lock, 1891, p. 171.
[9] Toussaint-Samat, Maguelonne, *History of Food*, Blackwell Publishers, Oxford, 1996 (Original French, Bordas, Paris, 1987), p. 120.
[10] Shoberl, Frederick, *Persia*, John Grigg, Philadelphia, 1828, p. 151.
[11] Morier, James, *Morier's Second Journey through Persia*, Longman, Hurst, Rees, Orme and Brown, London, 1818, pp. 369–370.
[12] Perry, Charles, *PPC* 14, Prospect Books, London 1983, p. 59.
[13] Aubaile-Sallenave, Françoise, '*Al-Kishk*: the past and present of a complex culinary practice', in *Culinary Cultures of the Middle East,* I.B. Taurus, London, 1994, p. 59.
[14] Ibid, p. 133.
[15] Perry, p. 59.

The Origins of the New York Dairy Industry

Andrew F. Smith

Spanish explorers and colonists first introduced Old World dairy animals into what is today the United States. Hernan De Soto probably brought cows with his expedition to Florida. Francisco Vásquez de Coronado transported cattle with him when he explored the American Southwest in 1540 and reportedly turned some loose in the Mississippi River Valley. Cattle herds accompanied Spanish settlers into Northern Mexico. These cattle thrived: beef and dairy products were among the more important staples of what is today the American Southwest.[1]

Other European colonists imported dairy animals to the eastern coast of North America. Richard Grenville took cows to Roanoke Island in 1585. Both the cows and the colony disappeared without a trace.[2] The first written record of cows arriving at Jamestown, Virginia, was a report of 100 head arriving in May 1611, but had been raised in the colony prior to this date. When the herd arrived, Lord Delaware reported that 'the Cattell already there are, much encreased, and thrive exceedingly with the pasture of the Country: the kine [cows] all this last winter, through the ground was covered most withy snow, and the season sharp, lived without other feeding than the grass they found, with which they prospered well, and many of them readie to fall with Calve; Milk being a great nourishment and refreshing to our people, serving also (in occasion) as well for Physicke [medicine] as for Food, so that it is no way to be doubted, but when it shall please God that Sir Thomas Dale and Sir Thomas Gates shall arrive in Virginia with their extraordinary supply of one hundred Kine.'[3] Seven years later, the colony counted over 300 cattle.[4] In 1620, the Virginia Company sent another 200 head to Jamestown. By 1634 all the 'better' plantations of Virginia were said to have plenty of milk, butter, and cheese.[5]

In New England, the Pilgrims' connection with dairy products predated their departure from England. According to William Bradford, the Pilgrims stored extensive quantities of butter for the voyage to the New World. However, before they left port, they were forced to sell off well over 3,000 lb of butter to generate money to buy other necessary provisions.[6] Despite their attraction to dairy products, the colonists failed to bring cows with them. For a time in New England, goat's milk was the only dairy product available. In

1624, the ship *Charity* landed one bull and three cows in Plymouth.[7] Subsequent deliveries were made and dairy farming was launched.[8]

When the Massachusetts Bay Colony established a settlement around what would become Boston in 1630, 30 cattle accompanied the first colonists. Other stock was sent subsequently. Increased availability meant expanded usage. Milk was drunk, churned into butter and converted into cheese. Milk was consumed with maize-based dishes, such as hasty pudding, samp, and suppawn. One favourite dish during the colonial period was baked pumpkin filled with milk and eaten with a spoon.[9] Curds and cream, a mixture of coagulated milk and cream, was also rated a delicacy. By 1650, dairy farming was so successful that the Massachusetts Bay Colony exported butter and cheese.[10]

The Dutch settlers who colonized Manhattan in 1624 were even more committed to dairy farming. They brought with them 103 head of livestock, including bulls and cows. Unfortunately, the population of the colony grew faster than the cattle herd. In 1628, Reverend Jonas Michaelius reported that fresh butter and milk were difficult to obtain, 'owing to the large number of people and small number of cattle and farmers'. The colony needed cows and 'industrious workers for the building of houses and fortresses, who could later be employed in farming, in order that we produce sufficient dairy products and crops'.[11] Subsequently, the Dutch specifically recruited dairy farmers to settle in New Netherlands. Despite the British conquest of the colony in 1664, the dairy industry remained largely in the hands of the Dutch. When Peter Kalm visited America in 1749, he reported that the New York Dutch drank their tea with milk. At breakfast they consumed either bread and butter or bread and milk, and quaffed buttermilk. At supper, they ate bread and butter, and downed milk infused with small pieces of bread. The Dutch served cheese at both breakfast and dinner.[12] It was not until after the American Revolution that New Englanders flocked to central New York and established a dairy industry based on British traditions.

Milk became a very common beverage throughout America in the eighteenth century. William Byrd in 1709 reported in his dairy that he 'ate milk for breakfast' almost every day. This milk was sometimes boiled and consumed with rice, potatoes, strawberries, rhubarb, apples, corn pone, and tea.[13] Landon Cater enjoyed marshmallow root boiled in milk and sweetened with brown sugar.[14] Shortly after the American Revolution, German-born Jacques-Pierre Brissot de Warville pronounced American cheese delicious and proclaimed that it was comparable with English Cheshire and French Roquefort.[15]

In the nineteenth century, milk products became even more abundant. Even small farms had one or more cows. In New York City, cows were tied to stakes and fed garbage. Property owners on Christopher Street rented out the privilege of staking cows in front of their homes. While the aroma generated by the cows might seem offensive to modern New Yorkers, to those of an earlier age this was a good financial investment. According to dairyman John Dillon, 'the disposition of the manure was a provision of the lease contract. At that time there was a demand for the manure on the farms cultivated on the land now covered by St. Patrick's Cathedral on Fifth Avenue and 50th Street, and the Empire State Building at Fifth Avenue and 34th Street.'[16]

For those who did not want to bother with keeping cows, women and men dispensed milk by ladle from hand-carried pails. As demand increased, distributors increased their carrying capacity to several gallons by employing a wooden yoke carved to fit over the shoulders and the back of the neck. Suspended from each arm were chains or ropes with hooks at the end. With this yolk placed over the shoulders, the carrier stood between two pails, stooped forward, attached the hooks, and then straightened up.[17]

Technology played an important role in increasing production of dairy products. For instance, constant experimentation improved the dash churn. Churning butter was a hard and lengthy endeavour. Creative solutions were devised to make this easier and less time consuming: pump handles were put on dashers, and rockers were put on churns; milk splashing against inside paddles converted milk into butter; the back and forth motion of swing churns made butter without paddles.[18] The hand-powered dash churns were also replaced by ones powered by dogs, sheep, horses, or oxen. The labour saved freed farmers to cultivate more land or churn more butter. The result was increased production of butter at a lower cost. Increased output enabled farmers to concentrate on one or two commercial crops which, when sold, yielded far more than subsistence. Between 1810 and 1850, Eastern farmers began specializing in dairy farming, 'partly because they could not meet western competition in producing small grains and livestock and partly because the increasing size of cities offered more profitable markets for perishable foods.'[19]

The growth of American dairying was observed by many European visitors.[20] Fanny Trollope, mother of British author Anthony Trollope, commented about the dairy business in Cincinnati, Ohio, which she visited in 1832:

> From the almost total want of pasturage near the city, it is difficult for a stranger to divine how milk is furnished for its supply, but we soon learnt that there are more ways than one of keeping a cow. A large proportion of the families in the town, particularly of the poorer class, have one, though apparently no accommodation whatever for it. These animals are fed morning and evening at the door of the house, with a good mess of Indian corn, boiled with water; while they eat, they are

milked, and when the operation is completed the milk-pail and the meal-tub retreat into the dwelling, leaving the republican cow to walk away, to take her pleasure on the hills, or in the gutters, as may suit her fancy best. They generally return very regularly to give and take through the morning and evening meal; though it more than once happened to us, before we were supplied by a regular milk cart, to have our jug returned home empty, with the sad news that 'The cow has not come home and it was too late to look for her to breakfast now.' Once, I remember, the good woman told us that she had overslept herself, and that the cow had come and gone again, 'not liking, to hanker about by herself for nothing, poor thing'.[21]

Her observations reflected accurately dairy conditions throughout America in the 1830s, but within 50 years, this life style had altered dramatically mainly due to increased demand for dairy ingredients in American cookery and the vast expansion of supply through the development of the factory system of dairying.

American dairy cookery

During the late eighteenth and early nineteenth centuries, several works published in the United States highlighted the importance of dairy to American cookery. In 1790, two editions of the British work, *Concise Observations on the Nature of our Common Food*, attributed to Thomas Hayes, were reprinted in New York. These volumes offered directions for improving milk, butter, and cheese production.[22] In 1793, *The Art of Cheese-Making, Taught from Actual Experiments, by which More and Better Cheese May Be Made from the Same Quantity of Milk* was published in Concord, New Hampshire. It was subsequently reprinted in Boston and Windham, Connecticut.[23] As the book contains recipes for making cheese, Eleanor Lowenstein included it in her *Bibliography of American Cookery Books, 1742–1860*. While the author is unknown, it was clearly written by an American.[24]

In 1796, three works were published in the United States, all of which had important sections on milk, butter, and cheese. The first was John Barlow's poem 'The Hasty Pudding', initially featured in *New York Magazine* and subsequently published in booklet form.[25] The American concept of hasty pudding was borrowed from the British and Scottish dish of the same name. Hasty pudding had been around Great Britain, according to the *Oxford English Dictionary*, at least since the late sixteenth century. Its main ingredient depended on where one lived: in England hasty pudding was based on wheat, while in Scotland it was based on oats; most recipes added milk, sweeteners, or spices. Hannah Glasse's *Art of Cookery Made Plain and Easy*, originally published in London in 1747, featured three recipes for hasty pudding, all of which contained milk and butter as ingredients.[26] Glasse's work was shipped to America and became one of the most popular cookbooks used in colonial times. When it was finally published in the United States in 1805, it included only one recipe for hasty pudding, but it also contained directions for selecting butter and cheese.[27]

The second work published in 1796 was Josiah Twamley's *Dairying Exemplified, or the Business of Cheese-Making*. It had been published originally in Warwick, England, in 1784. The American edition was based on the second British edition published in 1787. A major portion of the work focused on the 'business of cheese-making: laid down from approved rules, collected from the most experienced dairy-women'.[28]

The third was Amelia Simmons's *American Cookery*, which was the first general cookbook authored by an American. It reflected mainly English cookery practices and many recipes were borrowed from British works such as Susannah Carter's *The Frugal Housewife*. The first edition of *American Cookery* included an introductory section 'to procure the best viands', which featured tips on selecting butter and cheese. The author reported that 'tight, waxy, yellow Butter is better than white or crumbly, which soon becomes rancid and frowy.' The author recommended further, 'to have sweet butter in dog days, and thro' the vegetable seasons, send stone pots to honest, neat, and trusty dairy people, and procure it pack'd down in May, and let them be brought in night, or cool rainy morning, covered with a clean cloth wet in cold water, and partake of no heat from the horse, and set the pots in the coldest part of your cellar, or in the ice house.' As to cheese, the author pointed out that 'deceits are used by salt-petering the out side, or colouring with hemlock, cucumberries, or safron, infused into the milk; the taste of either supercedes every possible evasion.'[29] In the second edition of the work published in Albany, Simmons proclaimed that she had not written the introductory material, which had been added without her permission by the original publisher, but throughout the work numerous dairy products appeared in the recipes, including ones for pudding, custard, syllabub, creams, and cakes.[30]

Yet another dairy book was published in 1801 in Albany, New York: Joshua Johnson's *The Art of Cheese Making Reduced to Rules, and Made Sure and Easy, from Accurate Observation & Experience, Published for the Help of Dairy Women*. This book instructed dairy women how to 'make cheese which, when old, is sweet, full of spirit, and free from every nauseous flavor.'[31] Thirteen years later, John Nicholson of Herkimer County, New York, offered extensive directions for making butter and cheese, as well as for selecting proper churns.[32]

Dairying was an important component of all general cookbooks published subsequently in the United States. Three examples will suffice to illustrate the pervasive dairy content in cookbooks. H. L. Barnum, *Family Receipts,* for instance, includes fifteen pages on butter- and cheese-making in addition to individual recipes containing dairy ingredients. Barnum reported that most farmers let the cows roam around and did not feed them properly, particularly during the winter. This was a mistake, he proclaimed, for milk was 'not only the most nutritious but cheapest article of food'.[33]

The second example is N.K.M. Lee's *The Cook's Own Book,* which was essentially a compendium of recipes compiled from British and American sources. Lee's recipes were borrowed from other sources and were not necessarily reflective of culinary practices in New England during the early 1830s. Nevertheless, the popularity of Lee's cookbook was demonstrated by the dozens of editions it went through and by the frequency with which her recipes were borrowed by others. The sheer number of dairy-related recipes in Lee's work was amazing. She included directions to clarify butter and to make burnt butter, melted French butter, oiled butter, and butter sauce. She featured recipes for boiled, creamed, caked, pounded, roasted, toasted, and stewed cheese. She included recipes for creams flavoured with almond, barley, chocolate, coffee, lemon, orange, raspberry, ratafia, roseat, snow, and vanilla, as well as cream cakes, cream fritters, cream tarts, and whipped cream. Cream and milk also appeared in ice-cream recipes. Butter, milk, and cream appeared as ingredients in numerous recipes, including for alcoholic beverages.[34]

The third example is Eliza Call's *The Young Housekeeper and Dairymaids Directory* published in 1859 in Syracuse, New York. At the time of its publication Call was a 46-year-old widow who owned a farm in Fabius, a small community about twenty miles from Syracuse, New York. Dairy farming had been well established in the county. By 1860, the county had 35 dairy farms with from 25 to 125 cows.[35] Not much is known about Call. She was born in Cortland county, one of the more important dairy-farming areas at the time in New York. She married her husband, John R. Call, and they had five children. Her husband died in 1855, but she continued to manage the dairy farm. According to the 1860 Federal Census, her farm was valued at $5,400, and she had additional personal wealth of $1,175.[36] In the preface to the cookbooklet, she reported that she had been a housekeeper for 25 years. She claimed to have examined a number of cookbooks of the day carefully and had 'not in a single instance found one that has been written by an experienced Housewife'. She believed that a good cookbook 'written by one that was well acquainted with the science of House-Keeping was much needed'. *The Young Housekeeper* was intended to remedy that deficiency.[37]

Call's cookbooklet consists of 68 pages of miscellaneous recipes – most of which are unremarkable. The exceptions are the seventeen pages related to milk, butter, and cheese – which comes as no surprise as Call was a dairymaid for much of her adult life. In her cookbooklet Call proudly announced that her work was solely based on her 'many years of experience'. She did not believe that her dairy directions were 'the best in the world', only that she had 'never seen a single article or essay written on the subject but what in some respect might be bettered'. Her first rule was to make sure that 'everything appertaining to milk things should be kept perfectly sweet and clean. The milk should, as soon as it is brought in, be strained.' She recommended that the milk-room should be 'a distance from the fire–cool and airy' as heat caused milk to sour. The milk had to be carefully watched. As soon as it thickened, the cream was poured into stone jars or large covered tin pails. In summer, the pail was placed in cold water to cool the cream. The cream had to be 65° to churn. When the butter was well churned, it was placed into a wooden tray or bowl and cooled with a paddle. The butter then had to be washed and salted. To colour the butter, Call preferred yellow carrots to annatto. She had learned this mode of colouring 'from an Englishman, who used to get a penny or two a pound more for his butter made in this way, than his neighbor'. She believed that the juice of carrot gave a rich colour to the butter and also added 'sweetness to the taste'. She then packed the butter for long-term storage. Call believed that good butter was 'certainly one of the greatest luxuries of life'. It was 'a good substitute for meat and enriches many articles of food of different kinds, to say nothing of its accompaniments to our table in the form of delicious "bread and butter".'[38]

Call believed that the quality and quantity of butter and cheese depended on the conditions under which the cows were managed. For food, they required 'plenty of good, clean hay during winter. In the summer, cows needed 'good pasture and free access to water.' To increase the quantity of milk, she fed them pumpkins, carrots, and beets.[39] Call also believed in the importance of the dairymaid, who controlled 'the chief art of making cheese of the first quality'. The dairymaid needed 'skill, cleanliness and strict attention to her business'. Every operation needed to be performed precisely at the proper time, as 'hastening or delaying the execution' produced inferior cheese from milk from which the best might have been obtained.[40]

Call's dairy farm had nine cows. She milked them at half-past five in the afternoon, thus enabling her 'to have the milk run into a curd, and drained by a little after nine'. To make cheese, whey butter was used. As soon as it was churned, the whey was salted, and 'placed over the fire in an iron vessel and tried until the sediments come to a crisp and settle to the bottom'. After that, it was strained off, salted and placed in a cheese tub. Rennet was taken from the fourth stomach of a calf. It was prepared ahead of time and was carefully and thoroughly stirred into the tub. Curds formed after about an hour. The curd was cut by hand, bandaged, positioned in a hoop, and placed in a press. After removing the cheese from the press, the outside was coloured. Call rubbed the cheese with a solution of Venetian red in strong lime water until it was orange-coloured. The cheese was then stored, turned daily, and greased with fat.[41]

These cookbooks reflected this way of life that had existed with but few modifications for almost three centuries. Men mainly cared for the livestock and assisted with the heavy labour needed in dairying. Women milked the cow, churned the butter, and processed the cheese. It was extremely difficult work. As one observer reported cheese-making was an art which had to be learned. As most of its operations were performed by females, the dairy farmer was constantly training apprentices for 'when his dairymaid has been carefully taught the trade, she marries, and is at once lost to him'. This scarcity of skilled cheese-makers was felt severely throughout the whole dairy region. This required 'the farmer and his family, and more especially the female portions, to arduous labor; taxing their strength to a degree that tells heavily on health and constitution.'[42]

Dairying was conducted on the family-owned farm. Sales of surplus milk products were sold or bartered to neighbours or merchants in local communities. It was a way of life soon to disappear. During a fifteen year period from 1850 to 1865, New York dairy farms converted from small family-owned operations to a factory system that generated a multi-million dollar industry.

The beginning of the New York dairy industry

Even before the American Revolution, New England soil had begun to lose its fertility, causing a great decline in crop yields. After the Revolution, many farmers migrated to central New York, where land was inexpensive and had not been exhausted. Dairying became an important component of farming in central New York about 1800. One contemporary observer commented that 'its progress was slow, and the business was deemed hazardous by the majority of farmers, who believed that over-production was to be the result of those making a venture upon this speciality'. However, dairy farmers rapidly bettered their condition, 'outstripping in wealth those who were engaged in grain raising and a mixed husbandry'. This was particularly true when cheap grain from western states flooded eastern markets. Yet, the operations of dairy farms 'were rude and underdeveloped; the herds were milked in the open yard; the curds were worked in tubs and pressed in log presses. Everything was done by guess, and there was no order, no system and no science in conducting operations.'[43]

Several factors encouraged expansion of the dairy industry during the early nineteenth century. One of the more important was the rapid development of improved transportation. This relationship between dairy farmers and transportation was particularly significant in New York. The invention of the steamboat made it possible for dairy farmers along the Hudson River to ship their produce into the rapidly expanding New York City. The steamboat *Mary Powell* regularly brought milk to New York City from Newburgh and other points along the Hudson River.[44] The completion of the Erie Canal in 1825 meant that dairy farmers in central New York could easily ship butter and cheese to Albany and to New York City, and this further encouraged the growth of the dairy industry.

The expansion of the railroads northward from New York City meant that dairy farmers who lived north of the metropolis, but not on a navigable river, could ship fresh milk into the city daily. By 1842, the Erie Railroad had been extended as far as Goshen, Orange County. Two years later, the Orange County Milk Association was organized to ship milk to New York City. Unfortunately, the early operators of the Erie Railroad provided irregular service and the farmers were inexperienced with transporting milk. Eventually, these problems were overcome. Milk was transported in a tin can holding from 30 to 80 gallons. The cover had a flange six inches wide, which fit into the inside the can. An aperture in the top contained a cork, which acted as a vent to allow the escape of the air when the cover was placed over the can. This went within half an inch of the surface of the milk, so that the milk was not agitated while being hauled to the factory. Milk cans had spigots to draw off the milk at the bottom. Some factories manufactured their own cans, all of the same diameter, so that the quantity of milk contained could easily be determined. Farmers delivered the cans to platforms erected at crossroads: milk trains stopped and loaded them on the cars. The empty cans were then returned to the place of shipment. But accidents, carelessness, and theft caused extensive loss of the cans. To correct these abuses, the New York legislature passed the Milk Can Law, forbidding the use of branded milk cans by any person other than the owner.[45]

Milk trains that started in Goshen late in the afternoon might not arrive in the city for 36 hours. To transport milk for that length of time, particularly in hot weather, required special handling. At first, the milk was strained, put in long tin pails cooled by water, and carted to the train station where it was poured into fifty-gallon cans. During the 1850s, creameries were established within convenient distance along the route of the railroad. Farmers delivered the milk to the creamery, received credit for their delivery, then returned home. The creamery employees cooled the milk with ice and then shipped it to the train depot.[46]

The centralized factory rapidly escalated the production of dairy products. Dairy factories had first been developed in Switzerland about 1800. Experimental dairy factories had existed in the United States since the 1840s. Cheese factories were founded in Ohio. These establishments purchased unsalted curd from farmers and then pressed, cured, and marketed the cheese. George Hezlep, for instance, purchased curd made from the milk of 1,000 cows, and manufactured about 100 cheeses daily in 1850. Like other similar establishments, Hezlep's operation went under a few years later.[47]

Factory systems developed in New York during the 1850s. In the most successful system, farmers carried their own milk to a central factory, collectively paid a superintendent and other expenses and received a percentage of the profits. An early example of the factory system was the Cheese Manufacturing Association launched by Jesse Williams in Rome, New York. Williams was an experienced and skilful cheese-maker. When his son married and purchased his own farm, Williams contracted to make cheese from the milk of his son's cows. The son delivered the milk daily to his father's milk-house. Williams then converted the milk into cheese. This worked admirably. He then expanded his operation to take milk from his neighbours and erected buildings expressly for processing milk. Soon Williams manufactured cheese from the milk of about 1,000 cows, which greatly lowered the costs of production.[48]

On reaching the factory, the milk was weighed and credited to the sender. It was then run out into vats made of wood with an inner shell composed of tin. The space between the wood and tin could be flooded with cold water or hot steam, depending on what was required. A thermometer was suspended from a pulley over each vat so that the temperature could easily be determined. If the milk was less than 82°, steam was admitted. Based on experience, the superintendent added rennet and annatto to produce a rich cream colour. The intent was to induce coagulation within 45 minutes. When an hour had elapsed, the curd had the right firmness. It was then cut perpendicularly with curd knifes, which had five two-edged blades. Soon after this first cutting, the curd was placed in perpendicular columns, and the whey separated sufficiently to cover the curd. Steam was then piped in to raise the temperature to 86°. The curd was gently stirred to separate more whey, which was drawn off by a syphon into a trough and fed to pigs.[49]

The curd was then re-cut and reheated with steam for another hour. Again, it was cut and stirred regularly during the cooking until the particles were about the size of wheat grains. When the curd was judged to be sufficiently cooked, the remainder of the whey was drawn off. The curd was placed in a sink and drained. Salt was then added. The curd was poured into hoops and placed in screw presses. The newly-formed cheese was removed from the press, bandaged, and carted to a cheese house to cure. Cheeses averaged 21 inches in diameter and weighed 125 pounds. Cheeses were turned, rubbed and greased daily. After aging 30–40 days, they were packed into uniform round boxes, which gave a better appearance for market and afforded more protection in transportation. In 1860, cheese sold for about twelve cents per pound.[50]

The growth of factory system of cheese making was surprising. In 1851 only William's factory existed in New York. By 1855, there were seven factories. By 1860, there were 38. Five years later, 500 such establishments existed, and by this date similar factories had been established throughout the Northern states. The reasons for their success were several. First, private dairies had a greater hazard of souring of the milk from carelessness. The temperature of the milk could be controlled more easily in the factory system, which amassed larger volume. The factory system produced a uniform product, and each cheese was marked neatly with the date, its weight, and its producer. Cheese moved from a generic item to a name-brand product. Consumers sought out cheese that they particularly liked.[51] This increased sales and therefore expanded profits. Simultaneously, factories bought at wholesale and consequently reduced expenses. Finally, the factory system relieved 'the farmer and his family from the drudgery of the manufacture and care of cheese' and shifted the work to professionals who specialized in cheese-making.[52]

Due to these changes, the factory system rapidly expanded throughout New York, and the family-operated dairy farms declined in significance. By 1880, more than 1,500 dairy factories were in operation in New York alone and thousands of others were operating in other states. When Walter Gore Marshall visited America in the 1880s, he reported that milk was very popular, exclaiming: 'I have often been filled with wonder and admiration upon seeing the amount of milk an American will drink at one meal, without apparently getting bilious.'[53]

Adulterations
As New York City's population grew, demand for milk increased. The business of distributing milk became more profitable. As dealers multiplied, competition increased. Ambitious dealers became eager for even larger profits. To meet these ambitious, large stables were built near breweries in the 1820s. Mash or slop ran from the breweries to the cows' mangers through wooden chutes. By the 1830s, 18,000 cows in New York City and Brooklyn fed on this food almost exclusively.[54] The milk produced by cows fed on brewery waste did not produce milk rich in butter fat. Brewery cow milk was thin and light blue in colour. To correct these defects, the brewery dairies added water for volume, chalk and annatto for colouring, molasses for sweetness, and an occasional egg

for creaminess. The brewery dairies were generally ill-managed and badly kept. Simultaneous with their rise was the outbreak of cholera among infants. Infant mortality dramatically increased. By 1840, every fifth child in Manhattan died of cholera. Observers estimated total infant mortality rates almost reached 50 per cent.[55]

The first person to blame the brewery dairies for the high infant mortality was Robert Milham Hartley, a strong temperance advocate. In 1842 he published an exposé pointing to the abuses of the brewery dairies. He called the produce 'swill milk' or 'whisky-milk', which was doctored up with 'starch, sugar, flour, plaster of Paris, chalk, eggs, annatto, etc.'[56] This issue continued to surface. In 1857 the Common Council of Brooklyn created a committee to investigate the problem. What they found was shocking: 'The cows are tied in the stables when they are purchased, and kept there until they die or are sold to the butcher. They are fed three or four times a day with boiling swill, which remains steaming under their heads until it becomes sufficiently cool for them to drink.' The report asked: 'It is strange that lung disease prevails? And is it any wonder that their lungs become affected, and that they die at once when they dry up, and the milk ceases to carry off the poisonous secretions, which the lungs cannot throw off from want of fresh air?' After the report was issued, *Frank Leslie's Illustrated Newspaper* declared war on the brewery dairies in Manhattan and Brooklyn. A series of articles titled 'Startling Exposure of Milk Trade of New York and Brooklyn' reported that the stables surrounding the distilleries were 'dilapidated,' 'wretchedly filthy' and 'disgusting'. They were rude wooden shanties that were thickly hung with cobwebs. The cows were 'ranged in double rows, their heads to the swill troughs and their tails, or rather the remnants of their tails, towards each other: so close that sometimes one cow actually lies on the other.' The swill rushed foaming, and 'boiling hot and reeking with subtle poison it splashes into the troughs'. At first the cows revolted against the swill 'but after a week or two they seem to have a taste for it and in a short time we find them consuming from one to two or even three barrels of swill a day.' While developing the exposé, one of Frank Leslie's artists was attacked and a reporter was killed.[57]

The campaign eventually succeeded. One by one the brewery dairies were closed, but this did not end the concern with milk safety. Mary Ronald, editor of a 1897 cookbook, noted that milk was a proven disease carrier and urged mothers to boil it before giving it to their children.[58] However, the dairy industry had made great strides during the late nineteenth century. Pasteurization was introduced, as was an instrument developed by Dr S.N. Babcock at the University of Wisconsin that tested the richness of milk.[59] Henry E. Alvord's article, 'Dairy Development in the United States' published in the 1899 US

Department of Agriculture's *Yearbook,* proclaimed that: 'No branch of agriculture has made greater progress than dairying during the nineteenth century.' Alvord continued: 'it is now regarded as among the most progressive and highly developed forms of farming in the United States.'[60]

REFERENCES

[1] Juan N. Almonte, 'Statistical Report on Texas,' trans. by Carlos E. Castañeda, *Southwestern Historical Review* 28 (January 1925): 197; T. R. Pirtle, *History of the Dairy Industry* (Chicago, Illinois: Mojonnier Bros. Company, 1926), 16.

[2] Lewis Cecil Gray, *History of Agriculture in the Southern United States to 1860,* 2 vols. (New York, P. Smith, 1941), 1:19.

[3] Lord Delaware, as quoted in Ralph Selitzer, *The Dairy Industry in America* (New York: Dairy and Ice Cream Field and Books for Industry, 1976), 5–6.

[4] William Stith, *The History of the First Discovery and Settlement of Virginia* (New York: Reprinted for Joseph Sabin, 1865), 123.

[5] Richard J. Hooker, *A History Food and Drink in America* (Indianapolis & New York: Bobbs-Merrill, 1981), 32–33.

[6] William Bradford [Samuel Eliot Morison, ed.], *Of Plymouth Plantation 1620–1647* (New York: Alfred A. Knopf, 1952), 149.

[7] Ibid. 174.

[8] Edmund Sears Morgan, *The Gentle Puritan; a Life of Ezra Stiles, 1727–1795* (New Haven, Yale University Press, 1962), 128.

[9] George Francis Dow, *Every Day Life in the Massachusetts Bay Colony* (Boston: The Society for the Preservation of New England Antiquities, 1935), 98.

[10] Pirtle, op. cit. 17; Mary Caroline Crawford, *Social Life in Old New England* (New York: Little, Brown, and Company, 1914, 249; Dow, op. cit. 98; Bayrd Still, *Mirror for Gotham: New York as Seen by Contemporaries from Dutch Days to the Present* (New York: University Press, 1956), 25–26; Hooker, op. cit. 32–33.

[11] Jonas Michaelius, May 8, 1628, as quoted in Ralph Selitzer, *The Dairy Industry in America* (New York: Dairy and Ice Cream Field and Books for Industry, 1976), 10.

[12] Pehr Kalm, *Travels into North America* (London: Longman, Hurst, Rees, Orme, and Brown; and Cadell and Davies, 1812 [1772]), 346–347.

[13] William Byrd [Louis B. Wright and Marion Tinling, eds.], *The Secret Dairy of William Byrd of Westover, 1709–1712* (Richmond, Virginia: The Dietz Press, 1941), 3, 7, 8, 11, 32, 41, 57, 122, 298, 299, 300, 301, 302, 303, 354, 404.

[14] Jack P. Greene, *Landon Carter: an Inquiry into the Personal Values and Social Imperatives of the Eighteenth-century Virginia Gentry* 2 vols. (Charlottesville: University Press of Virginia, 1965), 1:211.

[15] J-P Brissot de Warville, *New Travels in the United States of America Performed in 1788* 2nd ed. (London, Printed for J. S. Jordan, 1794), 91.

[16] John J. Dillon, *Seven Decades of Milk: A History of New York's Dairy Industry* (New York: Orange Judd Publishing Company, 1941), 1.

[17] Esther Singleton, *Social New York under the Georges 1714–1776* (New York: D. Appleton, 1902), 358; Dillon, op. cit. 1.

[18] David Marshall Owen, '400 Years of Milk in America,' *New York History* 31 (October 1950): 451

[19] John H. Schlebecker and Andrew W. Hopkins, *A History of Dairy Journalism* (Madison: University of Wisconsin, 1957), 6–7.

20 John Leng, *America in 1876* (Dundee: Dundee Advertiser Office, 1877; reprint Arno Press, 1974), 56–57.
21 Fanny Trollope, *Domestic Manners of the Americans* Reprint (London: Penguin Books, 1997), 52.
22 [Thomas Hayes, attributed], *Concise Observations on the Nature of Our Food* (New York: T, and J. Swords, 1790); [Thomas Hayes, attributed], *Concise Observations on the Nature of Our Food* 2nd ed. (New York: T, and J. Swords, 1790).
23 *The Art of Cheese-Making, Taught from Actual Experiments, by which More and Better Cheese May Be Made from the Same Quantity of Milk* (Concord: Geo. Hough, 1793); reprinted (Boston: Benjamin Edes, 1797).
24 Eleanor Lowenstein, *Bibliography of American Cookery Books, 1742-1860* (Worcester: American Antiquarian Society, 1972), 1–8.
25 Joel Barlow, 'The Hasty Pudding,' *New York Magazine* New Series, Vol.1 (January 1796): 41–49; Joel Barlow, *The Hasty-Pudding* (New Haven: [Tiebout & Obrier], [1796]).
26 Hannah Glasse, *Art of Cookery Made Plain and Easy* (London: Published by the Author, 1747), 80.
27 Hannah Glasse, *The Art of Cookery Made Plain and Easy with an introduction by Karen Hess*, facsimile of the 1805 edition (Bedford, Massachusetts: Applewood Books, 1997), 7, 127–28.
28 J. Twamley, *Dairying Exemplified; or the Business of Cheese-making* (Warwick: J. Sharp, and sold by Messrs. Rivington's, and J. Taylor, London, 1784); 2nd ed. (1787); J. Twamley, American Edition (Providence: Carter and Wilkinson, 1796).
29 Amelia Simmons, *The First American Cookbook: A Facsimile of 'American Cookery,' 1796 by Amelia Simmons with an Essay by Mary Tolford Wilson* (New York: Dover Publications, Inc., 1984), 8–9.
30 Amelia Simmons, *American Cookery, Facsimile of the Second Edition with an Introduction by Karen Hess* (Bedford, Massachusetts: Applewood Books, 1996), 26, 28, 31–32, 38–40, 42.
31 Joshua Johnson, *The Art of Cheese Making Reduced to Rules, and Made Sure and Easy, from Accurate Observation & Experience, Published for the Help of Dairy Women* (Albany [N.Y.]: Printed for the Author, by Charles R. and George Webster, 1801).
32 John Nicholson, *The Farmer's Assistant Being a Complete Treatise on Agriculture in General, Also upon the Various Subjects of Gardening, Fruit-trees, Farriery and the Dairy, Arranged in Alphabetical Order* (Albany, [N.Y.]: Henry C. Southwick and T.C. Fay, 1814), 24, 33–34.
33 H. L. Barnum, *Family Receipts* (Cincinnati: Lincoln & Co., 1831), 251–266.
34 [N. K. M. Lee], *The Cook's Own Book and Housekeeper's Register* (Boston: Munroe & Francis, 1832), 30–31, 46–47, 54, 56–59, 66–67, 115–116, 161, 222, 251, 286, 196–297.
35 Dwight H. Bruce, ed., *Onondaga's Centennial. Gleanings of a Century* ([Boston]: The Boston History Company, 1896), 883–884.
36 1850 Federal Census; 1855 New York State Census, 1st Election District, by Nathan Abbott; 1860 Federal Census.
37 Eliza A. Call, *The Young Housekeeper and Dairymaids Directory Containing the Most Valuable and Original Recipes* (Syracuse: J. G. K. Truair & Co., 1859), iii.
38 Ibid. 5–9.
39 Ibid. 10–11.
40 Ibid. 12–13.
41 Ibid. 16–18.
42 X. A. Willard, *Willard's Practical Dairy Husbandry* (New York: The American News Company, 1872), 221.
43 Ibid. 214.
44 Dillon, op. cit. 4.
45 *Illustrated Annual Register of Rural Affairs for the Year 1864* (Albany, New York: Luther

Tucker & Son, 1864), 87; Dillon, op. cit. 2–3.
[46] Willard, op. cit. 247.
[47] *Illustrated Annual Register 1864, op. cit.* 86.
[48] Ibid. 86–87; Willard, op. cit. 215.
[49] *Illustrated Annual Register 1864,* op. cit. 87–88.
[50] Ibid. 90–91.
[51] Willard, op. cit. 216; *Illustrated Annual 1864,* op. cit. 91.
[52] Willard, op. cit. 222.
[53] W. G. Marshall, *Through America: or, Nine Months in the United States* Reprint of the 2nd ed. (New York: Arno Press, 1974), 98.
[54] Dillon, op. cit. 2.
[55] Pauline Arnold and Percival White, *Food: America's Biggest Business* ([New York]: Holiday House 1959), 174–176; Selitzer, op. cit. 35.
[56] Robert Milham Hartley, *An Historical, Scientific, and Practical Essay on Milk, as an Article of Human Sustenance; with a Consideration of the Effects Consequent upon the Present Unnatural Methods of Producing It for the Supply of Large Cities* (New-York: J. Leavitt, 1842).
[57] *Frank Leslie's Illustrated Newspaper* 5 (May 8, 1858): 359.
[58] Mary Ronald, *The Century Cook Book* (New York: The Century Company, 1897), 257.
[59] Owen, op. cit. 460.
[60] Henry Alvord, 'Dairy Development in the United States,' US Department of Agriculture. *Yearbook of Agriculture 1899* (Washington, D. C.: Government Printing Office, 1900).

The Wet-nurse

Raymond Sokolov

Speaking as a mammal, I wish to applaud, in a specific way, the choice of milk as the subject for the 1999 Oxford Symposium. Indeed, as I have reflected over the past year, it is altogether curious that it has taken so long for our august organizers to have picked milk as a topic for our discussions and research. After all, at the most general level, the Symposium is a conclave composed entirely of mammals interested in food, and milk is the food most central to our biological identity. It is, so to say, like mother's milk to us. And yet, curiouser and curiouser, if your personal food history is like mine, you have never tasted either your mother's milk or the milk of any other woman.

For my generation, and those born after World War II and on through the 1950s in America (and no doubt in other industrial countries) the norm was to substitute 'scientifically' tailored infant 'formula' for breast milk and to administer it in home-sterilized bottles with rubber nipples as interchangeable as other machine parts.

The history of this trend, and its rejection in the 1960s when alert mothers turned to nursing their infants as part of the spirit of those questioning times (or was it a special instance of the all-conquering natural foods movement?), is too wide a field to explore here.[1] My goal, instead, is to consider milk as a foodstuff like any other and to show how, throughout recorded human experience, even in our day, it has often been treated like a commodity, in effect bought and sold like bread, although usually delivered to the consumer in a package of such beauty and convenience (flexible, lightweight and easy to lift, washable in the home, with built-in climate control) that it seems divinely inspired. But first things first. What does human milk taste like and how does it compare to other milks?[2]

Since the overwhelming number of new mothers lactate from parturition and for months or even years thereafter, it would seem a simple task for the inquiring gastronomer to partake of human milk and gain a first-hand acquaintance with its organoleptic qualities, thereby correcting a gap in his gustatory (and olfactory) background that was largely the fault of his own mother, a bottle-boiling ewe among millions of others in the Spock flock.

To be fair, he had passed up another opportunity during the early years of fatherhood. And in recent months, after some cogitation, he refused an offer of a libation from a nursing diva.

Prof. Darra Goldstein, the distinguished Slavicist and author of standard works on Russian and Georgian food (as well as a mother with direct experience of our subject) was astounded to learn this spring at the annual meeting of the International Association of Culinary Professionals in Phoenix (Arizona) that the author had reached his current advanced age without tasting human milk. He conceded that it would be relatively easy, in an era of self-expression, rife with breast pumps and refrigeration, to find a fresh sample that could be imbibed in his test kitchen, perhaps together with a fig Newton or Bath Oliver.

Prof. Goldstein would have none of it. 'Not good enough,' she expostulated. 'You must take it from the breast. Context is crucial in one's experience of any food. As Georg Lichtenberg said, "Imagine eating steak with a scissors!"'

The author took this advice to his bosom. Like Panurge interrogating all and sundry about whether he should get married, he began asking friends and colleagues if and how he should proceed in arranging a dégustation à la mamelle. One freelance writer in the music area said he knew a coloratura soprano and nursing mom who just might countenance a field trip to her boudoir. The lady followed up with an encouraging e-mail.

The author decided not to respond, on the grounds that even if he did not feel obliged to conceal the matter from his wife, he could never publish the results of his research. It would be too scandalous. The tabloid press would have a field day. But he was still athirst with curiosity.

So he noted down testimonials from others: reports of auto-gustation, spousal suckling and the like. All informants agreed on two points: breast milk was sweet, and it was thin and bluish, like commercial (bovine) skim milk. The impression of sweetness squares with the data reported for human milk (from US samples) in Composition of Foods (United States Department of Agriculture, Washington, revised 1963). For every 100 grams, human milk contained 9.5 grams of carbohydrate (mostly lactose or milk sugar, $C_{12}H_{22}O_{11}+H_2O$). Human milk, therefore, had nearly twice as much carbohydrate as whole cow's milk (4.9 grams per 100 grams).

As to thinness, respondents were ostensibly reacting to a perception of low fat content, but the USDA found slightly more fat in human milk (4 grams per 100) than in cow's (3.7). Perhaps the apparent discrepancy between the anecdotal and the USDA data is the result of relative dilution in the human samples. A comparison with bovine milk by volume is essential for deciding this question.

The duration of nursing should also be taken into account. The first milk or colostrum, produced in the three or four days following parturition, is normally yellow and clear and jumpstarts the infant's immune system with antibodies. This is replaced by so-called early milk, bluish-white and thin.

Mature milk is white and creamy. Late milk, produced as breast-feeding winds down, is thin and white.[3] It would appear that informants reporting their milk had been thin and blue had not nursed for very long. Susan M. Love, in her popular account,[4] describes a different pattern in her own experience as a nursing mother: '...I found my own early milk to be white and the later milk to be blue. It was also interesting to me to find that milk left to stand in the refrigerator layered out with the cream at the top.[5] If I skipped a meal there would be less cream, 'skim milk'. When I ate regular meals, there was more cream.'

In general, moreover, human milk is low in protein (3.5 grams for cow's milk, 1.1 for human). This explains in large part why infants fed cow's milk often have to be provided with a special formula to prevent digestive and allergic complications from the overly rich protein content of whole cow's milk. By returning to breast milk, today's mothers have responded to the nutritional facts as well as to various other advantages of nursing, both tangible (disease immunity) and intangible (parent-child bonding).

Nursing, in human mothers, as in other lactating mammalian females, ordinarily can be continued for as long the mother stimulates milk production by regularly feeding her child or expressing milk mechanically. The author has had personal experience with a child kept at the breast until the age of four. Suckling stimulates the two milk-making hormones, prolactin and oxytocin. Regular suckling stimulates regular milk production, in principle indefinitely. Again, in principle, once a woman has lactated, she can resume lactation even after a lengthy lapse, by putting an infant to her breast for a few days.

This possibility of prolonged lactation is the basis of commercial dairying in non-humans, and it has also has been exploited commercially and socially in human life. From time immemorial mothers have farmed out their babies to other women, women still able to lactate because of a previous birth. These wet-nurses (as opposed to dry nurses who were really just nannies) took the burden of breast-feeding from the natural mother and fed her baby at their own breasts, usually for a fee or as part of their duties as a household servant or slave. The oldest and most vivid evidence for the universality of wet-nursing is literary.

The most venerable example is the nursing of Moses in Exodus 2:7-8. Pharaoh's daughter finds him abandoned in the river, because of her father's decree requiring male Hebrew children to be 'cast into the river'. While his sister Miriam secretly watches over him, Pharaoh's daughter rescues Moses. Miriam then comes forward and says: 'Shall I go and call to thee a nurse of the Hebrew women, that she may nurse the child for thee?' Pharaoh's daughter sends her off and she calls Moses's own mother.

So the wet-nurse in this story is actually the natural mother, a poignant resolution not only of the problem of Moses's survival, but also of the general emotional problem of parent-child separation created by the institution of the wet-nurse.

Chaucer raises the same issue implicitly in *The Clerk's Tale*, by injecting a wet-nurse into the catalogue of more explicit sufferings forced on poor Griselda: 'Whan it was two yeer old, and fro the brest /Departed of his norice.' (2.4)

Homer shows us more openly the divided emotions that flow from a child to his mother and to his nurse. Odysseus returns home in disguise, and the first person to recognize him is his nurse Eurycleia (Penelope's mind is elsewhere) when she recognizes an old scar while bathing him:

> This scar the old dame, when she had taken the limb in the flat of her hands, knew by the touch, and she let fall the foot. Into the basin the leg fell, and the brazen vessel rang. Over it tilted, and the water was spilled upon the ground. Then upon her soul came joy and grief in one moment, and both her eyes were filled with tears and the flow of her voice was checked. But she touched the chin of Odysseus, and said:
>
> 'Verily thou art Odysseus, dear child, and I knew thee not, till I had handled all the body of my lord.'
>
> She spoke, and with her eyes looked toward Penelope, fain to show her that her dear husband was at home. But Penelope could not meet her glance nor understand, for Athena had turned her thoughts aside. But Odysseus, feeling for the woman's throat, seized it with his right hand, and with the other drew her closer to him, and said:
>
> 'Mother, why wilt thou destroy me? Thou didst thyself nurse me at this thy breast, and now after many grievous toils I am come in the twentieth year to my native land.'
>
> (*Odyssey*, 19.468ff.)

Shakespeare, taking wet-nursing for granted, has Rosalind (*As You Like It*, 4.1) remark, in a highly condensed piece of bantering with Orlando, that a woman who can't turn the tables on her husband and make him pay for her misdeeds, 'let her never nurse her child herself, for she will breed it like a fool!'

In two other passages that subvert normal parent-child relationships, wet-nursing/natural mother confusions set up mistaken identities crucial to Shakespeare's plots:

> At three and two years old, I stole these babes;
> Thinking to bar thee of succession, as
> Thou reft'st me of my lands. Euriphile,

> Thou wast their nurse; they took thee for their mother,
> And every day do honour to her grave:
> Myself, Belarius, that am Morgan call'd,
> They take for their natural father. The game is up.
>
> (*Cymbeline* 3.3)

> Ay, there's the question; but I say, 'tis true:
> The elder of them, being put to nurse,
> Was by a beggar-woman stolen away;
> And, ignorant of his birth and parentage,
> Became a bricklayer when he came to age:
> His son am I; deny it, if you can.
>
> (*2 Henry IV.*4.2)

Defoe adopts a more prosaic, journalistic note in Moll Flanders' account of her infancy:

> ...nor can I give the least account of how I was kept alive, other than that, as I have been told, some relation of my mother's took me away for a while as a nurse....In the provision they made for me, it was my good hap to be put to nurse, as they call it, to a good woman who was indeed poor but had been in better circumstances, and who got a little livelihood by taking such as I was supposed to be, and keeping them with all necessaries....

Little had changed by Zola's time:

> ...Nana's greatest cause of distress was her little Louis, a child she had given birth to when she was sixteen and now left in charge of a nurse in a village in the neighbourhood of Rambouillet. This woman was clamouring for the sum of 300 francs before she would consent to give the little Louis back to her. Nana, since her last visit to the child, had been seized with a fit of maternal love...
>
> (*Nana*)

Not only poor women, but French urban mothers of almost every social level had traditionally farmed their babies out to wet-nurses in the countryside. As Emmanuel Le Roy Ladurie shows in 'Un phénomène biosocioculturel: L'allaitement mercenaire en France au XVIIIe siècle', many of the infants sent to the countryside perished, either during the journey, from neglect ('turkeys pecked out their eyes') or from starvation by cynical women unable or unwilling to carry out their charge. Le Roy Ladurie cites one particularly 'murderous' wet-nurse, Catherine Hiard, in a village near Rouen, who took in nineteen babies in 1789:

only two survived. At Rolleboise, Marie Bienvenu let 31 children die in 30 months.

Mortality rates of babies given to nurses averaged 52 per cent from birth to the age of six and a half. And the overwhelming trend was for natural mothers to send their offspring away to distant foster homes. One study of infant mortality between 1774 and 1784 in the suburban area to the south of Paris[6] reveals that of 24,000 babies born in that period, a mere 1,000 were nursed by their mothers, 1,000 more had wet-nurses living in the parents' home, 2,000 to 3,000 were given to nurses living in the Paris region, while the remaining 15,000 went far away.

Well before this time, public revulsion began to turn the tide. Rousseau published *Émile* in 1762, a scandalous but influential treatise on education. It begins with a ringing admonition to mothers to nurse their own children: 'If mothers would deign to nurse their children, our morals would repair themselves and natural feeling would reawaken in our hearts.'

Rousseau's pitch for family values sounds familiar today but he ignored the grim backdrop of poverty that encouraged the effective abandonment of children to women who were often little better than postpartum abortionists.

In our post-industrial age, the wet-nurse no longer plays a role in our lives or in our imaginations. James Joyce may have been the last important writer in English for whom the wet-nurse was a normal part of life, normal enough to crop up in casual conversation or in the stream of consciousness of a Dublin housewife, of Molly Bloom famously soliloquizing: '...there's the mark of his teeth still where he tried to bite the nipple I had to scream out aren't they fearful trying to hurt you I had a great breast of milk with Milly enough for two what was the reason of that he said. I could have got a pound a week as a wet-nurse all swelled out...'

REFERENCES

[1] In particular, there was the nutritional tragedy of third-world mothers encouraged by aggressive marketing to abandon breast-feeding in favour of commercial formula, which they couldn't afford long term. By the time they realized their predicament, they had no milk left.
[2] For a detailed breakdown of the milk of many species, see the attached table.
[3] *Dr. Susan Love's Breast Book*, by Susan M. Love M.D. with Karen Lindsey, Reading (Mass.): Perseus Books, p. 30.
[4] Ibid.
[5] This observation holds out tantalizing possibilities for the gastronome – human crème fraîche, cheese, rice pudding....
[6] Galliano, P., 1966, *Annales de démographie historique*, p 139 et seq.

Table: The milk of various species
(Adapted from Robert Bremel, University of Wisconsin)

SPECIES	FAT %	PROTEIN %	PROT./FAT	LACTOSE %	ASH %	TOTAL SOLIDS %
Antelope	1.3	6.9	5.3	4	1.3	25.2
Ass (donkey)	1.2	1.7	1.4	6.9	0.45	10.2
Bear, polar	31	10.2	0.3	0.5	1.2	42.9
Bison	1.7	4.8	2.8	5.7	0.96	13.2
Buffalo, Philippine	10.4	5.9	0.6	4.3	0.8	21.5
Camel	4.9	3.7	0.8	5.1	0.7	14.4
Cat	10.9	11.1	1	3.4	—	>25.4
Cow:						
Ayrshire	4.1	3.6	0.9	4.7	0.7	13.1
Brown Swiss	4.0	3.6	0.9	5.0	0.7	13.3
Guernsey	5.0	3.8	0.8	4.9	0.7	14.4
Holstein	3.5	3.1	0.9	4.9	0.7	12.2
Jersey	5.5	3.9	0.7	4.9	0.7	15.0
Zebu	4.9	3.9	0.8	5.1	0.8	14.7
Deer	19.7	10.4	0.5	2.6	1.4	34.1
Dog	8.3	9.5	1.1	3.7	1.2	20.7
Dolphin	14.1	10.4	0.7	5.9	—	>30.4
Elephant	15.1	4.9	0.3	3.4	0.76	26.9
Goat	3.5	3.1	0.9	4.6	0.79	12
Guinea Pig	3.9	8.1	2.1	3	0.82	15.8
Horse	1.6	2.7	1.7	6.1	0.51	11
Human	4.5	1.1	0.2	6.8	0.2	12.6
Kangaroo	2.1	6.2	3	Trace	1.2	9.5
Mink	8	7	0.9	6.9	0.7	22.6
Monkey	3.9	2.1	0.6	5.9	2.6	14.5
Opossum	6.1	9.2	1.5	3.2	1.6	24.5
Pig	8.2	5.8	0.7	4.8	0.63	19.9
Rabbit	12.2	10.4	0.8	1.8	2	26.4
Rat	14.8	11.3	0.8	2.9	1.5	31.7
Reindeer	22.5	10.3	0.5	2.5	1.4	36.7
Seal, grey	53.2	11.2	0.2	2.6	0.7	67.7
Sheep	5.3	5.5	1	4.6	0.9	16.3
Whale	34.8	13.6	0.4	1.8	1.6	51.2

Milky Medicine and Magic

Layinka M. Swinburne

After the death of his wife, the twelfth Lord Clifford fell into a grievous sickness and was so ill that he was laid out for dead under a velvet pall. Luckily one of his attendants noticed signs of life and he was resuscitated with cordials 'inwardly and outwardly'. After that, he sucked milk from a woman's breast for three or four weeks and continued the cure with ass's milk for several months. He made a complete recovery and lived another 23 years.[1]

Belief in the curative power of human milk and the medicinal strength of ass's milk goes back to Antiquity. Milk was a powerful, magical substance, one of God's gifts but sometimes unaccountably withheld. Its origin was mysterious. In medieval times menstrual blood was thought to be diverted from the uterus during pregnancy to nourish the foetus and conveniently turned itself into milk after delivery.[2] This ancient idea is illustrated in an extract from a medieval MS belonging to the monks of Bolton Abbey describing the development of the foetus in the uterus:

> From XXX daye ytt ys formyde in the forme of a mane…and froo the XL daye ytt begyns to be norcheyde w't' the bloode of the mothr by hys cowrs att the navylle…; and when that neyne monthes are fullfyllde, the blowde wherwythe that hee was norycheyde departs and assends uppe to the brests of the wooman, and ys theyre, as ytt wer, a thyke kreeme, and after hys byrthe hee ys norycheyde with mylke off his mothr.[3]

The belief persisted into early modern times and the connection between blood and milk remained strong. It seemed obvious since a woman's periods ceased whilst she was pregnant and suckling her baby, but returned when it was weaned. Blood and milk were sometimes combined in a medicinal remedy. A recipe for the stone in Lady Sedley's receipt book of 1628 required goat's blood and human milk.[4] At Elizabethan weddings, red and white ribbons tied to a rosemary bough were carried by attendants and used to decorate the table of the marriage feast. Amongst country folk in nineteenth-century France the two colours could be used to switch monthly periods on and off at will by sympathetic magic and were preferable to the old folk remedy of a toad hung round the neck which would also 'stay a flux'. The uterus was likened to the shape of a toad and metal toad amulets were used as votive offerings in France.[5] The white ribbon was a reminder both of milk and the absence of periods during childhood and pregnancy, whilst the red recalled the woman's 'monthly purgations' auguring future fertility.[6] 'Milk-white' connoted purity and innocence.

Witchcraft

Conversion of milk to blood in the reverse direction at the wrong time was much feared and was sometimes attributed to witchcraft. Killing a robin or pulling up the plant Herb Robert (*Geranium robertianum*), with its red angular stems, could both turn a cow's milk to blood, as was long believed in East Yorkshire as an explanation of this disaster.[7]

In *A Midsummer Night's Dream*, Puck, or Robin Goodfellow, was challenged by a fairy :

> Are you not he
> That frightens maidens in the villag'ry,
> Skim milk; and sometimes labour in the quern
> And bootless make the breathless housewife churn?[8]

Witches could also stop the butter from coming (another mysterious process) or stop cows giving milk.[9] A common accusation was that a witch would turn herself into a hare to perform these mischievous acts, a belief supported by hair-raising tales of goggle-eyed witnesses. To protect against witches, rowan – 'witten' – wood was to be hung at the byre door, along with elder to keep off the flies. The mark of a cross was a protection used on bread, butter or household objects. In Ireland, garlands of yellow flowers were hung on the cattle or strewn before the door of the dairy, but to bring good luck and good butter, it was important to tread on it as you entered. The golden garlands of the Greek first of May could have the same origin. Marsh marigolds and primroses would serve but it was unlucky to bring them inside before the first of May.[10]

Milk was a very vulnerable substance before the days of Pasteur and microbiology. All sorts of magical explanations were invoked to account for the way it 'went off' or changed its state in the dairy. St John's wort, another magical herb, put into ropy milk would cure it.[11] A hot iron plunged into the churn would counteract witchery in the dairy; whilst allowing milk to boil over could bring harm to the cow that had given it, unless the ashes were sprinkled with salt to avert it. Salt, especially consecrated salt that had been blessed on Palm Sunday, was a powerful agent for warding off evil. It was unlucky to use Friday eggs or butter in a Christening cake, whilst bewitchment of one's cows could be prevented by giving away Sunday milk to the poor.[12]

Hedgehogs – urchins – were believed to milk cows at night and the farmers' paranoia was quite as great as in the badger-TB debate today. When the

Reverend Josselin's bullock (?) gave blood in her bag, as he reported in his diary, 'we searched the field and found the hedgehog and killed him'.[13] There are remedies for urchin bites in country remedy books. Whether anyone ever got bitten is debatable but the remedy may have been to treat a prick by the spines – bites and stings were not clearly distinguished by country people. This was the case in 'A cure for a bite or sting by any venomous creature or worms' in a little pocketbook of remedies for the farm.[14]

No doubt surreptitious milking by dishonest neighbours in shared pasture-land, or vagrants looking for a free snack, did occur but witches got the blame in seventeenth-century English witch hunts.[15] Sometimes mysterious churning was to be heard at night and then in the morning strange butter was found smeared on the gate posts. The fungus named witches' butter (*Cladonia exidia*), which grows on rotten wood, indeed looks like black butter.[16]

Witches might bring inappropriate milk to the breasts of the new-born: known as witches' milk, due to infant lactation, a response to the mother's hormones transferred through the placenta. Accessory nipples, or anything that remotely resembled one, were one of the marks of a witch. In other societies, they were a sign of fecundity and possible future multiple births.

Human milk.

Human milk in medicine can be traced back to Ancient Egyptian times. The favourite use was for sore eyes, combined with rosewater and egg-white. In a remedy in *Arcana Fairfaxiana*, seventeenth- and eighteenth-century receipts of the Fairfax family, the mixture was to be put on a piece of bread and laid over the eyes. Robert Burton claimed that frontlets of human milk were known to every good housewife. A frontlet was a pack on the forehead for a headache. An Anglo-Saxon leech-book instructs: 'Take leek-seed and stampe it and temper it with woman's milk and the white of an egg; and bind it to the temple. And he shall sleep.'[17] Versions of this remedy crop up repeatedly. Milk occurs in the seventeenth-century Pharmacopoeia as *Lac mulieris* – to be distinguished from the *Lac Virginis* of the alchemists, which was composed of lead acetate in vinegar. It was one of the less horrific things that people put into their eyes, which might otherwise have been ground-glass, urine, or clary seeds to name but three. Perhaps the lysozymes, antibacterial agents which we now know are secreted in milk as well as tears, were beneficial in minor infections. Milk was also to be used for earache mixed with hare's gall. The ear was to be stopped with *black* wool which was always specified for ear problems.

If a woman's travail was slow, it could be speeded up by three remedies: the urine of a bitch, the urine of the father of the child, or the milk of another woman.[18] Dr Chamberlain, in his book on midwifery, commented on the latter 'that there is little reason for it and I am sure it is loathsome to most women'.[19]

Dr John Hall, Shakespeare's son-in-law, prescribed human milk as part of a cooling diet for a wealthy woman 'in a consumption' (suffering from tuberculosis), or any other wasting condition, along with lettuce, snails, river crabs, frogs, and *panatello* – bread moistened with milk and afterwards with almond milk.[20] Powerful restorative properties of the milk alone were used to bring back the Earl of Cumberland from the dead. It could also be used for determining the outlook for a patient: drop woman's milk into urine, 'If they mingle he shall be helped. If it float above he shall die!' In the Fairfax version, 'to know whether a sick man shall live or die certainly proved many times... Take a little of their water and put it into milk and if they die a dog will not eat it and if they live a dogge will eat it.'[21] Both may be likened to the old Egyptian usage, 'to know whether a woman can bear children or not', in which the milk of a woman who has borne a male child mixed with pumpkin or watermelon seeds is given a woman to drink, her response, whether vomiting or belching, giving the indication.[22]

Herbs to increase the flow of milk in a nursing woman naturally took their virtues via the doctrine of signatories, hence *Lactuca* (lettuce) with its milky juice; fennel; *Polygala* (milkwort) indicating lots of milk; and milk thistle with its leaves beautifully veined with white. Lentil posset was favoured by Eliza Smith.[23]

The character of the nurse, whether the mother or another woman, was believed to affect the character and temperament of the baby and many books gave advice on selecting a suitable woman and the way she should conduct herself, avoiding alcohol, strong tasting foods and concentrating on a light, easily digestible diet. In France it was the duty of Royal Physicians to taste the milk of prospective wet-nurses for the royal nursery.[24] In my training days in the Maternity Unit, the Registrar made our tea with breast milk on one occasion, but this was accidental and not part of our professional duties.

Animal Milks.

The most powerful remedy of all was the Alexeterial milk-water, good against the plague, surfeits, and almost everything else; and especially useful for the bite of a mad dog. It was a water distilled from numerous herbs and cow's milk. It occurs in many recipe books of the seventeenth century.

Animal milks were thought to have special virtues, derived from the animal's nature: from the ultimate luxury of dolphin's milk butter in Ben Jonson's *The Alchemist*,[25] to the precious remedy of an Italian Renaissance physician for infertility, a pill based on rabbit milk. Other milks were invoked for various reasons: Romulus and Remus gained power from their vulpine foster-mother; and for pure digestibility, ass's milk was favoured and had a special reputation for the weak and the consumptive.

In many remedies, such as those of Celtic origin used by the physicians of Myddfai and the old Anglo-Saxon leeches, the colour of the cow was important to the efficacy of the recipe. A one-coloured cow or a red cow was often specified, for example as a remedy for consumption sickness in *Arcana Fairfaxiana*:

> Take garden snails num. 5 break off the shells of them then boil them in a quart of new milk of a red cow till it comes to a pint and a half. drink of this first and last and at all times of the day.[26]

Also the sex of the calf that the cow had produced might determine the suitability of the milk for treating a child of the opposite sex.

Fernie, in *Animal Simples,* puts the order of digestibility as woman, sheep, goat, cow.[27] The school of Salerno arranged milks in the following order of virtue: goat, camel, cow, sheep, and ass.[28]

Whey from cow's milk had a special vogue as a preventative against the plague in 1650 (*The Rich Storehouse of Medicine*). It became the health drink of the eighteenth century and whey houses were set up where one could drink one's fill. Whole cow's milk was suspected of being indigestible. When drunk at all it was often advised straight from the cow or other beast to avoid souring. At one stage the preference for goat's and ass's milk was so marked in French orphanages that the infants were fed direct from the teats of a row of asses, brought into the wards or stabled alongside.[29] It avoided the problem of storage of milk and the intestinal upsets which killed so many nurselings. In Malta until the 1930s from 2,000 to 4,000 goats were driven into Valetta twice daily to supply milk at the door. Moreover, sick soldiers early in the century suffering from Malta fever (brucellosis) were encouraged to drink large quantities of it before it was discovered that the cure was the cause of the disease by transmitting the *Brucella* bacilli! Recently sheep have been implicated in transmitting *H. pylori* to the local population in Sardinia, where infection with this organism is endemic.[30]

For consumption or coughs, ass's milk was the favoured remedy. Asses in milk were loaned by the Verneys of Buckinghamshire to other local people needing their services, and they were regularly advertised in the *Leeds Mercury* by the wealthy Stanhope family of Cannon Hall and Horsforth.

One of the Constable family of Everingham in Yorkshire wrote of an acquaintance, 'he takes asses' milk and makes trips to Buxton and Bath for sciatica and rheumatism'.[31] However, the eighteenth-century medical author Quincy commented cynically that as it was often used as a last resort, 'his credit is usually safe who loses his patient in so creditable a manner'.

Herbal milks

Goat's milk had a good reputation for digestibility and moreover was a favourite as a facial cosmetic. Burton advised goat's milk whey and hellebore juice for melancholy and suggested that the goat should be fed on good herbs. Fernie mentions a practice of feeding cows deliberately with herbs to produce medicinal milks: pellitory of the wall for dropsy, madder for the treatment of rickets, nettles for piles and lettuce and purslane for costiveness. He did not say what happened to the cows or give examples of this practice but it would have imparted both a tint and a taste to the milk. Customary lore has it both that a milk for gout could be made by feeding goats dwarf elder and that a goat fed for a day with the leaves of dwarf elder will give a purging milk.[32]

This same principle can be seen operating in different cultures, although each society will use the milk of whichever animal is most convenient for it. In al-Jawziyya's text of Koranic medicine, the virtues of camel's milk are vaunted. The beasts were to be pastured on artemisia, achillea, camomile, anthemis, common grass and the like to get the desired medicinal effects. In one remedy for dropsy ordered by the Prophet, milk is to be taken warm from the camel, mixed with urine of an immature camel (rather than the blood of Western remedies).[33]

Today, even more complex techniques are being used to induce animals to secrete foreign proteins in their milk for therapeutic purposes, for example factor IX for the treatment of bleeding disorders. By contrast, nearly as much effort is employed to avoid unwanted effects of antibiotics and hormones given to milch animals.

Synthetic milks

Synthetic milks have been made since Antiquity, often with a clear attempt to imitate the qualities of animal milk. Almond milk was a medieval favourite and was used both in cookery and for women during childbirth and nursing. It did not curdle when kept and was said to be light of digestion. It was thought to thicken the blood and prevent untimely bleeding.

> To make gode almonde mylke[34] Take broken sugar or in default thereof take clarified hony and put it into fresh water and set hyt on the fyre and boyle hyt and skeme het clene and set hyt besyde the fyre and let het kele and then blaunce your almonds put them in a mortar and crush them small and temp them up with the same water.

Artificial ass's milk, 'the made asses milk', could be concocted from garden snails. Details are to be found in many a recipe book of the seventeenth or eighteenth century. Snail waters were particularly recommended for consumption. Quincy's

Dispensatory had six versions, of differing degrees of efficacy. Some elaborate brews he thought useless and he recommended above all others snails simply boiled in milk sweetened with sugar. The process started by putting the snails on a shovel and heating them over the fire till they frothed. Whether it was anything like the real thing I cannot say, not having tasted either. Candied eringo root, ox-eye daisies and barley were other important ingredients. Henry Prescott, Deputy Registrar of Chester Diocese in the early eighteenth century, took snail water nightly to alleviate pain and believed in its sedative effects.[35]

Do we still need cows?

The nineteenth century produced a host of milk-substitutes such as arrowroot, potato flour and cornflour. These were followed by von Liebig's composition, the *Suppe für Saüglinge*, a baked, malted flour with whey which was lacking in many nutritional essentials but was the forerunner of numerous better products and modified milks for babies and invalids.[36] The starchy alternatives to milk rapidly led to rickets in children fed exclusively on them. The early preserved milks were also devoid of vitamin C so that scurvy was another unfortunate sequel. Milk substitutes have burgeoned recently using modern technology providing good rice, oat, and soya milks for use by vegetarians and people with food allergies and intolerance. The two latter problems have also been circumvented by using sheep's and goat's milk: as milk, fermented to yoghurt, and as cheese.

For babies, mother's milk is still the best – as a convenient source of mineral salts and vitamins. Dairy products are important in many societies after weaning to prevent rickets in children and osteoporosis in later life. I do not subscribe to the current paranoia about animal fats and find that milk is still a useful source of essential fatty acids, phosphate, calcium and vitamins A and D. In the future we may, with great labour, make truly synthetic milks but in the past there have been mistakes in mimicking nature due to partial knowledge – such as von Liebig's concoction and other popular baby foods and early margarines which lacked vitamin D. Owing to some unknown quality of bitch's milk, the milk of dogs is hard to imitate and puppies are extraordinarily difficult to hand-rear. Elemental diets for humans applying current theory too strictly are constantly found to be deficient, and have to be modified or updated with unexpected ingredients in the light of new knowledge. Low-cholesterol 'Elmlea' and 'Benecol', 'Olestra' (composed of phytosterols from wood-chips), and their like should be used with caution. Cholesterol is an essential factor for brain and nerve tissue. Recently a growth factor has been described in whey which helps wounds heal in rats and abolishes wrinkles.[37] There are unknown perils of GM foods, and there may still be some magic in milk that is yet to be discovered.

REFERENCES

1. Williamson, George C. (1922), *Lady Anne Clifford: Her Life, Letters and Work*, p. 21.
2. Eccles, Audrey (1982), *Obstetrics and Gynaecology in Tudor and Stuart England*, p. 49; Beryl Rowland (1981), *Medieval Woman's Guide to Health*, p. 61.
3. Whitaker, T.D. (1878), *The History and Antiquities of the Deanery of Craven*, Vol. II, p. 477.
4. Guthrie, Leonard (1913), *Lady Sedley's Receipt Book,* Proc. Roy. Soc. Med., p. 151.
5. Gélis, Jacques (1991), *The History of Childbirth: Fertility, Pregnancy, and Birth in Early Modern Europe.*
6. Ibid, p. 12.
7. Grigson, Geoffrey (1975), *An Englishman's Flora*, p. 105.
8. Shakespeare, William, *A Midsummer Night's Dream*, Act II scenes 1 l. 34.
9. Atkinson, Rev. J.C. (1891, reprinted 1988), *40 years in a Moorland Parish*, p. 87.
10. Opie, Iona and Tatum, Moira (1981), *A Dictionary of Superstition*, p. 247.
11. Burton, Robert (1621), *The Anatomy of Melancholy,* Clarendon Edition 1997, II, 5. 1.5.
12. Kramer, Heinrich & Sprenger, James (1486), *Malleus maleficarum,* Arrow Books (1971), p. 400.
13. Josselin, Ralph, *The Diary of Ralph Josselin (1616–1683),* ed. Alan Macfarlane (1976).
14. Escrick, Thomas MS422 Leeds University Special Collections.
15. Atkinson, p. 53.
16. Phillips, Roger (1981), *Mushrooms,* p. 262.
17. Dawson, W. R., *A Leechbook or a collection of medical recipes of the fifteenth century,* p. 263.
18. Eccles, p. 102.
19. Eccles, p. 102.
20. Lane, Joan (1996), *Dr. John Hall and his Patients, p.* 39.
21. *Arcane Fairfaxiana Recipes of the Fairfax Family,* 1890, edited George Weddell, p. 45.
22. Dawson, Warren R. (1929), *Magician and Leech,* p. 141.
23. Smith, Eliza (1758), *The Compleat Housewife.*, p. 33.
24. Cressy, David (1997), *Birth, Marriage, and Death, Ritual and Religion in Tudor and Stuart England*, p. 91.
25. Jonson, Ben, *The Alchemist*, 131.
26. *Arcana Fairfaxiana,* p. 154.
27. Fernie, F.T. (1899), *Animal Simples,* p. 91.
28. Drummond, Jack and Wilbraham, Anne (1958), *The Englishman's Food,* p. 75.
29. Fildes, Valerie (1986), *Breasts, Bottles and Babies,* p. 268.
30. *Lancet* (1999), 354, p. 132.
31. *Letters of Constable of Everingham,* Yorkshire Archaeological Society Records Series, 136, p. 96.
32. Grigson, Geoffrey (1959), *A Herbal of all sorts,* p. 25.
33. Johnstone, Penelope (1998), *Ibn Qayyim al-Jawziyya, Medicine of the Prophet*, p. 33.
34. *Stere Htt Well* (C.15), ed. G.A.J. Hodgett, n.d., p. 32.
35. *The Diary of Henry Prescott LL.B., Deputy Registrar of Chester Diocese,* Records Society of Lancashire and Cheshire CXXXII (1994), pp. 388, 431, 543 etc.
36. Swinburne, L. M. (1995), 'von Liebig Condensed' in *Cooks and Other People,* Oxford Symposium on Food and Cookery 1995, p. 254.
37. Hawkes, Nigel (1997), *Times* July 30th, p. 14.

More on the Origin and History of the Ice-cream Cone

Robin Weir

'The ice-cream cone is the only ecologically sound package known. It is the perfect package.' United States Health, Education and Welfare official, quoted on the television show *60 Minutes* in 1969.

I have been fascinated by the methods used to eat ice-cream and sorbets for some time and this produced a paper at Oxford in 1991 on 'Penny Licks and Hokey Pokey: Ice-cream before the Cone'[1] and later a paper at Oxford in 1995 on 'Mrs. Marshall, Ice-creammonger Extraordinary'.[2]

In the latter I claimed that the first recording of an ice-cream being served in a cone was in *Mrs A.B. Marshall's Cookery Book*[3] published in 1888. Since then, at least, the US Post Office have produced a stamp that acknowledges as myth the popular notion that the ice-cream cone was invented at the 1904 St Louis World Fair and now advise that it was 'popularized' there.

But unfortunately these inaccuracies linger on and recent copies of *Bon Appétit* magazine (USA), the US Air in-flight magazine and the packaging of a major UK wafer-maker all repeat the myth that the origin of the ice-cream cone was the 1904 Fair. This is in spite of the above and the existence of a patent application at the US Patent Office for a multiple cone-making mould by a Mr Marchioni, 22 September 1903.

When sorbets, and later ice-cream, were first produced they were very expensive and therefore exclusive to the small number of wealthy people who had built ice houses on their estates and whose staff knew how to make them. It was not until Gilliers (1751)[4] and Emy (1768)[5] published their books with

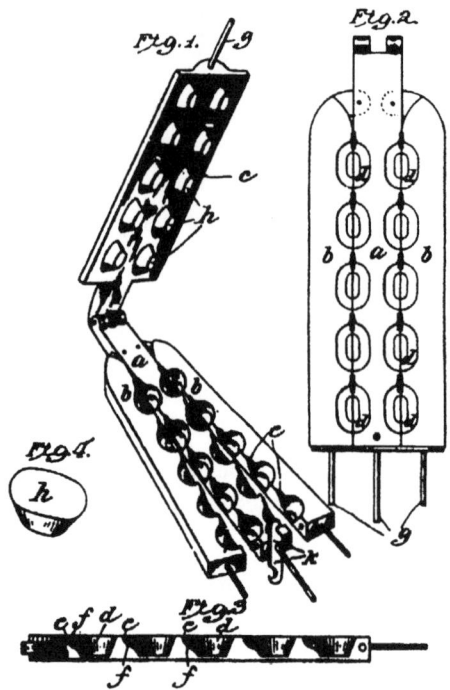

The Marchioni patent, 20 September, 1903.

MRS. A. B. MARSHALL'S COOKERY BOOK 403

Cornets with Cream.
(*Cornets à la Crème.*)

Mix together into a paste four ounces of finely chopped almonds, two ounces of fine flour, two ounces of castor sugar, one large raw egg, a pinch of salt, and a tablespoonful of orange-flower water. Put one or two baking tins into the oven, and when they are quite hot rub them over with white wax and let the tins get cool; then spread the paste smoothly and thinly over the tins (say, one tenth of an inch thick) and bake in the oven for three or four minutes; take out the tins and quickly stamp out the paste with a plain round cutter about two and a half to three inches in diameter, and immediately wrap these rounds of paste on the outside of the cornet tins which have been lightly oiled inside and out, pressing the edges well together so that the paste takes the shape of the cornet; then remove the paste and slip it inside the tin and put another one of the tins inside the paste so that it is kept in shape between the two tins; place them in a moderate oven, and let them remain till quite crisp and dry; take them out and remove the tins; these can be kept any length of time in a tin box in a dry place. Ornament the edges with a little Royal icing by means of a bag and pipe, and then dip the icing into different-coloured sugars; fill them with whipped cream sweetened and flavoured with vanilla, using a forcing bag and pipe for the purpose, and arrange them in a pile on a dish-paper or napkin. <u>These cornets can also be filled with any cream or water ice</u>, or set custard or fruits, and <u>served for a dinner, luncheon, or supper dish.</u>

The relevant text from Mrs Marshall's Cookery Book, *1888.*

Gilliers, 1768. 'Les goblets à neige'. *Emy, 1768. 'Goblet à glace'.*

exact details, quantities and methods, that the secret of freezing using ice and salt[6] and the making of sorbets, neiges and ice-creams became generally known. This produced an interesting range of serving vessels that were in keeping with the status and affluence of the household and the rarity of the frozen dessert.

As ice-cream became more popular and was available to the public it was sold either in small glasses, 'licks', that were then returned to the vendor and re-used, usually without washing, or sold wrapped in small pieces of paper,

A watercolour dating from ca. 1817 by Bartolomeo Pinelli (1781–1835).

Venditore di sorbetto a minuto – Franfelliccaro Napolitano. *Gouache on paper (21.6cm x 26cm), Savario Xavier della Gatta, Naples, 1820. (The Corning Museum of Glass.) The colour reproductions on the cover of this volume are taken from this picture.*

Margaret Cornets (above) and Christina Cornets (below) from Mrs Marshall's Fancy Ices, *1894.*

usually waxed, to take away and often referred to as 'hokey pokey'. There are numerous references to the glasses being used for ice-cream in early Neapolitan paintings and prints such as that by Bartolomeo Pinelli on the previous page. This shows clearly the vendor holding the glass by a base rim.

Mrs Marshall in *Fancy Ices*[7] suggests a number of ways of serving ice-cream cones. Margaret Cornets were made with half a pound of finely chopped or ground almonds, four ounces of caster sugar, four ounces of fine flour, two eggs a saltspoon of vanilla essence and one tablespoonful of orange-flower water. They were then filled with half ginger ice-water and half with apple ice-cream. Tricky but delicious.

Christina Cornets were the same recipe as Margaret Cornets but filled with ice-cream to which was added dried fruits – greengage, apricot, dried ginger, cherries etc., cut into very tiny dice shapes and 'as much ground cinnamon as would cover a three-penny piece, the same quantity of ground ginger and a tablespoonful of Marshall's Maraschino Syrup'.

However a recent purchase by the Corning Museum in Corning, New York State, of a gouache by Savario Xavier Della Gatta, painted in 1820, shows the ice being served in a glass cone. This differs from the lick in that it clearly has no base to stand on. The cones can clearly be seen by the way the purchaser is holding the glass and by the upturned cones resting on the tray at the vendor's feet. (This painting is reproduced in colour on the cover of this volume and in black and white opposite.) Despite numerous letters and a visit to Naples no original glass cone has been found to exist anywhere – so far.

This effectively moves the date of the first cone to 1820 and the first edible cone remains with Mrs Marshall at 1888.

The best-known ice-cream cone in the world is without doubt the Cornetto. This is now one of the best selling items that Unilever have in their range. The skill is to make a cone in such a way that it does not soften in freezing, no small feat.

Originally made in Italy by the Spica Company in Naples in the summer of 1959 and marketed under the Soave brand name (a cheaper brand) as it contained vegetable fat in the ice-cream. Sold without any advertising, the product was not an immediate success. It was sold under the Soave name in case it was not a success in which case it would not damage the more prestigious Algida brand.[8]

In 1960 it became clear that it was going to be a success and Spica registered the name Cornetto. In 1963 it contained just 46cc of ice-cream, 6g chocolate, 3g almonds and 3g cake crumbs. It was quite a small cone. In common with several other snack foods they are now much larger and today it contains 125cc by volume, nearly three times as much, a dietician's nightmare. It is now without question the largest selling branded ice-cream cone in the world.[9]

Unilever purchased 48 per cent of the shares of the Spica Company in 1962 and the balance of the shares in 1964. It is now made in factories all round the world.

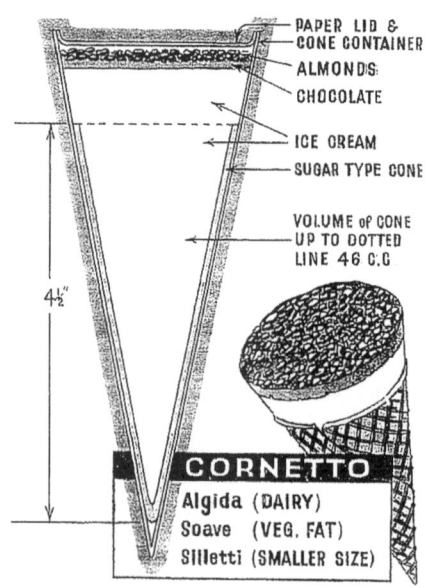

ACKNOWLEDGEMENTS

The painting on the cover of this volume is reproduced by kind permission Patricia Rogers, Head Librarian, The Corning Museum of Glass, Corning, N.Y. USA. The Juliette K. and Leonard S. Rakow Research Library. Cornetto, drawing copyright Unilever plc. All other illustrations in the author's collection. Mrs. A.B. Marshall, *The Book of Ices* and *Fancy Ices* are available in facsimile from Liz Seeber, Apple Tree Cottage, High Street, Barcombe, nr. Lewes BN8 5DH. Tel. 01273 401485. Fax 01273 401486.

REFERENCES

[1] 'Penny Licks and Hokey Pokey, Ice-cream before the Cone', *Public Eating: Proceedings of the Oxford Symposium on Food and Cookery 1991,* Prospect Books, 1991

[2] 'Mrs. Marshall, Ice-creammonger Extraordinary', *Cooks and Other People: Proceedings of the Oxford Symposium on Food and Cookery 1995,* Prospect Books, 1996. See also Weir, Deith, Brears & Barham, *Mrs Marshall, the Greatest Victorian Ice-cream Maker,* Smith Settle, Otley, 1998.

[3] *Mrs. A.B. Marshall's Cookery Book,* Marshall's School of Cookery, London, 1888.

[4] *Le Cannameliste français,* Gilliers, Nancy, 1751.

[5] M. Emy, *L'Art de bien faire les glaces d'office,* Paris, 1768.

[6] Caroline Liddell & Robin Weir, *Ices: The Definitive Guide,* Grub Street, London, 1995.

[7] Mrs. A.B. Marshall, *Fancy Ices,* Simkin, Marshall & Kent Ltd, London, 1894 (Facsimile of this book available from Liz Seeber, Apple Tree Cottage, High Street, Barcombe, nr. Lewes BN8 5DH. Tel 01273 401485. Fax 01273 401486).

[8] Pim Reinders, *Licks, Sticks and Bricks,* Unilever, Rotterdam, 1999.

[9] Ibid.

Use of Almonds in Late-medieval English Cookery

Caroline Yeldham

This paper is presented as part of this Symposium because almond milk was a standard ingredient in the late-medieval European kitchen. This paper sets out to investigate how this came to be so, and looks at some possible reasons why. Almond milk is very simple to make: blanch and grind your almonds as fine as possible, mix with a liquid (stock, wine, ale, rosewater and water are all mentioned in various sources) and drain. The resulting milky, bland mixture can be used just like milk.

Anyone reading cookery books of the period from England or, indeed, the rest of Europe, soon notices how frequently certain ingredients are mentioned: rice, sugar, spices, dried fruits and almonds. I shall concentrate on England in this paper partly because this is my area of interest, and partly because England is a more extreme example than the rest of Europe. Although some of these ingredients were grown in areas round the Mediterranean, all had to be imported into England. Their presence in English cookbooks, therefore, indicates the presence of strong arguments for incurring the expense involved. Why were almonds used so frequently, not only as a source for milk, but in their own right, in late-medieval English cookery?

We are usually offered a deceptively simple answer. Fasting was always part of the Christian church's ethos, beginning with Christ's 40 days in the wilderness, reproduced annually by the faithful in Lent. By the late fifteenth century up to half the days of the year involved some kind of restraint of diet – no animal flesh, sometimes also no animal products (eggs, milk or cheese – known as 'white meats'). Times of restriction included Lent, the Ember days, Wednesdays, Fridays and Saturdays, although the rules varied, as did individuals' degree of observance.

Despite Protestant propaganda about barnacle geese and beavers' tails, extant household accounts indicate a high degree of conformity with the Church's rules. For example, in Eleanor of Brittany's accounts for 1225, on Saturday 16 August the household purchased sole, almonds, butter and eggs; on Wednesday, herring, conger, sole, eels, almonds and eggs; and on Friday conger eels, sole, herring and almonds. In contrast the purchases on Monday were beef, pork, honey and vinegar; Tuesday, beef, pork, honey and vinegar; and on Thursday, pork, eggs and egret.[1]

As usual in the medieval world, our clearest evidence comes from the nobility. In the case of almonds this is not as much of a handicap as usual. Although reasonable compared to spices at about 4s per lb, almonds were not cheap, at 2d to 6d per lb and it is likely the market was limited to the wealthy. Also, as we shall see below, they were not a common item for sale outside London. Although almond trees appear on the 'Fromond' list of plants, dated to *c.* 1500,[2] it would seem they were mostly grown for their blossom, while the fruit was imported. Gerald's *Herbal* of 1633 refers to 'Divers sorts of Almonds: we commonly have 3 or 4 sorts brought to us a large sweet almond vulgarly called a Jordan almond, a lesser called a Valence almond, a better almond of the bignesse of the Valence almond and sometimes a lesse than it.'

Following the annual pattern of 'fast and feast' was a duty enjoined upon everybody by the Church. There was another duty which was of prime social importance – hospitality. As Matthew Paris indicates in this passage from 1250 about Henry III, failure in this duty was inexcusable:

> The King, shamefully deviating from the track of his ancestors, ordered the expenses of his court and the amusements of his usual hospitality to be lessened: an inexcusable act and bringing on him even the charge of avarice.

The duty of hospitality was two-edged, there was also a duty on the guests not to be too ostentatious in their restraint. In an instructive fable, a hermit and his disciple visited a monastery and dined with the monks. On the road, they passed a well and the disciple wanted to drink. He was stopped by his master, 'Today is a fast.' 'But, father, did we not eat today?' replies the disciple. 'That was love's bread, my son; but for us let us keep our own fast.'[3]

Restraint by the wealthy and powerful was admired: Thomas Becket, when Archbishop of Canterbury, was admired for his restraint. He kept a superb table, but curbed his appetite; he tasted the wine before his guests drank, but for his own use there was 'water used in cooking of hay'.

Both these cultural imperatives left the host, and his cook, with a social problem. How to obey the dictates of the Church yet provide hospitality to guests and household. The latter problem was particularly acute when the network of political influence and the need to impress contemporaries through ostentation is taken into account. Dinners, and feasts even more so, were never socially neutral events. They were always statements of power and influence, pride and the depth of a man's purse.

For the nobility and wealthy Northern Europe, the rice, sugar, dried fruits and almonds of Southern Europe, with spices from the East, provided a suitable answer. Their close association in household accounts, and the activities of the Grocers' Company, demonstrate their association in the minds of consumers, as in the recipes.

Some spices don't seem to have disappeared from the West after the retreat of classical civilization; pepper continues to be used, cloves are known in France from the time of the Merovingians. The trade routes from the East were complex, with overlapping stages and innumerable middlemen. One key port on the overseas route was Aden, at the southern end of the Red Sea. Indian and Chinese ships (from Fu-Kim) sailed to Arabia in 60 days from the Chinese, Indo-China and Malabar coasts. At Aden the spices were transferred to smaller ships whose pilots knew the Red Sea, thence by camel to the Nile, downstream to Alexandria the 'market of the two worlds'. To the Levantine markets came Western merchants, particularly Italians. By the eleventh century, Venetian ships came up the Channel to England and Flanders. An alternative route used in the thirteenth century was overland from Italy to the great fairs of Champagne. A major development that was initiated at the end of the fifteenth century was that Europeans began to explore the African coast, beginning to find their way to India and the Spice Islands.[4]

Most of this journey did not involve almonds. Despite the best type being called 'Jordan' almonds, more were coming from near Malaga in Spain than the Holy Land. 'Jordan' may be a mistranslation from *jardin*[5] but the almond was certainly associated with the Bible and Holy Land in the medieval mind, being the symbol of Divine approval. The almond is one of the fruit trees of Canaan, and the rod of Aaron grew buds, flowers and ripe almonds as a sign that God had chosen Aaron and his tribe of Levi to be the priests of Israel.

Knowledge of the almond's virtue had survived in medicinal/culinary texts. Anthimus wrote in the sixth century (but there are copies of his text through to the eleventh):[6] 'Almonds are good. Moreover if they are bitter they are suitable for those suffering from disorders of the liver. Eat almonds after putting them in hot water and removing their skins. They are recommended for those suffering from consumption after being skinned, ground finely and mixed with the finest honey: and as a remedy for catarrh when they are just beginning to ripen.'

A fuller statement is found in the thirteenth century in *Mensa Philosophia* by Michael Scott [Scotus]: 'Almonds are temperately heating, soothe the throat, lie long on heavy stomach, clear obstruction and allay burning micturition. If eaten with sugar they increase spermatic secretions. The oil of the nuts is exceedingly hot and relaxing. Avicenna says it is specific for shingles, erysipelas and for a style on the eye. Oil of almonds benefits chest, lung, bladder and kidneys, creates nausea and is evacuated slowly. Removes blotches from face, with honey it smooths down scars of wounds, clears vision and rids head of dandruff.'[7]

Mark Grant's commentary on Anthimus quotes a long list of classical medical experts praising almonds for use in remedies e.g. for kidney pains and stones (Gargilius Martialis, *On Medicine*), bladder pains (Marcellus Empiricus, *On Medicine*), liver complaints (Gargilius), for consumption and catarrh (Pseudo Theodorus Priscianus, *On Simple Medicine*) and for coughs and colic when ground and mixed with honey and sage (Gargilius).

Rice is dry in fourth degree and heating in the first. When boiled in water it is good for diurnal colic. But when cooked with almonds it loses its constipating quality and provides good nutriment – it generates good blood and augments the *sperma*.[7]

Mention of the *sperma* above may be one reason why almonds were so popular with those who could afford them. However, *Mensa Philosophia* and other medical texts may help to answer the question 'why almonds?' Why weren't other nuts used in the same way as almonds? Looking at medicinal comments on alternative nuts, almonds seem to have few faults.

There are walnut recipes, for example in *Two Fifteenth-Century Cookery Books*, where walnuts are the basis of a sauce for stockfish, with garlic, pepper, salt, breadcrumbs and fish broth,[8] but it was believed that fresh walnuts are cold, moist and slow to digest, giving little nourishment although Gerard says green walnuts boiled with sugar and eaten as succade 'are a most pleasant and delectable meat, comforts the stomach and expel poison'.[9] Older walnuts are very hot and dry, and best avoided by cholerics, cause headaches and giddiness.[10]

Chestnuts are strengthening and nutritious but hard to digest as they are dry and binding, they are easier to digest when roasted or boiled and less windy.[6] Phlegmatics should eat with sugar or honey. They are aphrodisiac, warm and make men fat.[10]

Hazelnuts are inferior to almonds, when fresh they are moist and windy, when older hard of digestion, passing slowly through the belly and may clog it. Nourishing when roasted but heating and tending to fatness; excess may cause headaches. Old nuts should be avoided by the elderly.[11] 'Hazelnuts are troublesome if mixed in a dish with other sorts of nuts.'[10]

As a final encouragement to choose almonds over other nuts, they are recommended for phlegmatics as they are warming and for the melancholic, the fussy eaters of the medieval world. Almond milk is good for consumption[11] and a few nuts taken before a meal will prevent drunkenness.[9]

If C. Anne Wilson[12] is correct and the use of 'spices' from the thirteenth century onwards is a result of contact with the Arab world, then certainly the way towards use of almonds was prepared by preexisting medicinal texts.

Use of almonds in England was dependent on a sophisticated trade system. The eastern end of the system was touched upon above. At the northerly end, the Venice and Genoa fleets brought all kinds of luxury goods to Southampton, London and Sandwich. However, this wasn't the only source of almonds. Trade records of Bristol show cargoes coming in from Bayonne including iron, steel (e.g. combs), pitch and rosin, bowstaves, beaver and Cordovan leather, honey, almonds, liquorice and saffron. Some of this was from Spain, but Bristol also received directly from Spain honey, almonds, liquorice and saffron, vinegar, lard, tunny-fish and rice.

Much of this luxury trade was dominated by the London Grocers' Company, which in 1316 drew up regulations to prevent fraud in the sale of spice, wholesale and retail. By 1386 the London grocers were trying to assert a monopoly of the retail trade. This was not solely for their own benefit, in 1400 is the first recorded prosecution for adulterated spice, powdered spices being particularly vulnerable to adulteration. This attempt to improve standards continues in the first half of the fifteenth century, and the scale of fines may indicate the potential profitability of the trade, as well as the importance of a good reputation. The fines go up to £10 with expulsion at the third offence.[13]

However, grocers were not the only people to buy from Italians. Records concerning Italians based in London show that between 1439 and 1444 draper Simon Eyre bought nearly £1,000 of spices, and fishmonger Stephen Forster bought 22 bales of pepper worth £529. The Grocers' Company tried to limit this intrusion on their trade: in 1455 it passed a regulation that no grocer to buy any 'sotill ware' from anyone outside the Company except from a merchant stranger, on pains of forfeiting half the value of the purchase. In 1471 this was amended to forbidding the purchase of spice outside the Company if it was as cheap within (imported saffron was exempt). This doesn't seem to have been successful, by the end of century all ordinances were cancelled.[13]

Inventories from Richard II's time show the great diversity of a grocer's stock, reminiscent of the dark, aromatic mysteries of an old-fashioned grocers:

> Honey, liquorice, dyes, alum, castle soap and nature soap, brimstone, paper, copperas, gall and gum arabic, vermilion and turpentine, coloured wax, lamp oil, painters oil, garden seeds, vinegar, salt, varnish, red lead, arsenic, books, firewood, canvas, Breton linen, blanket cloth, coloured cloth from Bristol, kersey, quilts and furs, varnish and vinegar, balances, candelabra and pot brass, nearly all had large stock of spices, pepper, ginger, saffron, anise, cumin, cinnamon, cloves, nutmeg, mace, almonds, raisins, figs dates, sugar, rice, flour of rice, confectionery and incense. [13]

However, it seems to have been difficult to buy almonds and spices outside London. Inventories of merchants in the provinces don't mention almonds. For example, the Chancery records of a bankrupt Leicester merchant – who seems to have been a draper, haberdasher, jeweller, grocer, ironmonger, saddler and dealer in timber, furniture and hardware – include notes of ready-made gowns in taffeta or silk, bowstrings, harp strings, writing paper, ink materials and vegetable seeds, purses of gold cloth, kerchiefs for nuns, cutlery, coal scuttles and horseshoes, but no almonds. In all that list, the only provisions mentioned are honey, raisins and salt.[13]

But there was an alternative source of supply, the seasonal fair. I have already mentioned the overland route from Italy to the great fairs of Champagne. In England, fairs were also important. Dame Alice de Bryene bought most of her almonds and spices at Stourbridge.[14] In 1412–13 she bought 40 lbs of almonds, at about 2 ½ d per lb and used about 38 lb over the year. To give an indication of usage levels, her household was quite small, usually feeding about 22 for dinner and the same for supper. Over the same period she had used 3 lb of rice, 1 lb of sugar, 15 ½ lb of various spices, excluding 1.5 bushels of mustard seed, 1 frail[15] and 4 lb of figs, 1 frail and 7 lb of raisins and 1 gallon 2 ½ quarts of honey. This is quite a highly flavoured diet, but not overly sweet.

The London grocers tried to control the extramural activities of their members. In 1420 members were forbidden to sell spices except at fairs, in 1423 'except at fairs' was struck out and fines of up to £4 were levied on members who went to fairs at Oxford, Salisbury or Stourbridge (the most commonly mentioned). Members fined included Wardens of the Company, so the profits obviously tempted even those chosen to support the rules of the Company. The aim behind the legislation was to try to get customers and chapmen to come into London, bring extra business into London, and also that they would carry the risks of transport.

There was another attempt in February 1487 when a Council ordinance forbid the Freemen of the City to send wares to outside fairs or markets for seven years. This was annulled by Parliament in November. The Statute of Repeal suggests that the ordinance would have ruined the fairs, especially those at Salisbury, Bristol, Oxford, Cambridge, Ely, Nottingham and Coventry – indication of the degree of dependence on the London merchants.[13]

Christopher Columbus is famous for his attempt to break the Italian monopoly on spices. However, there are other attempts recorded closer to home, which are less well-known. The late fifteenth century was a period of disturbance and change in the Mediterranean. In 1445 the Venetians were driven out of Egypt, threatening the monopoly of the Italians in the Levant.

In 1446 Robert Sturmy of Southampton, a prosperous citizen who had been bailiff in 1444, tried to seize this opportunity. In November 1446 a licence made out to carry tin (26,000 lb) and 40 sacks of wool (the vessel probably also carried cloth but no licence was needed) 'by way of the Straits of Morocco' to Pisa. The cog *Anne* left with 160 pilgrims and crew of 37. It went on Joppa where the pilgrims were safely landed, and it picked up a cargo (probably of spices but the details are unknown). Unfortunately the ship was wrecked off the coast of Modon (southern Greece).

Robert Sturmy's career didn't suffer too badly from this; he was Sheriff in 1451 and Mayor in 1453–4. However, political uncertainty grew in Europe. In 1457 English merchants were shut off from Gascony. In the Mediterranean, the Ottoman invasion progressively shut out Italians, in 1453 Constantinople fell and Genoa lost their colony at Pera, although Venice was in negotiation with the Sultan for free trade and self-government in Constantinople.

Robert Sturmy tried again. In March 1457 he received a licence to send 40 quarters of wheat to Italy with other goods. The *Katherine* sailed safely to 'divers parts of Levant' and picked up a cargo of green pepper and other spices. Then the Genoese came upon the *Katherine* near Malta and 'spoiled his ship' and another. There was political uproar. All the Genoese in London were arrested, put in the Fleet prison, and their goods confiscated.

A lawsuit was laid 'before the kyng and his conseile' from which the mayor and his brethren received 9,000 marks. On 25 July 1459 Sir John Stowton, Philip Mede, John Eyton, William Canynges, Richard Chok and William Coder received from the Treasury £6,000 to be distributed amongst those involved and the Genoese were released from prison. However, Robert Sturmy couldn't take advantage of the settlement, his will was proved on 12 December 1459.[13]

I gave a simple recipe for almond milk at the beginning of the paper, but this is not the only way almonds, or almond milk, were used, there is quite a large repertoire of recipes, often using almonds, rice, sugar, spices and dried fruits as their main ingredients. Almond butter, which has the advantage that it doesn't need salt to keep well, was often made with just sugar and rosewater, or might be made in more elaborate fashion in later, seventeenth-century, recipes (see, for example, one quoted in *The Art of Dining* by Sara Paston Williams).

However, almond cheese may be indicative of more ingenuity. The Viandier speaks of 'Flawns and tarts in Lent which will taste like cheese' made with roe and milk with almond milk. The Neapolitan collection gives recipes for almond cheese made with almond milk, pike broth and starch, and fake ricotta from almond milk and fish broth from pike or tench. The cook is recommended to mould the curds in wicker baskets like those used by ricotta pedlars.[16]

An important advantage of almonds over cow's milk was that the cook could be sure of the quality of the almond milk. The Menagier de Paris advises his wife to ensure that she watch the animal being milked, to prevent adulteration. Today we would also have concerns about the health of the cow and hence of the milk.

Having established that almonds were an important ingredient, for those that could afford them, and that contemporary medical opinion was in favour of using almonds, sometimes in preference to cow's milk, is this nutritionally sensible according to modern analysis? Almonds contain 160 calories per ounce, with 18.6 per cent protein and 54 per cent fat, of which 8 per cent is saturated, 68 per cent mono-unsaturated, and 19 per cent polyunsaturated. They are also high in amino acids (18 of the 23), and in essential vitamins: 15 almonds contain 4 mg Vitamin E, 14 micrograms folic acid and 41 mg magnesium. A quarter-cup of ground almonds contains 140 mg phosphorus. They are certainly a useful food source for vegans.[17].

Could this useful food source tempt the medieval noble to the sin of greed? Salimbene's description of the dinner given to Louis IX by the Franciscans at Sens might lead one to wonder; the dinner included fresh cherries, white bread, fresh beans cooked in milk, fish and eel pies rich with almonds milk and cinnamon, roasted eel in sauce and fruits with wine.[4]

Evidence from the Earl of Derby's records[18] shows that temptation existed even on pilgrimage. In 1392 he was on his way to Palestine, for which he had borrowed 8,888 ducats from the Alberti in Lombardy. The retinue no doubt included his chamberlain, treasurer, clerk of the household, and the officers of the stables, kitchen, buttery, pantry, poultry as well as a falconer – as it had done two years previously on crusade to Prussia. He stopped in Venice to purchase supplies including live poultry with cages and food, oxen for salting, 2,250 eggs, casks for water, cheese, oil, potted ducks, fish, vegetables, condiments, spices, 1,000 lb of almonds, sweet confectionery, sugar, choice wines, biscuits, fresh bread, four barrels of French fruit, butter and fuel. Some of these items are necessities, others would surely have been considered luxuries by the first generations of crusaders.

As European life grew more sophisticated, a complex trading system reaching half way around the world grew to satisfy demands for alternative flavours. The almond was particularly useful for this purpose, because it helped the cook and his master to provide a welcoming table whilst following strict ecclesiastical rules. Meat remained the focus of prohibitions, whilst cooks were lavishing care on 'sotill ware' to maintain the reputation of their masters for hospitality.

REFERENCES

1. *The Great Household in Late Medieval England*, C.M. Woolgar, 1999.
2. *Early Nurserymen*, John Harvey.
3. As quoted in *Fast and Feast*, Bridget Ann Henisch.
4. *Baronial Household of the 13th century*, M Wade Labarge.
5. *Modern Herbal*, M Grieve.
6. Anthimus, *De obseruatione ciborum*, trans. Mark Grant, 1996.
7. *Mensa Philosophia*, by Michael Scott. Trans. A.S. Way, *Science of Dining*, 1936.
8. Austin, Thomas, ed., *Two Fifteenth-Century Cookery Books*, EETS, 1888.
9. Gerard's *Herbal*, 1633.
10. Jane Huggett, *The Mirror of Health, Food Diet and Medical Theory 1450–1660*, Stuart Press, 1995.
11. Andrew Boorde, *A dyetary of helth*, ed. F.J. Furnivall, EETS, 1870.
12. 'The Saracen Connection: Arab Cuisine & the Medieval West', C. Anne Wilson, *PPC*, 7 & 8, 1981.
13. *Studies in English Trade in the 15th century*, ed. Power and Postan.
14. M.K. Dale and V.B. Redstone, eds., *The Household Book of Dame Alice de Bryene 1412–13*, Ipswich, 1931.
15. A frail is a wicker basket of variable size.
16. *The Art of Cookery in the Middle Ages*, T Scully.
17. Sylvia Fradley, private communication.
18. *Everyday Life of Medieval Travellers*, Marjorie Rowling 1971.

La Laiterie de la Reine at Rambouillet

Carolin C. Young

The *Laiterie de la Reine* (Queen's Dairy) was presented as a surprise gift to Marie-Antoinette by her husband, Louis XVI, in June, 1786. Today it stands, half-forgotten, in the woods behind the Château de Rambouillet, denuded of its furnishings, its elegant porcelains and most of its sculptures. It is a strange vestige of an unsustainable world, made as a play-dairy for a queen who lost her head on the guillotine before ever having a chance to use it. The Rambouillet dairy was the pinnacle of the development of the *laiterie d'agrément* (pleasure-dairy), and is one of the few examples of the form which remains today (figures 1 and 2). In creating it, France's leading Neo-Classical artists and designers melded the monumentality and mythology of Ancient Greece with Rousseau's love of nature to create the ultimate temple of milk in a harmony of luxurious materials, exceptional craftsmanship and purity of design. Paradoxically, this quintessence of *Ancien Régime* folly and refinement initiated a new Archaeological Neo-Classicism which became the decorative vocabulary of the republicanism that made it obsolete.

In 1783 Louis XVI purchased the Château de Rambouillet and its lands from the Duc de Penthièvre for use as a hunting lodge. He had a great deal of difficulty, however, in trying to convince his wife, Marie-Antoinette, to accompany him there; she found the fourteenth-century château gloomy and old-fashioned and was bored by hunting. Almost immediately, Louis XVI commissioned the Comte d'Angiviller, his *Directeur Général des Bâtiments du Roi*, and Governor of the Rambouillet estate, to oversee the creation of a *jardin anglais-chinois* (Anglo-Chinese garden), which was to include a *laiterie d'agrément* with which to entice Marie-Antoinette to visit Rambouillet.

Jacques-François Blondel, in his *Cours d'architecture* of 1773, defined a *laiterie d'agrément* as a place where ladies go to milk cows, churn butter and make cheese, and to relax with walks and country pleasures.[1] These pleasure-dairies grew enormously popular in France through the course of the eighteenth century, together with the *jardins anglais-chinois* and the *fermes ornées* (ornamental farms) which contained them. The most famous of these was Marie-Antoinette's own Hameau at the Trianon of Versailles, which Mique was building for her just as construction on the dairy at Rambouillet began (1784–

Figure 1. Jacques-Jean Thévenin, exterior of La Laiterie de la Reine, *built 1784–6 at the Château de Rambouillet (Yvelines, France)*

Figure 2. The rotunda of La Laiterie de la Reine *at Rambouillet, with a view into the* Pièce de Fraîcheur *containing 'Amalthea and the Goat that Suckled Jupiter', white marble, by Pierre Julien. The floor,* pietre dure *table, and heavier perimeter console were modifications at the order of Napoleon I.*

86). It seems quite appropriate that Louis XVI should have chosen to build a dairy for his queen, who had herself played a milkmaid in her theatre at the Trianon, singing:

> Voilà, voilà la petite laitière,
> Qui veut acheter de son lait?[2]

However, although the *Laiterie de la Reine* at Rambouillet was inspired by Marie-Antoinette, she was in no way involved in its planning or design. Similarly, since the Hameau was being built at exactly the same time as Rambouillet, it must be viewed as another expression of the popular motif of the *laiterie d'agrément* rather than an influence upon it.

The tradition of the aristocratic dairy in France can be traced back to the sixteenth-century dairy at Fontainebleau, decorated by Primaticcio. According to Johannes Langner, in his article on pastoral architecture under Louis XVI, this remained an isolated example for over a century, until the *Grande Laiterie* of the *Ménagerie de Versailles* was built from 1662 to 1664 on a design of Le Vau.[3] When Louis XIV gave the *Ménagerie* to the Duchesse de Bourgogne in 1698, he had Jules Hardouin-Mansart construct a second, more luxurious but smaller dairy, next to, but separated from, the first, specifically for her personal use. Years before Marie-Antoinette's pastoral entertainments at the Hameau the Duchesse de Bourgogne reportedly milked the cows herself and made butter for the table of Louis XIV, which the king was seen to enjoy and eat in her honour.[4] The separation of the *laiterie d'agrément* from the *laiterie d'utilité* (working dairy) which occurred there exemplifies its function, as a place for aristocratic recreation, rather than as the primary source of a household's dairy products.

The *laiteries d'agrément* which sprang up with increasing frequency in France during the second half of the eighteenth century were closely linked to the development of landscape gardening and the theory of the 'Picturesque'. The cult of the Picturesque can be summarized as a romanticized view of nature, based upon the seventeenth-century Arcadian landscapes of Claude Lorraine, which favoured irregularity and 'naturalness' in carefully constructed form. It evolved from the light-heartedness of the Rococo, epitomized by Watteau's *fêtes galantes* in the first decades of the eighteenth century and led into Romanticism in the early nineteenth century.

The ideal of the Picturesque was particularly strong in English landscape gardening and was exported to France after the end of the Seven Years' War in 1763. This grew more marked by the late 1770s and 1780s as the French developed their mania for all things English, the so-called *goût anglais*.[5] Gardens such as those at Stowe and Chiswick were held up as models for emulation in France.[6] The *jardin anglais-chinois*, which imitated the irregularity and exoticism of 'Chinese' gardens being built in England (the garden equivalent of Chinese Chippendale) and the *ferme ornée*, filled with ornamental pastoral follies, came into vogue. T. Wately's *Observations on Modern Gardening*, written in 1770, was translated into French almost immediately by F. de Latapie as *L'art de former les jardins moderne*. This was soon followed by Claude-Henri Watelet's *Essai sur les jardins* in 1774, which incorporated English ideas of picturesque gardening with the philosophy of Rousseau.

Dora Wiebenson has noted that the French interest in picturesque gardening differed from that of the English, although both found their roots in literature. In her words, 'where the English interest was a reflection of the importance of agriculture in the development of the national economy and the role of the individual within this context, French interest reflected a retreat of the individual to nature and to rural life in a period of political and economic decline.'[7] This distinction is apparent in the development of the dairy in these respective countries; the English dairy was usually utilitarian in nature, and located near the kitchen; by comparison, the French *laiterie d'agrément* was traditionally considered a place for meditation and increasingly set apart from other structures.[8] The first English dairy designed as an isolated pleasure pavilion was built at Richmond, sometime after 1736, and remained unusual.[9] However, in France the *laiterie d'agrément* was considered an appropriate, if not necessary, element to these gardens. Watelet included it in his list of suitably 'pastoral' embellishments for the garden of a country estate and, he built both a cow barn and a pleasure-dairy in his own *hameau* at Le Moulin-Joli in 1754.[10] Around 1770 the French pleasure-dairy then emerged as a separate, isolated structure, in the dairies at Monceau (1771–1773), Tivoli (1771) and Le Raincy (sometime after 1769).[11] These *laiteries d'agrément* were often designed as rustic huts, but also took on exotic forms, such as the Prince de Ligne's dairy at Beloeil, which looked like a mosque with minarets.[12]

The meditative quality attributed to the *laiterie d'agrément* grew out of the sentimental attitude of the French Eighteenth Century toward pastoral life embodied in the philosophy of Jean-Jacques Rousseau. He summarizes these ideas in *Émile*, when he states that, 'I would always stay as close as possible to nature, in order to indulge the senses I received from nature – quite certain that the more nature contributed to my enjoyments, the more reality I would find in them'. He then goes on to advocate a diet of simple and natural foods, and particularly mentions keeping cows for the dairy products he loves so much.[13] The French aristocratic predilection to relax in playing at the simple work of both peasant and servant anticipates Rousseau's writing, as exemplified not only by the Duchesse du Bourgogne's butter, but also in the Régence fashion for 'cooking' Sèvres used by people such as the Duc d'Orléans himself. Rousseau gave a voice and a philosophy to a preexisting desire to escape the pressures of court life, which immediately resonated with those caught in its hold. Marie-Antoinette herself enjoyed collecting engravings by Moreau le Jeune of scenes from Rousseau's *Émile*.[14] It is easy to see their appeal to a woman who complained to her sister that, 'people think it very easy to play the queen – they are wrong. The constraints are endless, it seems that to be natural is a crime'.[15]

Rousseau's writings also articulate the association between women and dairy. The idea that woman's primary function is to nurse her children is a cornerstone of Rousseau's philosophy, but he draws the further conclusion that women naturally love dairy products.[16] In addition, Rousseau defines cleanliness as one of the highest feminine virtues; a quality which Diderot's *Encyclopedia* considers central to a well-run dairy.[17] The *laiterie d'agrément* therefore provided aristocratic women with a forum from which to escape into a pastoral pursuit which was philosophically linked to the core of their womanhood as well as deemed suitably clean and proper to their social status.

Louis XVI visited both the Duc d'Orléans' dairy at Raincy and that of the Prince de Condé at Chantilly before commissioning the *Laiterie* at Rambouillet, and both were strongly influential to its final design. The *Ménagerie* at Chantilly was built from 1689–1694, under the direction of Jules Hardouin-Mansart by Daniel and Pierre Gitard. The dairy was destroyed in 1799, but surviving descriptions and drawings inform us that it was comprised of three rooms en suite leading to the principal rotunda made of white stone, paved in white marble, with a vaulted ceiling. A sideboard, set into the wall around the entire rotunda displayed an assortment of vases and faience. A small trench was cut into the marble floor for water which was accessible from four ram's head faucets. The room was hung with pictures depicting pastoral scenes and by paintings inspired by the cult of Isis, who was symbolized by a cow.[18] This arrangement became the model from which the building for the Rambouillet dairy was taken.

The dairy at Raincy, one of the first isolated *laiteries d'agrément*, was particularly noted for the set of cane-coloured Wedgwood dairy-ware commissioned for it. Although the order for it has been preserved in the Wedgwood archives no pieces are known to survive.[19] Wedgwood made numerous dairy-ware services in the late eighteenth and early nineteenth centuries, but the Prince de Condé was one of few, and certainly the first to acquire a set in France, and it was one of the most extensive services of its type ever made. The Wedgwood service made for the Countess Spencer's dairy at Althorp in 1786 is illustrative of the type of pieces which the Raincy set would have contained. This included: cream vases (large, footed ovoid vessels with a lid to keep cream cool, first made by Wedgwood in 1769), ladles, milk sieves, settling pans (large oval pans with a pouring lip at one end in which milk would stand to form cream), skimmers (shallow, perforated bowls with a pouring lip for separating and straining cream from the surface of milk), spoons, butter churns and tiles for lining the interior walls.[20] Raincy's isolated structure and extensive dairy-ware were both imitated at Rambouillet.

The *Laiterie de la Reine* went far beyond any of these previous dairies in luxury, design and iconography, largely due to the talents of the men assembled by the Comte d'Angiviller who gave the dairy the monumentality of a Greek temple. Selma Schwartz, in her thesis on the porcelain from the Rambouillet dairy, noted that almost all of these men had studied in Rome, and all were strong proponents of '*la vraie beauté*,' as the archeologically-based second phase of Neo-Classicism was known.[21] This style was based upon a purer adherence to classical antiquity than that of the 1760s and 1770s, which had retained much of the flourish of the Rococo, and is summed up by Winckelmann's famous statement that, 'the general and predominant mark of Greek masterpieces is noble simplicity and calm grandeur, both in gesture and in expression'.[22] The Comte d'Angiviller had been responsible for the reorganization of the French Academy in Rome prior to his appointment as the *Directeur Général des Bâtiments du Roi*, and there became an advocate of Winckelmann's ideas. D'Angiviller chose Hubert Robert, France's most important picturesque landscape gardener and painter, to oversee the building of the dairy. Robert, himself, had worked in an advisory role at the French Academy in Rome from 1768–72, and had already designed the classically-inspired *laiterie* at Méréville. Evidence indicates that Robert supervised the overall plan for the *Laiterie* at Rambouillet, with other artists and designers submitting their ideas for his approval.[23] They included the architect Jacques-Jean Thévenin, the sculptor Pierre Julien (who had won a *Prix de Rome* and was at the Academy from 1768–72), and the director of the Sèvres porcelain factory, Jean-Jacques Lagrénée (who had spent 1765–69 in Rome producing design books from antique wall-paintings). Robert himself designed the furniture, which was made by Georges Jacob (who later built the furniture for the painter J. L. David's studio).

The small but royal commission for the Rambouillet dairy provided these artists with an opportunity to practice their most radical innovations without the pressure to cater to conservative tastes frequently present in more public projects. Together they brought the aesthetics and iconography of antiquity to the *Laiterie*, melding simplicity of line with simplicity of spirit. What they created was a completely innovative work; the cows and goats which they employed in Rambouillet's decoration are not classically derived. Goats are known to have appeared as a Bacchic symbol in some Roman wall-painting and vases, but cows do not. Stylistically, the dairy appeared to imitate the classical past more archaeologically than prior Neo-Classicism, but the iconography of the dairy was invented to give form to Rousseau's thought.

The construction of the Rambouillet dairy probably began in August of 1784, although the original documentation has now been lost. A site was chosen to the west of the château, beyond its radiating canals. It stands near the *Chaumière des Coquillages*, another small garden folly whose interior is entirely covered in shells and mother-of-pearl. In 1785 the famous experimental farm was built on the other side of the dairy; the Merino sheep from Spain arrived in October of 1786.

A gate flanked by two semicircular pavilions precedes the *Laiterie* itself; the left-hand pavilion contains the '*Salon du Roi*', covered in *trompe l'oeil* relief-paintings of the 'Four Seasons' by Joseph Sauvage, on the right is the guard's house. Originally, a second gate was meant to demarcate the dairy from the courtyard in front of it, but this was dismantled, if it was ever built at all.[24] The *Laiterie* sits back beyond numerous exotic trees, which Robert imported from around the world.

The building is in the form of an early Greek temple with primitive, rusticated walls. Along the main axis it is a pure cube, rendered in large, severe sandstone bricks, relieved only by a simple cornice set one fourth of the way below the top. Three shallow steps lead up to the front door, which is flanked by two Tuscan columns supporting a semicircular pediment whose only ornamentation is a white-marble roundel of a cow suckling her calf, carved in relief, and the inscription, LA LAITERIE DE LA REINE.

The interior reveals two rooms: a sphere and a cube, the purest of Neo-classical schemes (figure 2). Through the door one enters into the Rotunda, whose domed ceiling is coffered with rosettes. An oculus lights the cave-like room, which is now dominated by a large, circular *pietre dure* table and multi-coloured marble floor, innovations of Napoleon. Originally, the Rotunda had a pure white marble floor, which matched the austerity of its design. Ten niches, alternately rounded and rectangular, are set into the walls beneath which runs a thin white marble console. The rectangular niches were added by Napoleon, as was the console, which replaced a less-ornate original. The niches were meant to hold vases filled with cooling milk, with other porcelain arranged symmetrically along the perimeter console. Napoleon's niches have replaced four bas-relief medallion sculptures by Julien, which depicted women in classical robes stoically milking cows, churning butter, and tending to other farm work. Jacob's dark mahogany furniture, purportedly covered in a nasturtium orange bordered with black, must have been a striking contrast to the cool whiteness of the room.[25] The Rotunda once held a round table

Figure 3. Pierre Julien (1731–1804), 'Nymph Milking a Cow', 1786–7, white marble relief sculpture created for the Rotunda of the Rambouillet dairy, (diameter: 65cm)

Figure 4. Pierre Julien (1731–1804), 'Nymph Churning Butter', 1786–7, white marble relief sculpture created for the Rotunda of the Rambouillet dairy, (diameter: 65cm)

surrounded by X-form Roman-style chairs and klismos-shaped milking stools, carved with rams' heads and other Etruscan motifs, some of which survive today at Versailles.

A pair of double-doors leads into the long, rectangular '*Pièce de Fraîcheur*' beyond it. The room is dominated by an enormous rock grotto, running from floor to ceiling, which contains Julien's white marble sculpture of 'Amalthea with the Goat that Suckled Jupiter', an unusual subject taken from Ovid.[26] The milk, which nurtured Jupiter through his infancy, is its unseen theme. This sculpture, considered a masterpiece in its day, was removed to the Louvre after the Revolution. Happily, it was returned to the *Laiterie* in 1953. Amalthea and her goat are surrounded by water, pouring from three fissures, which falls from a smaller basin into a larger one. A softly arched skylight, set into an octagonally coffered ceiling, lights the room. The surrounding walls are decorated only by a simple moulding. Originally, the side walls held long, frieze-like relief sculptures depicting 'The Education of Jupiter by Corybantes' and 'Apollo as Shepherd to Admetus', themes which set the Rousseauian pastoral ideal in mythic terms. A single, white marble medallion relief of a mother nursing her children was hung on the tympanum over the door. Guéridons by Jacob lined the walls, along with jugs of cooling milk. The effect was to create a sanctuary for communing with the mysteries of nature and reflecting upon her gift of milk. The idea for the grotto was inspired by the *nymphée* described in Longus' 'Daphnis and Chloe', popularly read in French translation during the second half of the eighteenth century.[27] The two rooms together, the first dedicated to work and the second to contemplation, play upon the belief that the nymphs themselves engaged in the simple occupations of pastoral life. The use of the most primitive Etruscan style suggests a time so ancient and mythic that it is inseparable from nature. The iconography throughout the dairy invites meditation upon milk.

The extensive Sèvres service created for the dairy at Rambouillet remains among the most elegant and sophisticated porcelain ever made, and certainly some of the most influential. Pleasure-dairies were typically equipped with dairy-ware services of faience, Chinese export porcelain, or, more rarely Paris porcelain, such as the LeBeouf set made for Marie-Antoinette's Trianon. The Rambouillet service was the only dairy porcelain made by Sèvres in the eighteenth century and, atypically for the factory, none of it is gilded, all except the '*jattes téton*' were made in hard-paste, and the Etruscan decoration emphasized the pure, white body of the porcelain itself. Sèvres held the royal monopoly for gilding on porcelain until the Revolution, and was renowned for its highly saturated ground colours, most of which could only be produced on a soft-paste body.[28] The white, ungilded body of the Rambouillet porcelain matched the cool austerity of the dairy, but also served to emphasize the purity and integrity of the material itself.

This 'truth to materials' attitude in the decoration of the Rambouillet porcelain was matched by its radical, archaeologically-inspired forms. The commission for the *Laiterie* presented Lagrénée, newly appointed director of Sèvres in 1785, with the opportunity to shift toward a purer imitation of the classically inspired shapes he had studied in Rome. He arranged to have Vivant Denon's 525-piece collection of Etruscan and Greco-Campanian pottery, recently purchased by d'Angiviller on behalf of the king, transferred to the studios at Sèvres. The Denon Collection, which survives at the Musée National de la Céramique at Sèvres, served as the primary design source for the Rambouillet porcelain; other ideas were taken from d'Hancarville's 1766-7 publication of William Hamilton's Greek vase collection, as well as the 1757 folio of the excavations at Herculaneum and Pompeii.

The pieces were painted with figures, goats and cows, alluding to milk, and most were edged with a simple black border. Many were painted with Etruscan designs, such as rows of lapets, palmettes linked by arches and a dotted, crosshatched motif, taken directly from one of the Denon pieces. Others had arabesque designs of scrolling foliage derived from Roman bath paintings. When used, the ground colours were '*grès*', the sandstone of the *Laiterie* itself and '*étrusque*', probably a terra-cotta colour.

There were 108 pieces originally ordered for the dairy, but this was subsequently reduced to 65 (with composite pieces, such as a bowl and stand, counting as one piece).[29] Very little of it has survived today, but drawings from the Sèvres archives help to illustrate what it looked like.

The most famous piece is undoubtedly the '*jatte téton*', a bowl in the shape of a woman's breast which sits in a tripod stand terminating in goat's head supports with goat's hoof feet (frontispiece). Rumours persist that it was modelled on Marie-Antoinette's breast, although no documentary evidence seems to support this.[30] This extraordinary form was derived from the ancient Greek '*mastos*' cup, which was also in the shape of a breast. Four *jattes tétons* were ordered for Rambouillet, but the factory continued to produce and sell the shape from the original model until 1885, when they replaced it with a new one. The Rambouillet service included four other types of *jattes*, (shallow bowls) which incorporated elements of Greek drinking vessels from the Denon collection into their design, but did not directly copy them.

Four designs for cups, two of which included a saucer, were produced for the *Laiterie*; all were classically derived. The '*gobelet à anses Étrusques*', for example, took its shape from two different vases from the Denon Collection, as demonstrated when they were shown together for the exhibition, *The Age*

Figure 5. Skyphos, *Etruscan drinking cup from the collection of Vivant Denon, used as the model for* gobelet à deux anses Étrusques et soucoupe *created for the Rambouillet dairy. Courtesy of the Musée National de Céramique, Sèvres, MNC 232.*

of Neo-Classicism.[31] The cup, with two upturned bracket handles, tapers gently to a foot of two stacked coils above a round plinth and is accompanied by a saucer. The shape is derived from a '*skyphos*', a common Greek drinking cup, and remained very popular through the 1790s, as did the '*gobelet à bandeau*' and the '*gobelet cornet*', which added a flaring concave profile to the traditional Sèvres '*gobelet à lait*'.

Two types of vases, based on Greek volute craters, were designed for the wall niches of the Rotunda. A footed centrepiece, derived from an Attic drinking bowl, was intended to sit on Jacob's central table. Two milk-pails and sieves, painted to imitate wood, with high-relief goat's head handles were ordered as centrepieces for the side consoles. Four types of '*pots à lait*', large milk jugs shaped like Apulian wine jugs, were designed, although only one was kept in the final service. These were the size of water jugs, without a basin. Similarly, the four '*sucriers ronds*' resembled water basins and were made without the cover usual to sugar bowls. Finally, four '*grandes terrines à pieds de chèvre*', large, low bowls on brazier-like stands, were made, two for each side of the rotunda. Several butter dishes and '*ecrémoires*', basins in which milk was stored to allow cream to form and be separated, were ordered and delivered, but later removed from the final service.

Louis XVI brought Marie-Antoinette to see the unfinished *Laiterie* on 20 June 1786. He had contrived to have it hidden behind a screen of branches, which was pulled back like a curtain at the last moment revealing the dairy to the unsuspecting queen.[32] However, the furniture and the first half of the porcelain were not delivered until May of 1787. The sculpture was only set into place the following month. The final delivery of porcelain did not take place until May of 1788, although the Queen never returned to see it.[33] It is unlikely that Marie-Antoinette ever had a chance to use her pleasure-dairy, but what might she have done there?

The symmetrical arrangement of the porcelain in the Rotunda plan with the milk-pails as centrepieces, the removal of the *ecrémoires*, together with the distinct absence of butter churns or a cow barn, indicate that the *Laiterie* was not intended as a space for the queen to actually play milkmaid. More likely, it would have been used as a pastoral setting for a light meal or dessert, similar to the one enjoyed by the daughters of Louis XV at the Chantilly dairy in 1777. They were rowed to the dairy in gondolas, followed by a sloop, carrying musicians who played as they ate, and a frigate firing cannons. Thousands of lanterns lit the gardens.[34] One can well imagine Marie-Antoinette taking a small group of friends out to the end of a canal in a gondola and walking the short distance to the *Laiterie*.[35]

Menon's 1774 edition of *La Cuisinière Bourgeoise* instructs readers that, 'les petits fromages à la crème, qui se mangent avec de la crème & du sucre,' as well as all other cheeses, except Parmesan and Brie, should be served at dessert.[36] He concludes his section on 'crèmes de dessert', which includes recipes for creams with strawberries and raspberries, 'fromage naturel à la reine' made with coffee and orange-flower water, and a variety of iced cheeses and ice-creams, with the advice that they be served in *gobelets*.[37] The *Laiterie* was replete with *gobelets*, as well as many *jattes*, suitable for serving these dessert creams. Milk could be cooled in the large vases, and poured from the oversized milk jug. The pails could hold *fromage blanc*, and be accompanied by sugar in the *sucriers ronds* and berries in the terrines, a treat which Marie-Antoinette was known to have enjoyed.[38] The Queen was, in fact, said to be so fond of her Trianon's creams and cheese that she would have them sent to her when at Fontainebleau or Paris.[39]

One wonders, however, if, for all its archaeological correctness and appropriateness to dairy, the Rambouillet porcelain would be enjoyable to use, and indeed, if this concerned its designers at all. Certainly, although simple and elegant, few cups are as uncomfortable to hold as an Empire teacup with an upturned handle, and few furnishings as unstable as a sabre-legged chair, both of which are derived from the archaeological neo-classicism of the *Laiterie*. Conversely, despite the decorative excesses of the rococo, nothing is as easy to pour from as an auricular-handled coffee-pot, and nothing more comfortable to sit in than a Louis XV *fauteuil*. Neo-classical perfection lay in reducing the object to its essential form not in simplifying its utility. Perhaps the true purpose of the porcelain is as a sculptural accent, the finishing touch to a space,

Figure 6. Gobelet cornet et soucoupe, *hard-paste porcelain, painted by Fumez after a design by Jean-Jacques Lagrénée, model by Louis-Simon Boizot, Sèvres, 1788, made for the Rambouillet dairy. The gobelet cornet was not originally paired with a saucer, although both pieces were original to the dairy. (gobelet: height: 11cm, diameter 10cm, soucoupe: diameter: 18.7cm) Courtesy of the Musée National de Céramique, Sèvres, MNC 6.795.*

Figure 7. Gobelet à deux anses Étrusques et soucoupe, *hard-paste porcelain, painted by Fumez, after a design by Jean-Jacques Lagrénée, shape attributed to Louis-Simon Boizot, Sèvres, 1788, made for the Rambouillet dairy. (height of cup: 8cm). Courtesy of the Musée National de Céramique, Sèvres, MNC 6.796.*

which, as Paul Guth suggests, would be perfect for contemplating Ovid and Rousseau at the same time.[40] The *Laiterie de la Reine* embodied the ideals of its own period within the immortal guise of the ancients; like all ideals part of its mystique lies in the knowledge that it was never used.

The Rambouillet dairy achieved that perfect balance of refinement and luxury that typifies the French decorative arts in the final days before the *ancien régime* collapsed. The *laiterie d'agrément* reached the apogee of its development at Rambouillet, where elegant, simplified design was achieved in sumptuous materials and mythic power was given to a queen's garden pavilion. The idea of the pleasure-dairy had no further place to go. The decadent world which brought it into being could not assume the guise of the Ancients without falling victim to the democratic ideals contained beneath its surface.

The porcelain had hardly been put into place when the French Revolution began, ensuring that the *Laiterie* would never see its intended use. Its fate was similar to that of the liberal aristocrats who helped initiate the events which led to their own destruction; the archeological neo-classicism brought into fashion at Rambouillet was adopted as the decorative style of the Revolution. Winckelmann stated that 'the independence of Greece is to be regarded as the most prominent of causes, originating in its constitution and its government, of its superiority in Art,' so it is logical that a hard-edged neo-classicism would bring republicanism with it.[41] The direct impact which the *Laiterie* had upon public taste can be gauged by Sèvres' immediate and continued sales of Rambouillet shapes. Even before the Queen's porcelain was completed, the painter Fumez was given two commissions for *gobelets de la Laiterie* an indication of the instant popularity of the service.[42] Sèvres continued to sell pieces made from the Rambouillet models through the 1790s and well into the Empire.

The *Laiterie* itself did not enjoy such sustained health. After the French Revolution the dairy sat neglected until state property was declared national on 10 August 1792. When an inventory of the Rambouillet estate was compiled, the only objects considered valuable were from the dairy, but much of the porcelain and furniture had already disappeared. The sculptures were then removed and taken to the Louvre, the furniture and porcelain packed up and put into storage. Joséphine eventually commandeered the porcelain for her own use at Malmaison and Napoleon replaced the white marble floor with a bombast of colour. The building itself remains, forever balancing a fragile union of luxury and idealism, with its roundel of the cow suckling a calf and the inscription LA LAITERIE DE LA REINE.

BIBLIOGRAPHY

Arts Council of Great Britain, *The Age of Neo-Classicism*, Catalogue to the Exhibition at the Royal Academy of Arts and the Victoria and Albert Museum, London, 1972.

Boyer, Marie-France & Halard, François, translated by Jenifer Wakelyn, *The Private Realms of Marie-Antoinette*, Great Britain: Thames and Hudson, 1996.

Conway, Peter V., *Jean-Jacques Rousseau*, New York: Twayne Publishers, 1998.

Cronin, Vincent, *Louis and Antoinette*, Great Britain: William Collins & Sons & Co. Ltd., 1974.

de Broglie, Raoul, 'Le hameau et la laiterie de Chantilly', *Gazette des Beaux-Arts*, Vol. 37, Oct.-Dec. 1950, pp. 309–324.

de La Borde, Alexandre, *Jardins de la France et ses anciens chateaux: Mêlée d'observations sur la vie de la campagne et la composition des jardins*, Paris: Delance, 1808.

Diderot, D. & d'Alembert, J., *Recueil de planches, sur les sciences, les arts libéraux et les arts méchaniques avec leur explication*, 3rd ed., Livourne: 1771.

Eriksen, Svend & de Bellaigue, Geoffrey, *Sèvres Porcelain*, Great Britain: Faber & Faber, 1987.

Feray, Jean, 'A Wedgwood Dairy in a French Collection', *Connoisseur*, American Edition, New York, Vol. 140, No. 563, September 1957, p. 21.

Guth, Paul, 'La Laiterie de Rambouillet', *Connaissance des Arts*, 75 (May 1958), pp. 74–81.

Honour, Hugh, *Neo-Classicism*, London: Penguin Books Ltd., 1991.

Langner, Johannes, 'Architecture pastorale sous Louis XVI', *Art de France*, 3 (1963), pp. 170–186.

Le Grand d'Aussy, Pierre Jean Baptiste, *Histoire de la vie privée des françois depuis l'origine de la nation jusqu'à nos jours*, Vol. 2, Paris: D. Pierres, 1782.

Loisel, Gustav, *Histoire des ménageries de l'antiquité à nos jours*, Vol 2, Paris: Octave Dion & fils & Henri Laurens, 1912.

Mauricheau-Beaupré, 'Un mobilier de G. Jacob dessiné par Hubert Robert', *Bulletin des Musées de France*, Vol 6, No. 4, April 1934, pp. 76–80.

Menon, François, *La Cuisinière bourgeoise*, reprint of the 1774 edition, afterword by Alice Peeters, France: Temps Actuels, 1981.

Pérouse de Monclos, Jean-Marie, *Le Château de Fontainebleau*, London: Scala Books, 1998.

Pinault, Raphaël, *Si Rambouillet m'était effeuillé*, France: L'Imprimerie Hérissey, 1987.

Reilly, Robin, *Wedgwood: The New Illustrated Dictionary*, Great Britain: Antique Collectors Club, 1995.

Reilly, Robin, *Wedgwood*, 2 Volumes, Great Britain: M Stockton Press, 1989.

Rey, Leon, *At Trianon: Wandering in Her Steps*, Savoy: Petit Savoyard, 1919.

Rousseau, Jean-Jacques, *Émile: or on Education*, published 1762, Introduction, translation and notes by Allan Bloom, New York: Basic Books, Inc., 1979.

——— *Julie: ou la nouvelle Héloïse*, published 1757, reprint, Paris: Editions Garniers Frères, 1960.

Schwartz, Selma, 'Sèvres Porcelain for the Rambouillet Dairy in Context', Master's Thesis, New York: Cooper Hewitt/ Parsons School of Design, 1993.

Villechenon, Marie-Noëlle Pinot de, *Porcelain from the Sèvres Museum: 1740 to the Present Day*, translated by John Gilbert, London: Lund Humphries, 1997.

Wiebenson, Dora, *The Picturesque Garden in France*, Princeton: Princeton University Press, 1978.

Winckelmann, Johann, *The History of Ancient Art*, Germany: 1764, translated. by G. Henry Lodge, Boston: Little, Brown and Co., 1856.

REFERENCES

[1] Jacques-François Blondel, *Cours d'architecture*, (Paris, 1773), quoted by Langner, p. 171.

[2] (Here, here is the little milk-seller,/ Who wants to buy her milk?) L. Anseaume, *Théâtre de M. Anseaume*, (Paris: 1766), quoted by Langner, p. 175.

[3] Langner, p. 172 – The Fontainebleau dairy and the *Ménagerie* have both been destroyed. Langner took his information on the dairy at Fontainebleau from A. Bray. *Le Château de Fontainebleau*, (Paris: s.d., p. 12). Primaticcio also created the *Jardins des Pins* at Fontainebleau organized around a grotto where François I enjoyed pastoral entertainments, well anticipating the eighteenth century pleasure dairies. This has been cited as the first rustic architecture in France. Only the grotto, in disrepair, survives (Pérouse de Montclos, pp. 200–222). Plans for the *Ménagerie* survive at the Bibliothèque Nationale (reproduced in Langner, plates 3 &4, p. 173).

[4] Le Grand d'Aussy, p. 52.

[5] The goût anglais helped spread and develop the French taste for picturesque gardening, but it by no means introduced the concept in France. One of the earliest examples of a picturesque garden in France was Lunéville, designed in 1737 by Stanislaus Lescynski, the exiled King of Poland, and father-in-law of Louis XV. The garden included a village of cottages for favoured guests. In 1742 Le Rocher, an imitation country village with 82 figures, arranged in vignettes, was added to the garden. The figures included a woman churning butter and some had mechanized movements and made sounds. Lunéville, with all its baroque artifice, was a forerunner in presenting a sentimentalized, ideal pastoral village in an entirely sanitized form for aristocratic diversion. Mme. de Pompadour's Hermitages, built in 1748, and Marie-Antoinette's Hameau at the Trianon of Versailles, 1784–86, are continuations of this tendency.

[6] The gardens of Stowe and Chiswick may both be visited today.

[7] Wiebenson, p. 1.

[8] Ibid, p. 92.

[9] Langner, p. 174, The dairy at Richmond has not survived, but is known through an engraving published in C. Campbell, *Vitruvius Britannicus*, (London: 1739), Vol. IV, plates IX–X, 'Plan of the House and Garden Park and Hermitage of Their Majesty's at Richmond, designed by J. Roque, 1736', reproduced by Langner.

[10] Wiebenson, pp. 17, 65, 99; Langner, p. 173, The cow barn and dairy (no longer extant) were later converted to court and kitchen, depicted in a wash drawing at the Bibliothèque Nationale (reprinted in Wiebenson, plate #26).

[11] Discussed by Langner, p. 173 and Wiebenson pp. 17, 65 & 99. All three dairies have now been destroyed and are known through contemporary descriptions and drawings: Monceau; separate dairy seen on the map from L. Carrogis (Carmontelle), *Jardin de Monceau, près de Paris*, (Paris, 1778, Prospectus) – reproduced in Wiebenson, plate #88, description from V. Thiéry, *Guide des amateurs et des étrangers voyageurs à Paris, ou description raisonnée de cette ville, de sa banlieu, et de tout ce qu'elles contiennent de remarquable*, (Paris, 1787), vol. 1, p. 67; Tivoli – illustrated in G.L. Le Rouge, *Détails des Nouveaux Jardins à la Mode*, (Paris: 1776–1787) and described by Baronne d'Oberkirch; Le Raincy – Hameau illustrated and described in Alexandre de La Borde, *Nouveaux jardins de la France et ses anciens châteaux*, (Paris: Delance, 1808), plate LXXII, pp. 132 & 154.

[12] Wiebenson, p. 94 and Langner, p. 175. Now destroyed, the dairy was described by the Prince de Ligne, *Coup d'oeil sur Beloeil*, (Dresden, 1795), Vol. I, p. 22.

[13] Rousseau, *Émile*, p. 345.

[14] Cronin, p. 211.

[15] Ibid, p. 136.

[16] Rousseau, *Émile*, pp. 361 & 395, and *Julie*, p. 103.

[17] Rousseau, *Émile*, p. 395; Diderot & d'Alembert Vol. 9, pp. 190–191 – definition of a

laiterie: 'Il faut qu'il soit voisin de la cuisine, ait un côté frais & non exposé au soleil, vouté s'il se peut, assez spatieux; & surtout tenu avec beaucoup de propreté; il faut qu'il y ait des ais, des terrines, des pots de différents grandeurs, des banquets, des barattes, des claies, des écliffes ou chazerats, des caferons des cases d'osier & en confier le soin à une servante entendue & amie de la netteté.'

[18] de Broglie, pp. 309–323; Langner, p. 171. Louis XVI would have also visited the new dairy built at Chantilly in 1774 by Jean-François Leroy. The description of the dairy is based upon Mérigot, *Promenades ou Itinéraires des Jardins de Chantilly*, 1791, p. 44 quoted by de Broglie, pp. 317–318.

[19] Feray, p. 21, attributed Wedgwood pieces in a private French collection as being from the Le Raincy dairy service. Their current whereabouts are unknown.

[20] Reilly, pp. 123–4, 128, 394.

[21] Schwartz, unnumbered.

[22] Johann Joachim Winckelmann, *Reflections on the Imitation of Greek Art in Painting and Sculpture*, (Germany: 1755), quoted by L.D. Ettinger, 'Winckelmann' in *The Age of Neo-Classicism*, pp. xxxii – xxxiii.

[23] Guth, p. 81.

[24] Langner, p. 179 – based upon the final group of Thévenin's drawings for the Rambouillet dairy.

[25] Boyer & Halard, p. 92. The source for this claim is undisclosed. The delivery invoice for the furniture for the dairy, transcribed by Mauricheau-Beaupré, pp. 78–79, includes a fee for the *tapissier* for ten cushions, although the colour is not specified. However, the claim for this coloration is in keeping with the appearance of the dairy and its porcelain and comparable to the cushions of red wool with black palmettes, also made by Georges Jacob in 1788, for the suite of furniture in J-L. David's studio. This furniture is no longer extant but can be seen in the artist's paintings, 'Paris and Helen' and 'Brutus', both at the Louvre.

[26] The same scene was used for one of plates from the famous 'Arabesque' porcelain service commissioned by Louis XVI and appears to have been a favourite of the king. See Cronin, p. 213.

[27] Langner, p. 185.

[28] 'Hard-paste' or true porcelain is made from a combination of kaolin and petuntse clays, fired at 1250–1350°C. 'Soft-paste' or imitation porcelain is made from a combination of white clays and silicas, usually fired well below 1250°C. Sèvres was producing quantities of hard-paste porcelain from 1768, but continued to produce more of the soft-paste because they were unable to achieve their famous colours in hard-paste. The Rambouillet service shifted taste toward the hard-paste, which was favoured after the Revolution.

[29] Schwartz, in her thesis for the Cooper-Hewitt, undertook the first complete analysis of the porcelain for the *Laiterie* at Rambouillet, including the commission, derivation, production, delivery and dispersal, which provided the foundation for the discussion of the porcelain herein.

[30] Guth, p. 78 and Boyer & Halard, p. 102 are among the publications which make this assertion. However, since the dairy was built as a surprise, and no evidence exists to support this, it seems unlikely.

[31] *The Age of Neo-Classicism*, No. 1418, plate 121a.

[32] Langner, p. 186, based upon an analysis of the king's journals & G. Lenôtre, *Le château de Rambouillet*, Paris, 1930, p. 110; Langner did not believe that Marie-Antoinette ever returned. The story of the branches being pulled back has long been part of Rambouillet mythology; no direct witnesses have described it.

[33] Schwartz reconstructed the delivery dates, based upon surviving receipts at the Archives National and the Archives de la Manufacture National, Sèvres. The furniture was signed for by Deshaies, concierge de la Garde-Meuble at Rambouillet on 29 May 1787, (AN O1 1919,

no. 18); Schwartz transcribed the delivery records for the porcelain from the MNS, '*Pieces de la Laiterie envoyées à Rambouillet le 25 Mai 1787*' and '*Livré une partie le 25 Mai 1787 et l'autre le 15 Mai 1788*' (MNS EB1 D12), and notes that a letter from Julien to d'Angiviller of 9 June 1787 (AN 01 1919-2 no. 134) expresses his delight that the sculpture was well received and considered a success. Schwartz concluded both that the king, the queen and court visited the dairy during the summer of 1787. This was based upon an analysis of the king's journals and comparison with the household expenses at Rambouillet examined in: Pierre de Janti, *Forêt, Chasses, et Château de Rambouillet*, (n.p., n. publ.,1947), pp. 99–100 and F. Lorin, *Rambouillet: La Ville, Le château et ses hôtes, 1768–1906*, (Paris: Librairie Alphonse Picard et Fils, 1907) p. 253. She records the last visit of the king to Rambouillet as August 1788, but does not think that the queen was present. No one has suggested a visit by Marie-Antoinette after 1787.

[34] This visit to the dairy at Chantilly, and numerous others are discussed in depth by de Broglie, in particular pp. 313–314.

[35] Schwartz records that gondolas were listed in the Rambouillet inventories (AN 01 3443).

[36] Menon, p. 328.

[37] Ibid, p. 449–454.

[38] Menu for the Queen's déjeuner (Paris: Archives Nationales), K505, no. 16, cited by Schwartz.

[39] Rey, pp. 17, 51–52.

[40] Guth, p. 79.

[41] Winckelmann, pp. 9–10.

[42] Eriksen & de Bellaigue, p. 350.

Other Papers Given at the Symposium

As in the previous year, it has not been possible to include all the papers presented in Oxford in this volume. This in no way implies that the papers excluded are of inferior quality; several of them will be or have been published elsewhere.

Rose Levy Beranbaum
Milk and Milk By-products in Baking and Desserts
A professional baker's description of and comments on the qualities of milk, butter, buttermilk and cheese.

Robert Chenciner
Red Milk
An essay on colour, especially red, with some references to milk.

John P. Gauder
Dessert Ices in Pillar Shape: A Uniquely British Phenomenon
Moulds for ice-cream are of many shapes but only in Britain between 1840 and 1940 was a classical pillar shape used.

K. Dun Gifford
Spilled Milk in the Canterbury Tales
In spite of the growth of the use of milk, particularly cows' milk, and other milk products in fourteenth-century England, Chaucer barely refers to it in *The Canterbury Tales*, even though there are many references to other foods. Perhaps he didn't like milk.

Juliet J. Harbutt
Cheese: An Endangered Species under Threat of Extinction
The paper describes the decline in artisan cheese-making in Britain and the need to encourage the demand for these local, seasonal, natural products.

Richard F. Hosking
Japanese Vegetarianism, Soya Milk and Yuba
The use of tofu and other products of soya milk in Japanese vegetarian – Buddhist – cuisine.

Eve Jochnowitz
Rotating Cows and a Sandwich Cake: Dairy Products of the Future at the New York World's Fair of 1939–40
The imaginative use by the growing American food industry of opportunities provided by the Fair to promote factory-made food products for the housewife of the future.

Jenny Macarthur
Curds and Yoghurt in India
A description of curd-making using the methods described in English-language Indian cookery books.

Colin Spencer
The Horrors of Milk Drinking
A survey of the use of milk for human food from early times with mention of the strangeness of this practice and some dangers involved.

List of those attending the Symposium

Dr Michael Abdalla, ul. Szydlowska 53/10, 60-656 Poznan, Poland.
Joy Adapon, c/o Dept of Anthropology, LSE, Houghton Street, London, WC2A 2AE.
Prof K. Albala, Univ of the Pacific, Dept of History, 3601 Pacific Ave, Stockton, CA 95211, USA.
Dr Joan P Alcock, 24 Queensthorpe Road, Sydenham, London, SE26 4PH.
Alice Arndt, 1821 Westlake Drive # 105, Austin, Texas, 78746, USA.
Hugo Arnold, 26 St Philip's Road, London, E8 3BP.
Dr Michael Ashkenazi, Gyosei International College, London Road, Reading, Berkshire, RG1 5AQ.
Sara Baer-Sinnott, Oldways Preservation Trust, 25 First Street, Cambridge, MA 02141, USA.
Priscilla Bain, 7 The Norton, Tenby, Dyfed, SA70 8AA.
Anne Bamborough, 18 Winchester Road, Oxford, OX2 6NA.
Chitrita Banerji, 9 Chauncy Street # 31, Cambridge, MA 02138, USA.
Ann Barr, 36 Linton House, 11 Holland Park Avenue, London, W11 3RL.
Rosemary Barron, 12 Centenary Way, Cheddar, Somerset, BS27 3DG.
Najmieh Batmanglij, 1408 35th Street NW, Washington DC, 20007, USA.
Rose Levy Beranbaum, 110 Bleecker Street, Apt. 7d, New York, NY10012, USA.
Dr A Blake, Firmenich SA, 1 route des Jeunes, CH 1211 Geneva 8, Switzerland.
Fritz Blank, Deux Cheminées, 1221 Locust, Street, Philadelphia, PA 19107, USA.
Daniel Block, 2609 W Farwell, Avenue, Chicago, IL 60645, USA.
Prof. Phyllis Pray Bober, Box 589, Harpswell, Maine, 04079, USA.
Lynne Bradshaw, Lychwood House, Caldy, Wirral, L48 1LP.
Una Bray, Dept Mathematics, Skidmore, College, Saratoga Springs, NY 12866, USA.
Marilyn Bright, 11 Sion Hill Avenue, Dublin, 6W, Ireland.
Catherine Brown, 30 Hamilton Park Avenue, Glasgow, G12 8DT.
Deirdre Bryan-Brown, Rose Cottage, 14 Henley Road, Shillingford, Wallingford, OX10 7EH.
Peter Burt, 26 Bushy Mead, Widley, Portsmouth, Hants, PO7 5DY.
Prof John & Mrs Moira Buxton, Pillar House, Needham Market, Suffolk, IP6 8DG.
Charles Campion, Old House Farm, Church Road, Crowle, Worcester, WR7 4AT.
Lisa Chaney, 13 Upper Price Street, York, YO2 1BJ.
Robert Chenciner, 11 & 12 Lloyd Square, London, WC1X 9BA.
Dr Jeremy Cherfas, Crossways Cottage, West Stoughton, Wedmore, Somerset, BS28 4PW.
Janet Clarke, 3 Woodside Cottages, Freshford, Bath, BA3 6EJ.
Dr Albert Coenders, Prof. Regoutstraat 77, 5348 AA - Oss, Netherlands.
Caroline Conran, Flat 1, 17 Holland Park, London, W11 3TD.
Andrew Dalby, Le Bourg de Saint-Coutant, 79120 Lezay, France.
Jane Bak Andersen & Lone Lysdal, Karoline's Kitchen, Frederiks Alle 22, 8000 Aarhus C, Denmark.
Silvija Davidson, 12 Lords Close, West Dulwich, London, SE21 8EZ.
Caroline Davidson, 5 Queen Anne's Gardens, London, W4 1TU.
Alun & Gilli Davies, Glebe Farm, St. Andrews, Major, South Glamorgon, CF6 4HD.
Joy Davies & Gareth Spencer Jones, 501 Cinnamon Wharf, 24 Shad Thames, London, SE1 2YJ.
Tricia Dawson, 56a Penton Road, Staines, Middlesex, TW18 2LD.
Tamasin Day-Lewis, Splatt Mill, Splatt Lane, Spaxton, Bridgewater, Somerset, TA5 1DB.
Rachel Demuth, 30 Belgrave Crescent, Bath, BA1 5JU.
Judith Dern, 2907 Swest Crockett Street, Seattle, Washington, 98199, USA.
Mrs Carol Déry, Department of Classics, University of Wales, Lampeter, Dyfed, SA48 7ED.
Gentilissima Signora June di Schino, Via Orazio 31, 00193 Roma, Italy.
Anne Dolamore, 10 Chivalry Road, London, SW11 1HT.
Jennifer Donovan-Pyle, 38 Ordnance Hill, London, NW8 6PU.
Fuchsia Dunlop, 13 Sandringham Road, London, E8 2LR.

Hugo Dunn-Meynell, 14 Avenue Mans., Sisters Ave, London, SW11 5SL.
Joy Durston, PO Box 1055, Elsternwick 3185, Victoria, Australia.
Anastasia Edwards, OUP (China) Ltd, 18/F Warwick House East, Taikoo Place, 979 King's Road, Quarry Bay, Hong Kong.
J Audrey Ellison, 135 Stevenage Road, Fulham, London, SW6 6PB.
Michael Erben, 21 Dorchester Court, Ferry Pool Road, Oxford, OX2 7DT.
Rianna Erker, 3100 Briarcliffe Road NE, 517 Atlanta, Georgia 30329, USA.
Rachael Evans, Kiln Cottage, Culham, Abingdon, Oxon, OX14 4NE.
Sarah Jane Evans, Crescent Wood, Cottage, 6 Crescent Wood, Road, London, SE26 6RU.
Prof Doreen G Fernandez and Maria Elena G Besa, 3 First Street, Acacia Lane, Mandaluyong, Metro, Manila 1501, Philippines.
Elizabeth Field, 46 St Helen's Road, Booters Town, Co. Dublin, Ireland.
Wendy Fogarty, Far Outlook, Old Road, Shotover Hill, Headington, Oxford, OX3 8TA.
Ove Fosså, Parkveien 11, N-4307 Sandnes, Norway.
Svein Fosså, Bioddgaten 17, N-4890 Grimstad, Norway.
Sarah Freeman, 47 Onslow Gardens, London, N10 3JY.
Mrs Jean Freemantle, Hill House, Waddesdon, Bucks, HP18 0JF.
Susan Friedland, Harper Collins Publishers, 10 East 53rd Street, New York, NY 10022-5299, USA.
Elizabeth Gabay, Flat 3, 16 Belsize Park Gardens, London, NW3 4LD.
Brenda & Silvia Garza, PO Box 124, San Pedro Garza Garcia, NL, Mexico, 66230.
John P Gauder, 3306 April Lane, Stevens Point, WI 54481-5523, USA.
K Dun Gifford, Oldways Preservation Trust, 25 First Street, Cambridge, MA 02141, USA.
Sally Grainger, Timberua, Glen Road, Grayshott, Hindhead, Surrey, GU26 6NB.
Henrietta Green, 17 Hopefield Avenue, London, NW6 6LJ.
Juliet Harbutt, Old Woolman's House, Hastings Hill, Churchill, Oxon, OX7 6NA.
Jane A D Hedges, Fulscot Manor, Didcot, Oxfordshire, OX11 9AA.
Victoria F Hingley & Ruth Hingley, 49 Plater Drive, Waterside, Oxford, OX2 6QU.
Ruth-Hege Holst, Sch. of Hotel Mgmt, POB 2557, Ullandhaug, N-4004 Stavanger, Norway.
Geraldene Holt, Le Village, St Montan, 07220 Viviers, France.
Richard F. Hosking, 55E Lamont Road, London, SW10 0HU.
Jeffrey Hyman, 32 Chandos Way, London, NW11 7HF.
Philip & Patsy Iddison, 3 Upper Grotto Road, Twickenham, Middlesex, TW1 4NG.
Jan Krag Jacobsen, Kasatnieallee 39, 3520 Farum, Denmark.
Eve Jochnowitz, 21 East 10th Street #2a, New York, NY 10003-5924, USA.
Rosemary Joekes, The Hermitage, St. Catherine, Bath, BA1 8HE.
David Karp, 633b Palms Boulevard, Venice, CA 90291, USA.
Karen Karp, 42 Lancaster Park, Richmond, Surrey, TW10 6AD.
Cathy Kaufman, 718 Broadway #10A, New York, NY 10003, USA.
Mary Wallace Kelsey, Dept Nutrition & Food Mgmt, Oregon State U., Milam Hall 108, Corvallis, OR 97331-103, USA.
Lidia & Sotiris Kitrilakis, 27 Audella Street, Thessaloniki, 55131, Greece.
Diane Kochilas, Kehagia 29, 152 37 Filothei, Athens, Greece.
Mark P Lake, Park Farmhouse, Sandford Saint, Martin, Oxford, OX7 7AH.
Rachel Laudan, c/o J B Thatcher, Ley Farm, Teffont, Evias, Salisbury, Wiltshire, SP3 5RW.
Gilly Lehmann, 25620 Bonnevaux-le Prieuré, France.
Paul Levy & Penelope Marcus, PO Box 35, Witney, Oxon, OX8 8BF.
Audrey Levy, 60 Gloucester Road, Kingston-upon-Thames, Surrey, KT1 3RB.
William & Yvonne Lockwood, 2210 Struthers Road, Grass Lake, Michigan, MI 49240, USA.
David Lockwood, Neal's Yard Dairy, 17 Shorts, Gardens, London, WC2H 9AT.
Elisabeth Luard, Brynmeheryn, Ystrad Meurig, Dyfed, SY25 6AH.
Fiona Lucraft, The Limes, 40 Station Road, Haddenham, Ely, Cambridgeshire, CB6 3XD.

LIST OF THOSE ATTENDING THE SYMPOSIUM

Jenny Macarthur, 13 Wavell Road, Maidenhead, Berkshire, SL6 5AB.
Dr Jeremy MacClancy, 14 High Street, Eynsham, Oxon, OX8 1HB.
Laura Mason, 4 Saint John Street, York, YO3 7QT.
Stephen W Massil, 138 Middle Lane, Crouch End, London, N8 7JP.
Carolyn McCrum, 57 Oakthorpe Road, Oxford, OX2 7BD.
Christine McFadden, 71 Prior Park Road, Bath, BA2 4NF.
Tessa McKirdy, Cooks Books, 34 Marine Drive, Rottingdean, Sussex, BN2 7HQ.
Susan McLellan Plaisted, PO Box 1162, Morrisville, PA 19067-5979, USA.
Richard C Mieli, Flat 1, 85 Cornwall Gardens, London, SW7 4AY.
Janny de Moor, Ulco de Vriesweg 29, 8084 AR 't Harde, Netherlands.
Dr H & Mrs F Morrow Brown, Highfield House, Highfield Gardens, Derby, DE3 1HT.
Caroline Morrow Brown, 49 Eggington Road, Hilton, DE65 5JG.
Lizabeth Nicol, 16 rue du Saussoy, 77515 Saint Augustin, France.
Jill Norman, 1 Rosslyn Hill, London, NW3 5UL.
Shirley Olivier, The Coach House, Eaton Road, Hove, East Sussex, BN3 3PP.
Sonia Ortega, Spain Gourmetour, Paseo de la Castellana, 16, Madrid, 28046, Spain.
Sri & Roger Owen, 96 High Street Mews, Wimbledon Village, London, SW19 7RG.
Gaitri Pagrach-Chandra, Bulkstraat 48, 4196 AX Tricht, Netherlands.
Rupert & Vola Parker, 151 Wilberforce Road, London, N4 2SX.
Helen Peacocke, Rose Cottage, 43 Acre End Street, Eynsham, Oxford, OX8 1PF.
Dorothea A Pelham, 6 Portland Road, Oxford, OX2 7EY.
Charles Perry, 12912 El Dorado Avenue, Sylmar, CA 91342, USA.
Elia Petridou, 12b Medley Road, London, NW6 2HJ.
Maya Pieris, 17 Grays Lane, Hitchin, Herts, SG5 2HG.
Gae Pincus, PO Box 59, Glebe, NSW 2037, Australia.
Hannah Rapport, 4 Harbour Road, London, SE5 9PD.
Anjan K. Ray, 16 West 85th Street, Apt. 3A, New York, NY10024, USA.
Gillian Riley, 11 Kersley Road, Rottingdean, N16 0NP.
Alicia Rios, Avenida General Peron 19 - 8ºC, 28020 Madrid, Spain.
Cherry Ripe, c/o Abraham, 42 Tavistock, Road, London, W11 1AW.
Joe Roberts, 31 Brock Street, Bath, BA1 2LN.
Ann Rycraft, 1 Mill Mount, The Mount, York, YO24 2BH.
Helen J Saberi, 75 Haldon Road, London, SW18 1QF.
Alice Wooledge Salmon, 14 Avenue Mansions, Sisters Avenue, London, SW11 5SL.
Camille Savory, 86 Queensbridge Court, Queensbridge Road, Haggerston, E2 8PA.
Dan M Schickentanz, Pakenham, Park Lane, Long Hanborough, Witney, Oxon, OX8 8JU.
Liz Seeber, Apple Tree Cottage, High Street, Barcombe, Near Lewes, East Sussex, BN8 5DH.
Dr Colleen Taylor Sen, 2557 West Farwell Avenue, Chicago, IL 60645, USA.
Maria José Sevilla, 26 Holland Park Gardens, London, W1Y 8EA.
Regina Sexton & Shane Lehane, The Yellow House, Vicarstown, Co. Cork, Ireland.
Margaret Shaida, Teulades V, Apt. 201, Els Vilars, Escaldes, Andorra.
Sue Shephard, 7 Leigh Road, Clifton, Bristol, BS8 2DA.
Helen J Simpson, Burton Court, Eardisland, nr. Leominster, Herefordshire, HR6 9DN.
Andrew F Smith, 135 Eastern Parkway #11A, Brooklyn, New York, NY 11238, USA.
Diane Sokolofski, Kraft Foods, 1 Kraft Court, Glenview, Illinois, 60025, USA.
Raymond Sokolov, 34 Barrow Street, New York, NY 10014-3735, USA.
Colin Spencer, Winchelsea Cottage, High Street, Winchelsea, East Sussex, TN36 4EA.
Rosemary Stark, 6 Chamberlain Street, London, NW1 8XB.
Jeffrey L Steingarten, 29 West 17th Street, New York, NY 10011, USA.
Dr Layinka M Swinburne, 16 Foxhill Crescent, Leeds, LS16 5PD.
Anne Tait, Ridgeway Cottage, Glanvilles Wootton, Sherborne, Dorset, DT9 5QF.

Malcolm Thick, 2 Brookside, Harwell, Oxon, OX11 0HG.
Jane Thrift, 3 Beechcroft, Dorchester on Thames, Wallingford, Oxon, OX10 7LS.
Linda Tobey, 26 Byne Road, Sydenham, London, SE26.
Pat Van Den Wall Bake-Thompson, La Cuisine Francaise, Herengracht 314, Amsterdam, 1016CD, Netherlands.
Linda Vijeh, 21 North Street, Crewkerne, Somerset, TA18 7AL.
Harlan Walker, 294 Hagley Road,, Birmingham, B17 8DJ.
Jennifer Walker, 10 Whitley Park Lane, Reading, Berks.
William Woys Weaver, Box 75, Devon, PA 19333-0075, USA.
Kathie Webber, Framewood Manor, Framewood Road, Stoke Poges, Bucks, SL2 4QR.
Robin Weir, 104 Iffley Road, London, W6 0PF.
Margaret Willes, 17 Appleby Road, London, E8 3ET.
Faith Willinger, Via Della Chiesa 7, Florence, 50125, Italy.
Bee Wilson, 16 Priory Road, Cambridge, CB5 8HT.
Mary Wondrausch, The Pottery, Brickfields, Compton, Nr Guildford, Surrey, GU3 1HZ.
Caroline Yeldham, 356 Ripon Road, Stevenage, SG1 4NQ.
Carolin Young, 57 Mudge Avenue, S Hamilton, MA 01982, USA.
Sue Young, La Toque d'Or Gourmet, 55 rue de Varenne, 75007 Paris, France.
Sami Zubaida, 2 Avenue House, Belsize Park Gardens, London , NW3 4LA, .